Visual Basic Programmer's Guide To Serial Communications

4th Edition

By Richard Grier

Visual Basic Programmer's Guide To

Serial Communications

4th Edition

Richard Grier

PUBLISHED BY:

Hard & Software
12962 West Louisiana Avenue
Lakewood, Colorado 80228 USA

Phone: 303-986-2179
Fax: 303-593-9315

Internet: dick_grier@msn.com
WWW: http://www.hardandsoftware.net

Fourth Edition, Second Revision
Date of Printing: June 1, 2006

1-890422-28-2

Library of Congress Catalog Card Number: 2002111093

This book is dedicated to my wife, Peggi. Thanks for putting up with me for the last 38 years.

— Dick, June 2006

The 2nd, 3rd, and 4th Editions also are dedicated to my mother, Joyce.

— Dick, June 2006

Exceptionally knowledgeable, creative and intuitive, Richard Grier is truly a programmer's programmer. Through his precise methodology, scientific approach and keen attention to detail, ideas and concepts are transformed into outstanding software achievements. When others are lost in a technical quagmire, Grier is certain to find the solution.

—Barry Bittman, MD
Neurologist and CEO
TollFree Software Solutions, Inc.

Dick Grier's thorough understanding of serial communications and his unique ability to clearly communicate the subject to others is only surpassed by his willingness to help programmers solve real world problems. This is definitely a book you want in your reference library.

— Don Brown
President
Lighthouse Software, Inc.

Table of Contents

Forward to the Fourth Edition

Second Revision: June 1, 2006.

Ten years ago I decided to write a book on serial communications using Visual Basic. Looking back, my initial ideas seem a little naive — though that may be somewhat harsh. I have learned so much in the succeeding years that I now know, well, more! The 2nd, 3rd, and now 4th Editions of this book have led me to the realization that I, as is true of most authors, never will know everything there is to know about a subject, and that I may not be as adept at communicating what they do know as I might like to be. Actually, I have been very pleased with the response to earlier editions. That has motivated me to add this one to the mix.

The Fourth Edition includes additional information, corrections, and elimination of some items that have lost relevance since the 1st, 2nd, and 3rd Editions were published. In the four years since the 3rd Edition of the book (February, 2002), Visual Studio .NET has gone through three releases (VS .NET 2002 and VS.NET 2003 and Visual Studio 2005 released in November of 2005).   Several of the companies featured earlier are no longer in business, and others have entered the marketplace.

I would like to thank all of the people who have contacted me with comments about and criticism of the 1st, 2nd, and 3rd Editions. I have been pleased that most of the feedback that I have received has been favorable. This has led me to devote the time required for this newer work. Much of the new example code that is included on the accompanying CD ROM was written at the behest of several readers. I hope that the improvements in the Fourth Edition meet my reader's needs.

This printing is the first undertaken without the help of Mabry Publishing, who published all previous versions. I certainly want to express my appreciation to James Shields and Ruth James of Mabry for their many hours of effort through the years.

I have updated the Introduction to reflect the additions and other changes that I have made in the book content.   In fact, all areas of the book have received attention in the Fourth Edition; some will have noticeable changes, others are only cosmetic.   All are designed to help you out.

Introduction

I have worked for over 35 years as an engineer, and for the last 22 years I have worked designing modems and data communications systems hardware and software. I have also designed data acquisition and other electronic hardware and software. In almost every case, an integral part of the designs of these other systems was a data communications subsystem that frequently used RS-232 or similar serial methods.

I started designing software for Windows at the same time that VB 2.0 was released. I had looked at VB 1.0 but felt that it was not quite "ready for prime time," although there had been a lot of useful software written using VB 1.0.

As soon as I started using VB 2.0, I became a regular visitor to the CompuServe MSBASIC forum. I found the MSBASIC forum to be a valuable source for answers to questions and I quickly became familiar with the expert programmers that frequented the forum. When I felt competent to contribute to the online discourse, I did so. An area that was insufficiently covered was serial communications and, more specifically, the MSCOMM.VBX. Fortunately I already had a background in serial communications and modems (which generate more questions than any other subject in the area of serial communications). So, I started writing example code and answering questions online. I was satisfied that my efforts were beneficial; many correspondents sent me messages of thanks and other positive feedback. I also felt that I would like to do more. Microsoft has moved their developer tools onward with the release of the various incarnations of Visual Studio .NET. While serial communications may appear to be less important with the emphasis on networking and the Internet, it still has many important applications.

In early 1994, the Microsoft engineers in charge of the MSBASIC forum activity recognized my online assistance by naming me a Microsoft Developer MVP (Most Valuable Professional). The MVP program was started by Microsoft as a way to recognize those people who provided accurate and valuable online help to other developers. The MVP program was a way for Microsoft to say "Thanks" and to encourage future contributions from the MVPs. See http://mvp.support.microsoft.com/default.aspx for more information about the Microsoft MVP program.

I have continued my participation in the Microsoft MVP program with Microsoft's move to the Internet.

So, Why Did I Write This Book?

I found that I could never spend as much time online as I wanted. And, I could not put as much detail into responses as I would have liked. Frequently, a simple response to a question brings up many subsequent questions. One of the most frequent questions asked of me was, "What book can be used to learn what you have told me?" There was none that I could find. I do not think that I can kill all questions with this one tome (puns and mixed metaphors are a favorite pastime of mine) but it gives me a chance to get everything down in writing that I think may be useful. It also has kept me out of my wife's hair for several months.

A common complaint that I have seen is that the documentation on MSCOMM that is furnished with Visual Basic is incomplete and borders on useless. Well, let me be charitable and say that it is not as practical as I would like, that it does not cover most of the issues that I'm tackling here. I also am able to furnish information on commercial serial communications products.

When Visual Studio 2002 and 2003 were developed, serial communications support was **not included.** Thus, resources were even more limited.

I have had plenty of help gathering information and resources for this book. Please see the Acknowledgments section for a list of the people who have helped me out. If I make an error conveying to you, either in fact or by implication, something that they have provided to me, please let me know and I will try to rectify the problem.

Please help me out by letting me know when you find something that should be changed or included.

What Is In This Book

The working title for this book was *An Engineer's Guide to Practical Serial Communications in Visual Basic*. The title has been changed but the approach remains the same. I have tackled the problem of serial communications as an engineer would. That is, I have included background information on serial communications and details on how the various parts of a serial communications system act and interact. This "systems approach" should be useful to engineers and non-engineers alike.

I have tried to include detailed information on the use of the Windows API (Application Programming Interface) for serial communications, the use of the MSCOMM custom controls that are furnished with VB Pro and Enterprise versions and Visual Studio.NET, and commercial serial communications add-ons. In addition to general-purpose serial communications, I have included special purpose information on Alphanumeric Paging and a variety of other specialized serial communications topics.

WindowsCE devices are growing in popularity and utility and they present some unique problems that need to be addressed.

I have expanded the discussion of checksum and CRC calculation and have included working example code.

New to the Fourth Edition is a discussion of various Virtual Serial port implementations, including how some might be used by Visual Basic Programmers.

I have placed the discussion of XMCommCRC.ocx and associated example programs in a separate chapter. XMCommCRC.ocx is an ActiveX control that I wrote using VB6. XMComm wraps the functionality of MSComm32.ocx and adds XMODEM/Checksum and XMODEM/CRC error-corrected file transfer capability. It may be used freely in Visual Basic, Excel, Access, Visio, LabView, or other environments that support ActiveX controls. There is no license restriction on its use. The source code for XMCommCRC.ocx is the starting point for this new chapter. There I discuss its design and implementation. The accompanying examples illustrate the use of the XMCommCRC control.

I have devoted one full chapter to VB.NET, the Visual Basic portion of Visual Studio.NET. VB.NET examples cover the same general subjects that are in other chapters for earlier versions of Visual Basic. Several of these examples used the Upgrade Wizard that is part of the VB.NET development environment. The code generated by the Upgrade Wizard may not be as intuitive as one might hope plus it sometimes needs modification to work well. However, the Wizard does a fair job. These examples allow me to illustrate some of the built-in functions that the .NET framework offers to make a developer's life more productive. This productivity comes at a cost — I discuss that, too.

With the release of Visual Studio 2005 serial communications is included for the first time. The System.IO.Ports namespace provides this support. I have added detailed information on the use of this namespace, along with example programs that provide practical starting points for its use. I developed a native VB .NET class that encapsulated the Windows communications APIs and the FileStream and other .NET methods for use with Visual Studio 2002 and Visual Studio 2003. This class was in the 3rd Edition, and still is valid for those development environments, which had no built-in serial support. However, to make space for Visual Studio 2005 in this edition, I have moved the associated text to the CD ROM.

The 3rd Edition of this book included a chapter covering serial communications using eVB for Windows CE and the Pocket PC. However, with the release of Visual Studio 2003, Microsoft moved practical application development to the Compact Framework. The Compact Framework, like the desktop .NET Framework, did not include serial communications, so it was left to developers like me to create that functionality. The eVB chapter has been replaced by a Compact Framework chapter in the Fourth Edition. All of the text in the 3rd Edition Chapter 8 is retained on the CD ROM.

Occasionally, I write C# code. Some C# will creep into the examples on the CD ROM. While I'd like to cover C# programming in more detail, I will have to leave that to another time.

Included on the CD ROM are code examples to illustrate as many of the things that I talk about as possible. To use this example code, you will need to have installed the version of Visual Basic that is specified along with any custom controls that are mentioned. But, often, the example code will be portable to other versions of VB and other custom controls. The example programs are designed as applets. That is, each stands alone and implements one or more features that are discussed in the book. None of the applets depends on another, although there is, naturally enough, some overlap in details.

I have included a folder on the CD ROM for X10 communications. These controls, examples, and information are freeware, downloaded from the Internet. I do not have the space to go into any detail on this subject in the book.

The CD ROM that accompanies the book includes custom controls and products from a variety of vendors. Please see the documentation that accompanies each of these sets of software for ordering or registration information. Some of these vendors will offer readers of this book discounted software or other benefits. See the README or similar file in each associated folder on the CD ROM for details on these offers.

I have used this icon in the text where I want you to refer to the CD ROM for a complete program listing or for information and services directly from the publisher of the product being discussed.

Also included on the CD ROM are several freeware DLLs and utility programs.

I have included as much information on debugging serial communications problems, both hardware and software, as I can. Debugging is critical and requires an understanding of the interaction of several disparate elements. These elements are the software that you have designed, the computers used and any multitasking that they must do, and any modems that are used. The tools that I mention are invaluable but there is no substitute for uncommon sense and for a methodical approach.

An appendix is devoted to resources. There are addresses, both mail and online (if available) and phone and fax numbers for all of the companies that are mentioned. A short description of the products that they offer, that may be valuable, is included. I have included as many other products as I could find, even if I do not discuss them explicitly in the book. If any product that should be included is not, let me know.

OK, What Is Not In The Book?

I have included nothing on serial communications via the Internet. This is a subject that should have a book devoted to it. There are several such books available and there will be many more coming. I have also decided NOT to include the use of fax add-ons or Microsoft Exchange. Time and resources have left this as a subject for the future.

I have included information on the features that are included with the MSCOMM32.OCX that is furnished with VB 5/6 and now VB .NET. However, some of the new features of VB .NET (and there are many that will be useful) will not be discussed in detail.

I do not try to cover synchronous serial communications. Windows has no built-in support for synchronous communications while Windows support for asynchronous communications is substantial. Synchronous hardware is uncommon but it is available. So, if this is your need, you will have to do some research or get some consultation in this very specialized area. However, one product that is included in Chapter 7 (see LUCA) includes synchronous communications support.

I do not cover the control of PLCs (Programmable Logic Controllers). It is possible to implement the more common PLC communications protocols using MSCOMM or some other add-on. But perhaps a more practical solution is to use one of the custom controls designed specifically for this purpose. I have listed some of them in the Resources appendix. Also, refer to the LUCA product description in Chapter 7 and on the CD ROM. LUCA includes support for several PLC related protocols. At the time of this writing, I had plans to include a variety of commercial add-ons that support PLC communications in the Resources portion of the CD ROM.

The Universal Serial Bus (USB) is not discussed. USB, as its name suggests, is a bus designed to provide "Plug and Play" capability to peripheral devices. It requires a device driver for supported devices on the bus, similar to printer, network and video adapters, or other conventional PC hardware that may be used. Unlike conventional hardware buses like PCI, USB devices are "daisy-chained" together using special serial cables. So, while the name says "serial", that is a physical description of the way that it connects. In other practical ways, the "bus" part of the name is what is important while serial is not. There are USB serial devices, however. These provide a way to add serial ports to a PC without opening the box and inserting a card. Use of these serial ports is the same as those on an internal card (with minor caveats that will be discussed in the Debugging chapter). Appendix A provides a list of pertinent vendors. Jan Axelson's book *USB Complete Third Edition, Everything You Need to Develop Custom USB Peripherals* (ISBN 1931448027) is a good source of information on USB. It has example code written in Visual Basic and VB .NET.

Bluetooth is a serial communications technology that might be of interest. I do not attempt to cover this in detail. Why not? First, often it is designed for networking computers and compatible devices. It may not be oriented toward raw serial communications but may use an extensive protocol stack; this depends on the actual Bluetooth device with which you are communicating. There are some Bluetooth devices that may be treated as standard serial ports (an example is most GPS receivers that provide a Bluetooth interface). Second, in general, Bluetooth will not use a conventional serial port, though there are serial to Bluetooth adapters. See the Resources Appendix for Bluetooth resources. Last, I have not had the need to develop any Bluetooth enabled applications, so there are more things that I do not know, than there are things that I do know about it. There is an old saying, "Those who can, do. Those who cannot, teach." While there is some small truth in this statement, I prefer to admit the limitation.

Most of the discussions in the book and all of the code samples on the CD ROM use the North American (English) versions of Windows and modems designed for North American use. Users in other areas of the world may find differences that invalidate one or more things that are covered. Unfortunately, this is a fact of life. Microsoft TAPI attempts to alleviate this problem and to date it has been only somewhat successful.

Parts of earlier Editions of the book have been moved to the CD ROM to make space for new text. These cover a variety of topics that may still be of use, such as the 16-bit Windows serial communications API, eVB, and various communications products that may still be present in legacy systems. Refer to the CD ROM 2ndEdition and 3rdEdition folders for .doc files that contain the deleted text.

If I learn something new and useful or if you let me know of something that I can mention, specifically, about serial communications in other parts of the world, I will add it to the next revision of this book. An important aspect of this book will be the continuous upgrades and improvements to the text and software.

How to Use This Book

This is not a "Dummies" book. I have no quarrel with that highly successful genre. But I do not think that serial communications is a subject for novices. I have assumed that you know how to program in Visual Basic. I do not go into how to edit code, how to place and use standard VB controls, or any details of UI (User Interface) design, except where those affect the subject at hand, serial communications.

I do not attempt to teach you a programming style or to use programming techniques that are specifically designed to encourage code reusability. If you want to learn how to program, how to design class modules, or to design a User Interface, see the Resources appendix. There are several books listed there that cover these subjects.

I do not assume that you have done any serial communications programming or that you have an intimate familiarity with modems or other serial communications devices. I will ask you to read the manuals that are furnished with the communications device that you are going to use. As painful as the fact may be, there is no better source for accurate information than that furnished by the manufacturer.

I will offer opinions on programming techniques and approaches to problem solving. These opinions are my own and contrary opinions may well exist. I will offer them to you because I have found them to be valuable to me. See Conventions and Style for more information.

A Quick Preview of the Book and CD ROM Content

Chapter 1 covers serial communications, Windows, the PC hardware that is used, and flow-control. Error-corrected file transfers and terminal emulations are covered next.

Chapter 2 covers modems and serial standards. More than 40 of the most common questions about modems are answered.

Serial hardware standards are discussed. Various RS-232 null-modems are illustrated. RS-422 and RS-485 are discussed, with code fragments presented for RS-485.

I have added a new section to this chapter for a more detailed discussion of checksum and CRC calculation in Visual Basic. Several examples are included.

Chapter 3 discusses the details of the Windows communications API, including the Telephony API.

Chapter 4 discusses the MSCOMM custom control. The versions for VB 2.0 through VB 6.0 are covered along with programming concepts for VB6. Examples for earlier versions of Visual Basic will be found on the CD ROM. This chapter is where we start to see some practical program examples to illustrate the use of MSCOMM.

There is a program included called DMM/LOGGER. This program illustrates how to use MSCOMM to communicate with a (DVM) Digital Volt Meter. The ideas that are conveyed with this program include how to implement a communications protocol. In this case, the protocol has been designed by an electronic instrument manufacturer.

Two MSCOMM programs implement Global Positioning Satellite receiver interfaces. These illustrate implementation of a serial protocol called NMEA-0183. One receiver program synchronizes your computer to the GPS receiver time and date and provides highly accurate location (latitude and longitude). The other decodes more NMEA data such as altitude, speed over the ground and the number of satellites in view but does not synchronize computer time.

A simple magstripe and barcode programs each are provided to illustrate the use of magstripe, barcode and similar scanning devices.

Several other programs (applets) are included to illustrate various points. See the CD ROM for example programs that did not make it into the text. Chapter 4 also covers XMCommCRC.ocx. XMComm is an ActiveX control written in Visual Basic 6.0 that adds XMODEM/checksum **and** CRC error checked file transfers to the underlying capability of MSComm32.ocx. It can be used in a variety of development environments that support ActiveX controls (the only exception is browsers where it is not appropriate to host an ActiveX control that accesses client-side hardware such as the serial port). The source code for XMComm illustrates the design of the ActiveX control. The XMTerm, Remote/Host and Flashlite example code illustrate the use of XMComm in actual VB applications.

Chapter 5 explores some of the new .NET territory. I provide four example programs that illustrate the opportunities and issues in Visual Basic.NET. I have ported three VB6 programs to .NET. One is the XMTermNET program which uses the XMCommCRC.ocx ActiveX control for terminal emulation and error-checked file transfer. The second is a GPS program that uses native .NET methods for converting UTC (Universal Coordinated Time) to local time and date information, and which automatically compensates for Daylight Savings time. The third program is a straightforward port of VBTerm, using MSComm32.ocx, to VB.NET. These programs illustrate the use of .NET COM Interoperability. That is, Visual Basic .NET and other .NET framework languages allow you to continue to use ActiveX controls and other ActiveX objects in .NET programs.

I have included one native VB.NET program that illustrates the .NET built-in support for serial communications using the FileStream class. Unfortunately, FileStream does not furnish all of the "hooks" needed for a complete implementation of a serial communications object. We have to use unmanaged code from the Windows communications API to do some of the heavy lifting. Thus, some of this code will look quite a bit like some of the equivalent code in Chapter 6. However, .NET does offer some real power that I mention. Method overloading is something new to VB programmers. It provides an important tool that we will use.

MS.NET is an evolving program for Microsoft. Developers will have to add new techniques to their toolset in order to work in this area. Visual Studio 2005 adds the System.IO.Ports namespace. This furnishes a native serial communications class. I discuss this and provide examples, including XMComNET (XMODEM for .NET).

Chapter 6 goes into the Windows 32-bit serial communications API. 16-bit API information and examples have been moved to the CD ROM.

Next is TAPI (the Telephony API). A simple Windows 95/98 dialer is shown plus other uses and limitations of TAPI are discussed.

Chapter 7 is an overview of a variety of commercial communications add-ons for Visual Basic. Sax Comm Objects, Greenleaf CommX, MagnaCarta CommTools, LUCA, amComm, SuperComm, and Crystal CrystalCOMM are discussed.

One example program is provided. This program uses Sax Comm Objects to create a program that provides a combination Host (BBS) and Remote (client) communications program that implements automatic file transfers. This program illustrates TAPI modem configuration, dialing, post-connection user validation, and automatic file transfers using the Zmodem file transfer protocol.

Previous editions of this book provided example programs that used PDQComm from Progress/Crescent Software. PDQComm is no longer available. However, the example code and accompanying text has been placed on the CD ROM in the PDQComm folder.

Chapter 8 discusses paging. Numeric paging is shown and its limitations are discussed.

Alphanumeric paging (AlphaPaging) is the real forte of this chapter. One commercial Visual Basic add-on is discussed: the Logisoft Page/X ActiveX control.

A program is included that uses Ron Tanner's PowerPage DLLs (the DLLs are included on the CD ROM) and illustrates the AlphaPaging process. This DLL is no longer available, so this has been moved to the CD ROM.

Chapter 9 covers serial communications using VB .NET and the Visual Studio 2003 Compact Framework 1.x. Serial Programming for Visual Studio 2005 and the Compact Framework 2 is equivalent to that in Chapter 5, and an example is included on the CD ROM.

The 3rd Edition of the book included eVB for Windows CE and discussed actual coding using the ceComm control, various design, development, and debugging issues, along with performance considerations in a WindowsCE hand-held or embedded PC environment. The accompanying eVBTerm and ceVoltmeter examples are used to illustrate this area. This text has been moved to the CD ROM.

Chapter 10 examines the use of direct I/O port manipulation to do things that cannot be done using more conventional API methods. This requires that you understand the physical I/O structure of your PC. Four methods are discussed. The first uses VBASM.DLL (included) to access PC I/O ports on a system using a 16-bit version of Visual Basic. The second uses WIN95IO.DLL to do the same using a 32-bit version of Visual Basic.

Two programs are included that monitor the status of the Carrier Detect bit in the UART. This allows a Visual Basic program to record the total connect time of another Windows program. A third program is used to overcome the maximum speed limitation (19.2k bps) of MSCOMM32.OCX furnished with VB 4.0. Included are examples that are mainly oriented toward Internet applications but also use the Windows and RAS APIs to do some of the actions listed.

Scientific Software Tools, Inc. DriverLINX PortIO software is a freeware DLL and kernel mode driver that permits direct access to standard I/O ports under Windows 9x/Me and Windows NT/2K/XP. It is included on the CD ROM. An example program is provided that allows a program to force DTR false, thus causing a modem to disconnect on a comm port that has been opened by another application. Another example illustrates sending data **directly** to the serial port UART, which bypasses the Windows serial API. While not everyone's "cup of tea," these examples may help solve a problem that otherwise is intractable.

Chapter 11 is important. Debugging communications applications can be difficult. Here are discussed both hardware and software methods for debugging your applications.

Techniques for optimizing your code and tips to increase its reliability are presented. The use and utility of telephone line simulators are discussed. Last, debugging serial port hardware problems are discussed.

Appendix A is a list of resources. Contact information is provided for all of the products that were mentioned in the book. Lots of additional contact information is provided for companies that offer products or services that may be useful but that could not be discussed in detail.

Appendix B is a VT100 terminal emulator written using VB6. Earlier editions of the book included 16-bit VB2 code (still on the CD ROM in the 2ndEdition folder).

Appendix C is the complete NMEA-0183 protocol used in the GPS receiver program.

Appendix D details the I/O port description of 8250 and 16550AF UARTs.

Appendix E is a chart of the ASCII character set.

Appendix F is the basic AT modem command set.

Appendix G is a description of the Telocator Alphanumeric paging protocol.

Visual Basic and Windows Versions

Visual Basic is now in its ninth version and Windows in its umpteenth version. I'll discuss some of the issues with various versions here and in more detail in other sections of the book.

Windows 3.0 was furnished with a device driver (COMM.DRV) that implemented interrupt driven serial communications. This and later drivers are "virtual device drivers." That is, they provide a controlled interface to the services provided to support serial communications and isolate the application from the actual hardware. The Windows 3.0 COMM.DRV did not provide support for 16550 AF UARTs (discussed later) so it was not reliable at speeds in excess of 19200 bps.

Windows 3.1 was furnished with an improved version of COMM.DRV. It provided built-in support for 16550 AF UARTs, an improved notification scheme, and was capable of reliable communications up to 57600 bps.

VB 1.0 had no built-in support for serial communications. If you needed serial communications, you had to rely on the Windows API. This was a fairly complex process, as can be seen from the API chapter in this book. VB would work with Windows 3.x in either Standard or Enhanced mode.

It would be a mistake to do any serial communications under Standard mode Windows. True multitasking is needed to avoid loss of data when a serial communications application runs in the background. So, do not try to write any serious serial communications application that runs under Windows 3.x Standard mode.

There were a few DLLs (Dynamic Link Libraries) that encapsulated the API but these were not too common nor did they work too well.

It is possible to encapsulate the Windows communications API functions in a VBX or OCX and to make those functions easily available to VB programmers. VB 2.0 Professional Edition was furnished with MSCOMM.VBX that provided a simplified communications interface. MSCOMM.VBX also provided event driven communications. Event driven comm is desirable because it means that communications routines can be written in an analogous way to the routines that are associated with other VB events, e.g., CommandButton Click events.

VB 2.0 Standard Edition, and later Standard Editions of Visual Basic, had no built-in support for serial communications. You could use the Windows API functions or purchase a commercial communications add-on.

At the time of the VB 2.0 release, commercial vendors started to offer VBXs that encapsulated the communications API and that offered other enhancements. Some of these enhancements were built-in error-checked file transfers and terminal emulation.

VB 3.0 followed VB 2.0 by only six months. VB 3.0 Professional Edition was also furnished with MSCOMM.VBX. The new MSCOMM.VBX used a feature of the Windows 3.1 API that permitted communications event notification of the control. However, it was soon found (within a few days!) that this event notification did not work reliably at speeds higher than 9600 bps. A new version of MSCOMM was released to deal with this fault. See the section on VB 3.0 and MSCOMM for more details on this problem.

Windows for Workgroups 3.11 was released in this same time frame. WFW 3.11 introduced a new problem. The communications driver (COMM.DRV) that was furnished with WFW 3.11 had a fault of its own. It did not properly identify 16550 AF UARTs. This meant that it would attempt to enable the FIFO (First In First Out buffer) on the UART when, in fact, that FIFO did not exist. This caused unreliable processing of receive data. There were several solutions for this problem. The first involved replacing COMM.DRV with the COMM.DRV from Windows 3.1. The second solution involved editing the SYSTEM.INI file to disable the FIFO for any port that was known to not have a 16550 AF UART. The third possibility was to replace COMM.DRV with a commercial replacement. You can see the Resources appendix for TURBOCOM.DRV from Pacific CommWare. This driver is my preference for solving this problem because it works and because it is a higher-performance driver than the Microsoft-furnished COMM.DRV. Another such driver is HiCom/9 from Cherry Hill Software Corporation.

Windows NT (later Windows 2K and XP) and Windows 95 (later Windows 98 and Me) are 32-bit operating systems. These various operating systems improved serial port handling with each new version (with a few warts here and there). TAPI became viable only under these OS's, and serial port drivers have been updated to improve "Plug and Play" operation with a variety of serial devices. The chapter on Debugging will discuss some of the problems that may be encountered under these OS's.

VB 4.0 introduced several new features and issues. This version of VB came with two development environments. VB 4.0/16 was for Windows 3.x (although, of course, 32-bit versions of Windows would also run VB 4.0/16 programs) and VB 4.0/32 was for Windows 95/98 and Windows NT 3.51 or later. Accompanying the Professional and Enterprise Editions of VB 4.0 was MSCOMM16.OCX for use with VB 4.0/16 and MSCOMM32.OCX for use with VB 4.0/32. From the standpoint of the VB programmer, OCXs (OLE custom controls, where OLE stands for Object Linking and Embedding) were just like the earlier VBXs. These new controls offered no new features but they did cause some new problems that were not seen in earlier versions. See the section on VB 4.0 and MSCOMM for more specific details. To solve some of the problems, new versions of both controls were released on 1/26/96. These updated controls were part of the unadvertised VB 4.0a release of VB 4.0.

VB 4.0 introduced a new issue with respect to binary data and the use of the String data type for Binary data. See the section on VB 4.0 and MSCOMM for an extended discussion on this area of concern. I discuss the best solution for this potential problem also.

VB 5.0, VB 6.0, and VB.NET are 32-bit development environments, only. A new version of MSCOMM32.OCX was furnished with VB 6.0. The VB5 and VB6 versions had some features that were new and very useful. I go into detail on some of these in the VB 5.0/VB 6.0 section. The VB 6.0 version of MSCOMM32.OCX does not offer any features that were not in VB 5.0. So, discussion of VB 6.0 will be combined with that for VB 5.0.

The Visual Studio .NET 2002 and 2003 Frameworks introduce many new functions that are designed to improve programmer productivity. As its name implies, it is largely oriented to development for the still emerging Internet-centric applications. Many of the new features offer value – but not too much that is directly applicable to serial communications. The GPS example application that is included in the VB.NET chapter illustrates the use of a .NET built-in library to perform calculations involving Daylight Savings time. The XMTermNET application illustrates the use of COM interoperability with the XMCommCRC control and the use of ActiveX controls in .NET. A similar VBTerm example, using MSComm32, also is provided on the CD ROM.

The NetTerm example illustrates.NET methods for implementing serial communications without using an ActiveX controls. Instead it uses the .NET FileStream class to read and write serial data using a communications port that has been opened using Platform Invoke (P/Invoke) to call a variety of underlying Windows APIs. The DesktopSerialIO dll that I provide on the CD ROM is an improvement on these techniques.

Visual Studio .NET 2003 Professional and higher Editions include the Compact Framework for PocketPC 2000 and higher devices, and other Windows CE systems using WinCE 4.x and higher. The Compact Framework does not provide any serial communications support, so I provide a complete, practical, serial communications class (DLL). The source code is included, so any modifications that are vital for a specific project may be made. I also furnish example code that illustrates the use of this class.

Visual Studio .NET 2005 includes native serial communications. Explanation and examples are included. See the CD ROM for Compact Framework 2.0 examples for the Pocket PC.

Conventions and Style

There are so many decisions to make when you write a book. One of the most important is how much humor to display? I have found (many, many times) that things that amuse me are not even slightly funny to others. So, be forewarned. If I say something stupid, write it off to my strange world-view. If you give me the benefit of the doubt, I will not have to work so hard.

The word "data" is a plural noun. However, common usage assigns a singular meaning to the word. Grammar dictates that the plural form of a verb should accompany plural subjects in a sentence. The choice made in this book is to follow common usage, rather than the grammatically correct form. So, the word data will be used in sentences with a singular form of the verb.

Microsoft and other knowledgeable authorities suggest that Hungarian notation should be used for variable and objects. For example, a TextBox might be named txtReceiveData. A string variable might be named sBuffer, and an integer variable might be named nTotalCount (or iTotalCount, depending on whom you ask).

The rationale behind these naming conventions is that one can tell from the name what the variable type is or what kind of object is named. This can simplify debugging of your own code, simplify maintenance, and make the code more readable by a casual viewer. All of these arguments, and perhaps others, are valid.

However, I often do not use Hungarian notation. I try to use descriptive names for my variables and objects. Habits of twenty years or so are hard to break and the extra prefixed characters that describe type or function are not natural to me. I apologize in advance for this failure and hope that it will not be too great a limitation. I do not use the default property of controls for an assignment. If one were to do so, readability would be enhanced by using Hungarian notation. However, statements like ReceiveData.SelText = Buffer is rather unambiguous. I do use Hungarian-like notation for private variables inside class modules and at various other times – I have tended toward verbosity as the years have gone on. This is my own idiosyncrasy. It reminds me of their use and allows me to use similarly named property Let and Get names.

Occasionally, I use a variable type suffix. For example, I might use Ret% to designate an integer or Buffer$ to designate a string. When writing code, I do this on an ad hoc basis (and have abandoned it in recent years). I suggest that you decide on your own coding style or use the standard that your organization has adopted. However, these suffixes are not supported in VB.NET. For that reason, most of the code that has been modified for the 3rdand later editions of this book removes these suffixes.

Some of the code examples in this book use Chr$(13), the Carriage Return character, instead of the equivalent VB intrinsic constant vbCr. The reason is that versions of Visual Basic earlier than 5.0 did not provide this and other built-in intrinsic constants. I like to use vbCr and vbCrLf when possible. However, you may see either form in the book text. VB.NET changes the syntax for these constants, so code written for .NET may use syntax like ControlChars.Cr for the Carriage Return character. VB .NET still supports the VB Constants vbCr and others, so you can choose the syntax that you prefer. See the .NET Help system for more information.

I always enable Option Explicit (Require Variable Declaration) and I suggest that you do so, too. I explicitly type variables. If a variable must be converted from one type to another, I use an explicit conversion and do not use any form of variant conversion.

Variants and conversion of types to variants can be troublesome. Variant variables require more memory, are slower to access, and their use can cause unpredictable errors. VB.NET no longer supports Variant data types. This reinforces the idea that Variants should not be used. VB.NET adds a compiler directive called "Option Strict." Option Strict restricts implicit data type conversions to only *widening* conversions. This explicitly disallows any data type conversions in which data loss would occur and any conversion between numeric types and strings. I suggest that you employ Option Strict On. It means that your code might be more verbose. However, the added safety is worth the effort.

I use Public or global variables only when they make sense. Artificial data hiding that requires extra parameter passing can make code maintenance easier but, contrary to some popular opinion, it does not make code more "modular." However, the use of global variables can cause side effects. So it is best to limit their use as much as is practical.

The sample code that is furnished with this book is as compact and concise as I could make it, within reason. I have kept the user interface as simple as possible. I have limited the number of forms any project uses to the minimum number that would work while providing the functionality that I desire.

One of the decisions that I had to make was what to call the sample apps that I furnished. I decided to call them applets to indicate that they are complete but that their functionality is limited. I hope this decision will not cause them to be confused with the applets that Microsoft furnishes with various operating systems and programming environments.

In some of the simpler programs, I hard-code some variables that a production application design would allow the user to configure. For example, the NIST Automated Time Program and RingDetect programs are hard-coded to Com1. Both of these programs run minimized, without normal user interface.

A production application will most often have a user interface that will allow the user to configure all critical variables. Such a production application will persist configuration variables in INI files, the Registry, or in a database. These configuration parameters will be read and used when the program is run subsequently. Several of the programs in this book illustrate that technique while others do not have this feature.

When I discuss step-by-step approaches to problem solving, usually I will do one of two things. I may present "pseudo code." This is a code-like expression of the steps that might be used. It is not real code. The other approach that I will use is to present code, based on MSCOMM, which could be compiled. Pseudo code allows me to present a concept without getting into the issue of making certain that the syntax is exactly correct. While real code is concrete, it takes more work to make certain that it will actually work as described. If I present code based on MSCOMM, it should work unchanged with all commercial communications custom controls. However, if you use a DLL-based product, you may have to make some changes in syntax but the logic will often be the same.

When I had a choice about placing subroutines or functions in a form or in a .BAS module, and if I did not need a .BAS module for other reasons, I placed the code in the form. The reason for this decision is twofold. First, fewer files are required to make a functional program, thus distribution of the source code is easier. Second, it is easier to view code on screen and on the printed page when associated procedures are close together.

A good programming technique is to use numeric constants with easy-to-understand names. Visual Basic includes CONSTANT.TXT with a number of these pre-defined constants. For example, an OnComm receive data event can be tested for using Const MSCOMM_EV_RECEIVE = 2. If MSCOMM1.CommEvent = MSCOMM_EV_RECEIVE Then (do something). By all means, use constants to make your code easier to read and maintain.

When later versions of Visual Basic were introduced, the name of some constants changed but not their meaning or value. For example, the MSComm CommEvent constant that indicates receipt of data in an OnComm event is comEvReceive for MSComm32.OCX (version 5 and later) while it was called MSCOMM_EV_RECEIVE in earlier versions. The actual value, 2, was retained. You may see either variation of these constants in the example code furnished with the book.

Public Enums extend the concept of providing easily interpreted constants, such as those used by MSCOMM and all of the classes in the .NET Framework.

Simply said, some of my code examples have no class. By that I mean that many of my examples do not use class modules. Classes enhance code reusability and maintenance. However, the reason that they enhance reusability and maintenance is that they can obscure details of the underlying design and implementation. These details are exactly what I want to emphasize in this book. However, I include a number of examples that do use class modules. One notable example is the source code for the ActiveX custom control XMComm. ActiveX projects must use class modules for their public interfaces so they are a natural to illustrate modular, object-oriented design. Visual Studio .NET is truly object-oriented. Use class modules when programming for .NET. This is a natural and inevitable element of design/implementation. When you create your own applications, I encourage you to use classes where appropriate. The Resources appendix lists a couple of books that discuss this in detail.

I should make one comment on the typographical conventions that I have adopted. I have decided to paste code from my Visual Basic projects into the examples sections of the book. Some of the sample code "wraps" from one line to line to the next. If it is typed in exactly as it appears, it will not run in 16-bit versions of VB. I could format it with artificial underscore characters; the line-continuation character that is used in VB 4.0 and later versions, but that is not available in earlier versions. The editing decision has been made to add underscores to indicate a line-continuation in code but not in comments, regardless of the version of VB in use.

On the CD ROM that accompanies the 2nd, 3rd, and 4th Editions of this book, I have included a number of example programs and other files that are not described in the text. This was done in order to keep the book as affordable as possible. I have included a text file that accompanies each of the extra examples that gives a short explanation of their use and utility. I hope that you find them to be useful. Several of these have been furnished by readers and are included with their approval. Here are some of the topics that are included in this supplementary set of files:

- **ShareCom**. An ActiveX EXE (OLE Server) that allows a single serial port to be shared between several Client applications.

- **RASConnect**. A simple VB/32 project similar to CDMonitor in Chapter 8. The difference is that it uses API methods and is limited to DUN or RAS modem connections.

- **RASHangup**. VB/32 code similar to the Hangup programs in the book. This code will only work with RAS connections. However, it does not require an add-on DLL.

- **GPS4VB3, GPS4VB4, and GPS4VB5-6**. GPS projects that further illustrate decoding NMEA-0183 sentences. Included are displays of Universal Time, Latitude, Longitude, Speed, and Course over the ground, Altitude, and Satellites in use.

- **CommProtocol**. An example of a simple communications protocol that might be appropriate for a dedicated application. This sort of protocol might be appropriate for controlling a device or devices on a dedicated serial network where you control the software implementation for all devices. For example, this might be an RS-485 network that communicates with embedded controllers. If you do not have to adhere to any specific standard protocol, this example shows how to implement one that has several desirable features.

- **Comm Spy**. A project written by Leon Kenison, a student in the Computer Engineering Technology programs at New Hampshire Technical Institute. See the accompanying files for more information

- **TapiClass**. A set of class modules, and an example VB5 project that substitutes class objects for Crescent's PDQTapi. TapiClass was written by Will Fookes. These class modules improve on certain characteristics of PDQTapi. These class modules are designed to be used with PDQComm.

- **XMODEM.TXT**. This file describes the XMODEM and Ymodem error-checked file transfer protocols.

- **9BitData**. A short discussion of how to tackle the 9-bit data problem (PC serial port hardware supports data bits up to 8).

- **Access**. Several Access databases that employ the NETComm.ocx to add serial communications.

- **AppendBinaryArrays**. A project that illustrates how to append binary data in arrays, in a way similar to the way one string might be appended to another.

- **ASP**. Illustrates the use of the NETComm.ocx serial communications ActiveX control in a simple web client (browser) application.

- **CreatePacketSend**. A simple communications protocol for a specific serial hardware system (laser table positioner).

- **Excel**. Use of NETComm.ocx to add serial communications to Excel worksheets.

- **FileCompression.** Various file compression methods, including ZIP.

- **FindTAPIModem.** Use TAPI to identify connected modems.

- **IntelToMotorols.** Convert Intel floating-point values to Motorola floating-point.

- **Scales.** VB and VBA examples that interface with various scales (industrial weighing devices).

- **WindowsMobile5.** Compact Framework code for Visual Studio 2005.

- **GSM INTERFACE SPECIFICATION.** Cellnet's Short Message Service Centre (SMSC) TAP interface provides external companies with the facility to submit short messages of up to 160 characters to GSM mobile subscribers and subsequently to determine the status of those messages. I have added a folder specifically for the 3rd and later Editions that have additional files, folders and utilities. Here are some of these:

- **Zmodem.txt**. This file contains a description of the Zmodem file transfer protocol. It could be used to help a reader implement this protocol. However, anyone really interested in this should consider the comments that I make on this subject in the chapter that deals with file transfer protocols

- **Breakout** is a complete setup program for a utility that I have written. It uses direct port I/O to read the serial port UART modem control and modem status registers. The resultant data is displayed both as a state indication (similar to the LEDs on a breakout box) and as a time-domain logic scope display of the status of the RTS, DTR, CTS, DSR, CXD, and RI lines of the serial port. The Breakout program works with all 32-bit versions of Windows (9x/Me and Windows NT/2K/XP). See the README file that is included for a complete description.

- **SDA322** contains source code for two projects that work with the B&B Electronics (see Appendix A) SDA line of serial data acquisition modules. One project is an ActiveX control that provides an easy way to interface to these modules for PC-based applications. The other is an ActiveX EXE that provides a similar interface that might be called from a Visual Interdev (ASP) or equivalent application so that the same SDA data can be accessed by a browser via the Internet.

- **mComm** contains a class module that encapsulates MSComm32.ocx. The unusual aspect to this code is that it allows MSComm to be used **without** the need to place an instance of MSComm on a form. I have had lots of people ask for this (though, I admit, I have never found a need to do this).

- **mCommTLB** is a class module that encapsulates MSComm32.ocx (see above), but that uses a Type Library implementation of the ClassBuilder to eliminate licensing issues seen when using MSComm without a form.

- **CalculateEvenParity** illustrates parity calculation in code.

- **ConversionRoutines** illustrates numeric conversions that may be needed when dealing with external systems.

- **EnumPorts** enumerates installed hardware ports. These include serial, parallel, and network ports.

- **SendMail** is an illustration of various network email functions. This code was written by Monte Hansen.

- **SimpleMAPI** is an illustration of MAPI written by Michael Kaplan.

- **X10** contains information and examples covering the popular X10 home networking/control system.

- **NETComm** is the source code form an ActiveX control that wraps the functionality of MSComm32.ocx, so that it may be used with on license restrictions in Visual Studio .NET, Access, Excel, Visio, or other ActiveX clients. Simple Access database and Excel examples are included. NETTerm is a terminal example that illustrates using NETComm.ocx in VB .NET.

- **VT100** is a VB6 terminal emulation that employs MSComm32.ocx.

- **VBTerm.net** is a port of the VBTerm example to .NET. This port used the Upgrade Wizard, though I made a few manual changes.

- **DesktopSerialIO** provides a simple but powerful serial object for Visual Studio 2002/2003 (located in the Chapter 5 folder).

- **XMCommNET** is the XMODEM file transfer protocol implemented using calls to the Windows serial API and is based on the serial code in DesktopSerialIO (located in the Chapter 5 folder).

- **VirtualSerialPort and DataMonitor.** This is a utility that I wrote to use a pair of virtual serial ports **and** a hardware serial port to facilitate testing and debugging of serial applications. Find it in the Serial Communications Products folder under HardAndSoftware on the CD ROM.

If there is a feature that you want to use in your application that is not implemented in one of the programs in this book, it may be in another. So, look around.

Acknowledgments

I have had a number of people who have added considerably to the technical content of this book. I would like to thank them all, in no particular order.

Daniel Appleman provided the original code for the 32-bit Windows API chapter. The API32Term program appeared first in his book *Visual Basic Programmer's Guide to the Win32 API*, ZD Press. I highly recommend his book for anyone who needs to program using the API. This book follows the popular 16-bit version *Visual Basic Programmer's Guide to the Windows API*, ZD Press. I took the liberty to modify the program to match my preferences and prejudices but I appreciate the knowledge that Dan has shared with the VB community.

To compile a practical application that incorporates file transfers or emulation, you may want to purchase a commercial communications add-on. A number of such products are described in the book. Any of these might be used, with appropriate modification of the code that I have furnished, which rely on Sax Comm Objects. Often these modifications will be minor and I will outline the differences between add-ons in each section.

Ron Tanner furnished the 16 and 32-bit Power Page DLLs that are included on the CD ROM for alphanumeric paging. However, these DLLs are not supported.

Jonathan Wood furnished the VBASM.DLL for 16-bit applications. This DLL complements the WIN95IO.DLL that I have included on the CD ROM for 32-bit access to I/O ports under Windows 95/98. VBASM.DLL is freeware. Thanks to SoftCircuits and Jonathan Wood.

Jim Stewart of JK Microsystems furnished a Flashlite Single Board Computer embedded PC so that I could develop the Flashlite Control program. The quid pro quo was that JK Microsystems could include the program that I developed for the Flashlite on their Utilities disk and on their Web site, to benefit all of their customers. Fair all around, I think.

I want to thank Jim Mack of MicroDexterity for the BuffToHex functions in the RGUTIL16.DLL and RGUTIL32.DLL files that I have included on the CD ROM (see the Debugging chapter). MicroDexterity sells the outstanding product named Stamina. It features many add-on functions that can be invaluable for serial communications programs.

Karl Peterson, a friend and fellow MVP, has provided dozens of code examples and ideas on his web site (www.mvps.org/vb) that are valuable. I have incorporated one or two of his ideas in the code in this book.

I want to repeat my appreciation to James Shields and Ruth James at Mabry Software. Without their support I would not have been able to publish this book.

Last, I thank all readers of the First, Second, and Third Editions of this book who took the time to contact me with improvements and corrections. I would like to give special thanks to Ronald Frakes, Terrance Simkin, Will Fookes, and Jeffrey Schnell. I received valuable comments on the Second Edition from Dan Karmann and Fabio Varriale. Valuable input from John Kozee was used in the Fourth Edition. Thank you.

How to Contact Me

Richard Grier
Hard & Software
12962 West Louisiana Avenue
Lakewood, CO 80228

303-986-2179 (voice)
(303)593-9315(fax or voice mail)
Dick_Grier@msn.com (email)
dick_grier@hotmail.com (email)
www.hardandsoftware.com
www.hardandsoftware.net

Chapter 1   Serial Communications

1.1   Background

There are two basic ways to move data between computers or other digital systems. One way is by parallel transmission; the other is by serial transmission.

Parallel transmission sends four, eight, or more bits at the same time. This means that the number of wires needed is equal to the number of bits that are sent. Additional wires are needed to control the flow of data and to synchronize the two systems.

Parallel data transmission can be very fast because a number of bits arrive at the destination at once. However, it can also be expensive because of the number of wires used to convey the data. And, transmission of parallel data is not inherently compatible with modems. See Appendix A for information on Jan Axelson's book *Parallel Port Complete.* Axelson's book is a good resource for VB programmers interested in parallel I/O.

Serial transmission sends data between computers one bit at a time. In general, only a single wire is needed to transmit the data. Actually, two wires are required because a complete electrical path must exist for actually transmitting the signal. In an unbalanced system, the return or ground wire can be shared for all signals. (For a discussion of balanced systems, see the section on RS-422 and RS-485.) If data is to be sent bi-directionally, three wires are needed. And, there may be additional wires required for other functions.

Serial data is also inherently in a form that can be used by a modem or equivalent device to send data over long distances, often by telephone lines.

Serial data is sent one bit at a time. A mechanism is needed to determine when a data bit is in a high or low state. This is done by electronic circuitry on the receiving side that samples each bit position to determine if it is a zero or a one. This sampling needs to be as close as possible to the center of the bit cell so that any possible jitter in the timing of the bit does not cause the sample to move to an adjacent bit. If the hardware samples the wrong bit, an error occurs.

There are two ways to make certain that the sampling of a bit is correct ----- ways to assure that jitter does not cause an error.

The first technique is with synchronous serial data. Synchronous data is sent with a clock signal that is used by the receiving hardware to decode the serial data. See *Figure1.1*. If these were oscilloscope traces, the upper trace would represent serial data and the lower the accompanying clock signal. The decoded data is (LSB to MSB), 1011010010001110101100.

To extract data from a synchronous data stream, it is necessary to define a **unique** data pattern that represents the start of a series of bytes and a **unique** data pattern that represents the end of the series of bytes. Various patterns are used. That discussion goes beyond the scope of this book. However, those interested can look up SDLC (Synchronous Data Link Protocol), HDLC (High level Data Link Protocol) or Bisync, three popular such protocols, in other references on serial data communications.

It is also possible to encode asynchronous data in a synchronous data stream. This is what modems in asynchronous mode do. Then data is decoded bit for bit from the synchronous data stream with the asynchronous nature of the data restored after decoding.

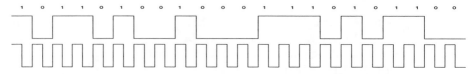

Figure 1-1. *Synchronous Data and Clock*

When a modem is used to send serial data, the clock is combined with the data to modulate an analog signal. The clock is derived from the analog signal that is received and used to decode the synchronous data stream. If the modem is in synchronous mode, that clock is often output on the modem serial port and looks like that shown in *Figure 1-1*. If the modem is in asynchronous mode, the clock that is derived from the receive analog signal is used to decode the data but is not output on the modem serial port. See Chapter 2 for additional information on how modems work.

I will not discuss synchronous serial data in any more detail because Windows has no built-in support for synchronous serial hardware. And, synchronous serial hardware and applications for the PC are uncommon.

The second way to assure that data is correctly extracted from the serial data stream is to use asynchronous serial data. This method adds a pair of synchronizing bits to each data character that is sent. The first bit of a serial character is termed a **START BIT**. After the start bit, several data bits are sent. Depending on the word length and parity that is used, this data is from 5 to 9 bits in length. Then a **STOP BIT** is sent. A start bit is a digital **0** and a stop bit is a digital **1**. See *Figure 1.2* for an example of asynchronous serial data. The three asynchronous characters that are illustrated are 01101001 (&H69), 10110010 (&HB2), and 11001000 (&HC8), assuming no parity bit. **S** represents the start bit and **s** represents the stop bit.

You may also see the terms **Mark** and **Space** to represent the signal levels on an RS-232 port. Mark = logic 1 or a stop bit. Space = logic 0 or a start bit.

Note that these characters are not shown to be "connected" between start and stop bits. While asynchronous characters may be sent back to back with a start bit immediately following a stop bit, this sequence is not required. Each asynchronous character stands alone. The actual state between characters is a 1, the same as a stop bit. This assures that a transition from 1 to a 0 is recognized as the start bit of the next character. As the name "asynchronous" implies, there is no specific time interval between characters.

Figure 1-2 *Asynchronous Data*

One note should be made about the diagrams of serial data in *Figures 1-1* and *2-1*. These figures show a logic 1 as "low" and a logic 0 as "high". The actual electrical levels that represent 0 and 1 in serial communications depend on where the data is measured. At the logic levels within the PC, a logic 1 is represented by a high level while a logic 0 is represented by a low level. RS-232, on the other hand, inverts these levels. So make sure that you know what you are looking at, before you assume that you know what it means! The levels illustrated in this diagram are RS-232 levels.

The serial port hardware detects a start bit (a transition from Mark to Space, or from 1 to 0). When a start bit is detected, the hardware starts a sample clock that samples the serial data at 16 times the selected bit rate. If the Start bit does not stay in the same state long enough, the hardware decides that it was a false transition, perhaps caused by noise. If the start bit is validated, each subsequent bit is sampled using the same 16X clock and is shifted in to a shift register to convert it to parallel form. The hardware also samples the stop bit to verify that it is the correct state (Mark, or 1).

If the serial data does not meet the expected configuration, an error is generated. The error might be a framing error if the start and stop bits are not detected with the correct time relationship. A parity error is generated if the parity of the receive data does not match the selected parity (see the discussion of parity later in this section).

Figure 1-3 shows a more generic representation of an asynchronous data character.

Again, **S** represents the start bit, **Data** is the actual character, **P** is a parity bit (if used), and **s** is the stop bit. From now on, this is a mental picture that will be valuable. We often are not concerned with the actual electrical signals that represent serial data but with the bit pattern of 1's and 0's that represent the data in binary form. This diagram also lets us represent the data more naturally with the LSB as the right-most bit and the MSB as the left-most bit.

S	P *	Data	S

Figure 1-3 *Asynchronous Data Character*

The start, stop, and parity bits are added by the serial port hardware to the data character that you send. When a character is received, these extra bits are discarded and the serial port returns only the data bits of the complete data character.

The parity bit provides a rudimentary form of error detection. If parity is used, the serial port hardware automatically places the parity bit between the data character and the stop bit.

Parity is a term that refers to the number of bits in the data character that are a 1, excluding the start and stop bits. If the number of 1's in the character is even then the character has even parity. If the number of 1's in the character is odd then the character has odd parity.

If parity is enabled, the serial port hardware counts the number of 1's in the character. The hardware then adds a parity bit of 1 or 0, such that the total number of bits (including the parity bit) is even or odd, for even or odd parity respectively. The complete asynchronous character, with start, parity, and stop bits added, is then sent.

Mark or Space parity is a variation of parity that occasionally is used. If Mark parity is selected, the parity bit is always set (1). If Space parity is selected, the parity bit is always reset (0). This contrasts with the normal parity in which the number of bits in the character is used to set or reset the parity bit.

If parity is enabled, the serial port hardware receives each asynchronous character and calculates the parity of the received data. If the data bits and the parity bit do not match the selected parity, a parity error is detected. This error then is reported to your software. Both computers in the communications link must use the same parity. If parity is set to **None**, the parity bit is not used. As I noted previously, the number of bits for each character can vary from 5 to 8 bits. The serial port hardware must be configured with the correct number of data bits per character. Both computers in the communications link must be configured to use the same number of data bits per character.

Simple arithmetic shows that the total number of bits in an asynchronous character can be as few as 7 bits and as many as 11 bits. A character can have more than 11 bits if more than 1 stop bit is selected. See the paragraph below concerning stop bits. Also, see the section on Windows and the modem chapter for other restrictions in asynchronous character length.

The break signal is a state where the serial output sends continuous logical 0's. A normal break signal consists of 20 (or more) consecutive logical 0's. Each asynchronous character consists of at least one bit set to a logic 1 (the stop bit). So, a break signal is not an asynchronous character. It is a special purpose signal. A break signal can be interpreted in several different ways that depend on the design of the communications system. The serial port hardware sends and detects break signals. The break signal is not commonly used in PC communications systems.

The stop bit is used to signal the end of an asynchronous character. It is possible to configure the hardware for 1, 11/2, or 2 stop bits. One and one-half and 2 stop bits are needed only for teletypes and similar electromechanical systems. Modern systems should not need to use more than 1 stop bit. Both systems should be configured for the same number of stop bits, although there usually is enough flexibility in the design of serial port hardware that a difference in this area will be ignored. Using more stop bits than required slows data throughput.

1.2   Half-duplex And Full-duplex Transmission

Serial data might be sent in only one direction — from one device to another but not back. An example of unidirectional transmission is from a PC to a printer.

Usually, however, data is sent in both directions. When data is sent in both directions, say between two computers, there are two methods that can be used. The first is called half-duplex transmission. Half-duplex transmission sends data in one direction at a time. Full-duplex transmission allows data to be sent in both directions at the same time.

Most modems that are used with Windows PCs support full-duplex transmission. Data can be sent and received at the same time. However, there are some file-transfer protocols that are half-duplex protocols. Both half-duplex and full-duplex file-transfer protocols are discussed later in the book

Some hardware that you may need to use also requires half-duplex communications. An example of this is RS-485. RS-485 will be discussed later in the book.

1.3   Windows And Serial Communications

Serial communications in a multitasking environment is challenging. When serial data is received by the computer hardware, your application may not be executing. It may be suspended, waiting for Windows to return control so that it can continue. Other programs and Windows itself all require processing time so that each can run. Windows must allow for the possibility that your application will not be ready and buffer the data until it is running.

Also, Windows must provide "virtual access" to the communications hardware. It would be a poor idea to allow applications to directly access the hardware. If applications directly access the hardware, one application could interfere with the operation of another that attempted to use the same hardware. At the least, this would cause data errors. At worst, it might cause a fatal error that halts your application or even Windows itself.

1.3.1   Interrupts (IRQs)

Each time you boot your PC, the system BIOS checks your COM ports to see what serial devices are installed on your PC and copies this information to the system BIOS data area (BDA). Windows "reads" the BDA and uses the default interrupt request line (IRQ) for each device that is registered there.

Usually, each COM port requires a unique IRQ to communicate with your PC. Certain PCs and some I/O cards support IRQ sharing. **However, although this feature is supported by some hardware and some versions of COMM.DRV, it should not be used.**

Some early BIOS versions may not check for devices on COM ports 3 and 4. If you have devices installed on COM port 3 or 4 and your system BIOS does not recognize these ports, you need to register the addresses for COM3 and COM4 by using the Windows Control Panel.

To register devices in Windows 3.x, select the Control Panel icon in the Program Manager Main Group. Then select the Ports icon and click on the port to be registered. Click on the Settings and then the Advanced button to enter the correct I/O port address and IRQ in the list boxes that are displayed. For comparable registration in Windows 95/98, select Settings from the Start menu and then the Control Panel icon. Next select the System icon and click on the Device Manager tab. From the Tree that is displayed, select the communications port. Then click on the Resources tab. If the information that is displayed needs modifying, select the port and click on the Change Setting button. These instructions will vary slightly depending on the version of Windows that you are using. Nonetheless, the steps will be similar to those outlined here.

Generally, PCs come with built-in ports COM1 and COM2 preset to the following values:

Port	Address	IRQ
COM1	03F8	4
COM2	02F8	3
COM3	03E8	4
COM3 (PS/2)	3220	3
COM4	02E8	3

Because COM1 and COM3 both use IRQ4 as the default, and COM2 and COM4 both use IRQ3, you may need to reassign the IRQ if you use serial devices (such as a fax card or modem) on COM ports 3 or 4. You can reassign IRQs by using Control Panel as described above. Of course, you must configure the actual serial port hardware to agree with whatever change you have made.

The Windows 95/98 and Windows NT 4.0 (or later versions of these operating systems like Windows XP and Windows 2003) "Plug and Play" feature may require that you use the Windows Add Hardware Icon to add or change hardware. Some wags, never I, have suggested that this feature should be called "Plug and Pray." It is not as reliable as it should be so it is often better to configure hardware for a specific I/O address and IRQ rather than depending on "Plug and Play." Each succeeding version of Windows has made "Plug and Play" more reliable. However, it still has limitations that may require that you adjust some settings manually. **Always** check your hardware manufacturer's web site for up-to-date drivers.

The following table shows the most common IRQ settings. You can use Control Panel to specify an IRQ from this table if it matches your hardware requirements.

IRQ	Description
0	Timer
1	Keyboard
2	Link to IRQs 8-15
3	COM2, COM4
4	COM1, COM3
5*	LPT2, or Reserved
6	Floppy disk controller
7*	LPT1, LPT3
8	Real time clock
9	Redirected IRQ2
10*	Reserved
11*	Reserved
12	PS/2 mouse
13	Math coprocessor
14*	Hard disk controller
15*	Reserved

The IRQs marked with an asterisk "*"are often available. (This does not guarantee they are currently available on your PC.) If you encounter problems with IRQ conflicts, see the section on Debugging.

1.3.2   COMM.DRV

COMM.DRV is the serial communications driver that is furnished with Windows 3.x. This driver virtualizes the serial communications port hardware. It has the interrupt service routines that manage the hardware, the routines that provide data buffering, and routines that interface to the Windows kernel. These routines are the lowest level (closest to the hardware) embodiment of the Windows serial communications API (Application Programming Interface).

In this book, COMM.DRV is a somewhat generic term. There are commercial products that replace the Windows 3.x COMM.DRV with a higher performance driver. One example is TURBOCOM.DRV (see the section on TurboCommander Pro). Other versions of Windows may use other files that have essentially the same functions that are in COMM.DRV, e.g., SERIAL.VXD and SERIALUI.DLL in Windows 95/98 or Windows NT/2K/XP. For the sake of simplicity, I will speak of COMM.DRV with the understanding that I mean all such virtual device drivers.

The serial communications drivers that have been used since Windows 3.1 have supported the FIFO (First In First Out buffers) in 16550 AF UARTs (Universal Asynchronous Receiver Transmitter). These FIFO are important hardware features that help avoid receive data overruns and improve throughput in a multitasking environment.

Commercial replacements for COMM.DRV often provide a program that allows optimization of the FIFO settings. A 16550 AF UART has a programmable interrupt trigger level that determines how COMM.DRV uses the UART FIFO. There are two FIFOs, one to receive data and one to transmit data. Each FIFO can hold up to 16 bytes.

The interrupt trigger level setting for the receive FIFO is fixed in COMM.DRV but may be programmable in replacement drivers and the drivers used with Windows 95/98 or Windows NT. The receive trigger level is 8 bytes. An interrupt is generated when there are 8 or more bytes in the buffer. This requests COMM.DRV to empty the FIFO. If more than 8 additional bytes are received before COMM.DRV empties the FIFO, an overrun error will be generated. The default in Windows 95/98/NT/2K/XP is 14 bytes.

If you experience data overrun errors using Windows 95/98/NT or a commercial replacement for COMM.DRV, you may want to adjust the trigger level. You might reduce the level by 50%, from 8 to 4 on Windows 3.1 or from 14 to 7 or 8 on Windows 95/98/NT/2K/XP, to see if this improves the reliability.

Although there is a 16-byte transmit FIFO in the 16550 AF UART, the FIFO itself has no programmable transmit threshold. An interrupt is only generated when the transmit FIFO is empty. COMM.DRV places only a single byte in the FIFO for each transmit-empty interrupt. As soon as it is sent, an interrupt is generated that must be serviced by COMM.DRV. The default in Windows 95/98 is to add as many as 16 bytes to the FIFO (if available). Most commercial replacements for COMM.DRV on Windows 3.x allow you to increase this buffer level. Such an increase can result in a marginal improvement in multitasking performance.

1.3.3   UARTs

UART is an acronym for Universal Asynchronous Receiver Transmitter. The UART is the heart of a PC serial port. Internal modems incorporate a UART identical in function to that in a standard serial port. So, internal modems will be discussed as though they were external and were connected to a conventional serial port with a cable. There are a couple of important distinctions between internal and external modems. The differences will be addressed in the modem chapter.

The UART performs serial-to-parallel conversion of data received from a peripheral device such as a modem and the parallel-to-serial conversion of data. The parallel data consists of bytes (8 bits) and is what the CPU (central Processing Unit) of the PC can handle. The CPU can read the complete status of the UART. Status information includes the type and condition of data transfer operations being done by the UART as well as error conditions (parity, overrun, framing, or break).

The UART includes a programmable bit rate generator (also known as a baud rate generator) that divides the timing reference clock by divisors ranging from 1 to 65535. This bit rate generator produces a sample clock that is 16 times the actual serial bit rate.

The UART includes modem control capability. That is, modem control signals like RTS (Request To Send), CTS (Clear To Send), DSR (Data Set Ready), CD (Carrier Detect), DTR (Data Terminal Ready), and RI (Ring Indicate) are provided.

Two types of UARTs are commonly found in PCs. The original PC used an 8250 UART. In addition to the serial shift registers, this UART has just a single buffer for receive and transmit data. It is also limited to a maximum serial rate of 57.6k bps. The other common UART is a 16550 AF UART. This UART has a serial shift register and 16-byte FIFOs to buffer transmit and receive data. The maximum speed supported by the 16550 AF UART is 256k bps.

There are a few manufacturers of newer UARTs (16650 and 16750 UARTs, to name two) that have deeper FIFOs than those in the 16550 AF. There are UARTs with 32 and 64 byte FIFOs and are available from different manufacturers. These extra-deep FIFOs are not supported by any version of Windows at this writing. However, there are some commercial communications replacement drivers available that do support these UARTs. These UARTs will provide improved performance and reliability under some circumstances. These newer UARTs add some new features such as programmable transmit FIFO trigger level and hardware support for flow control. Hardware support for RTS/CTS and Xon/Xoff flow control can improve reliability and remove that task from the communications driver, resulting in enhanced multitasking performance.

The actual maximum speed that your serial port hardware will support is also a function of the design of the clock circuitry used by the UART. Even though a 16550 AF UART supports speeds up to 256k bps, often the hardware design of the clock limits the actual speed to 115k bps. Some of the code in this book assumes such a limitation. If you are interested in testing hardware for its maximum speed, a program is also included for that purpose.

See 1.3.6 for information on Virtual Serial Ports and software UARTs.

Refer to Appendix D for information on the internal registers of 8250 and 16550 AF UARTs. You can use this information to manipulate the UARTs manually and thereby overcome some of the limitations imposed by Windows and MSCOMM.

1.3.4   Flow Control

Computers can send serial data faster than connected devices can receive and process them. If this happens, data is lost. This is troublesome at least, and fatal at worst. If possible, we should attempt to assure that this data loss does not happen. That is where flow control comes into play.

If a serial device (perhaps a computer or some other device in the communications link, e.g. a modem) detects that data is approaching a point where it might overflow the receive buffer, it can request that the sending system stop transmitting data for a while. This gives the receiving device time to process data already received. When the receiving device has processed some or all of the data previously buffered, it can signal the sending system to continue to send data. This signaling system is called flow control.

Flow control comes in two flavors. These are commonly called hardware and software flow control. Sometimes, for reasons that will become obvious, hardware flow control is called out-of-band flow control and software flow control is called in-band flow control.

1.3.4.1   Hardware Flow Control

A complete discussion of hardware flow control requires the introduction of two additional terms. These terms are DTE (Data Terminal Equipment) and DCE (Data Communications Equipment). These are old-fashioned terms that almost serve to confuse as much as to elucidate. However, they are what we have to work with — so work, we will.

DTE implies a terminal (or computer), and DCE implies a modem. A DTE often connects to another DTE using an intervening DCE. Serial communications links are not bi-directional; you can only send or receive a signal on a single wire. DTE and DCE indicate which wires are used to send and receive signals. Diagrams of the various connections for DTE and DCE devices are shown in the sections on Null Modems, Cables, And Adapters. Your PC and all computers almost always are treated as a DTE. The following discussion will assume that is the case.

Hardware flow control uses a separate pair of wires on the serial port to perform the signaling between the connected devices. Usually the two signals are called RTS (Request To Send) and CTS (Clear To Send). Occasionally DSR (Data Set Ready) and DTR (Data Terminal Ready) are used instead of CTS and RTS, respectively. In rare cases there is a mix of these two pairs of signals. I will show one such oddball connection in the section on Null Modems, Cables, And Adapters.

When a receiving DTE needs to signal the sending device that data flow should stop, it lowers the RTS line. A DTE raises the RTS line when it is able to receive data. Likewise, a DTE monitors the CTS line. If CTS is lowered, the connected device is signaling that it cannot receive much more data so the DTE is obligated to stop sending data. When CTS is raised, the DTE can resume sending data.

On the other hand, a DCE reverses the meaning of these control signals. If a DCE needs to halt the flow of data, it lowers CTS. It then raises CTS to permit data to flow. Likewise, a DCE monitors RTS. If RTS is lowered, no data should be sent by the DCE while RTS high indicates that data may be sent.

The sections on RS-232 and Null Modems, Cables And Adapters cover some additional details on hardware flow control and the associated signal lines.

1.3.4.2   Software Flow Control

One of the features of serial data transmission is that it requires only one wire to send data in each direction (ignoring signal ground). However, hardware flow control adds an additional path for each control signal. Perhaps there is another way. There is such a way, of course. That's software flow control. When a device needs to control the flow of data, it might be possible to define a special data character that it can send to halt or restore the data flow. There have been a couple of different pairs of characters defined to do this. The most common form is called XON/XOFF. An XOFF character (also called DC3 or Device Code 3) is sent to signal that no more data should be transmitted. XON (also called DC1 or Device Code 1) is sent to signal that data can flow again.

Software flow control is also known as in-band flow control because this control acts just like other serial data; it is sent on the same signal wires as other data. Software on each end of the path must respond appropriately to these special characters to suspend and restart the data flow.

1.3.4.3   Which Flow Control Method Should I Use?

Like most simple questions, this one has no absolute answer. Both methods have advantages and disadvantages. COMM.DRV has built-in support for both types of flow control.

The advantage to software flow control is that it requires no extra signals to operate. The disadvantage is that it requires software overhead to execute so it can be slower and less reliable than hardware flow control. More importantly, the software flow control is limited to situations where the characters that are used are available, that is, they are not data. So, software flow control is not appropriate when transferring binary data. In fact, one file-transfer protocol (Kermit) goes to great lengths to allow software flow control when transferring binary data. This results in a substantial reduction in performance.

An advantage to hardware flow control is that it is fast. The UARTs themselves interpret changes of state in the input signals and can generate an interrupt that COMM.DRV can react to immediately. Hardware flow control is out-of-band, so it is inherently compatible with binary data. But, the extra control signals required need special handling when it comes to using them with data sent over the telephone system.

So, the answer is to use software flow control only if the system that you are communicating with requires it. Otherwise, use hardware flow control. The advantages of hardware flow control are significant and, as we will see in the chapter on modems, there are ways to make it work over the telephone system.

1.3.5   Multiport Serial Cards

Most PCs come with two standard serial ports. These RS-232 ports are designated Com1 and Com2. What do you do when you need more than the basic two ports?

You can add additional ports, often two at a time, using conventional serial add-in cards. You can also add one or more internal modem cards. The only restriction is that each new port should have an independent IRQ so that one serial port does not interfere with another.

However, this may not be a practical solution. Most PCs have too few expansion slots to use them indiscriminately. If you need more than four serial ports, you may want to consider the use of a multiport serial card. Several vendors of multiport serial cards are listed in the Resources appendix. Cards with 4, 8, and 16 serial ports are fairly common.

If the multiport card is to be used under Windows 3.x then the card manufacturer should furnish a replacement for the Windows COMM.DRV. COMM.DRV is designed to work with only Com1-4 so is not compatible with multiport cards. Also, the MSCOMM custom control allows the assignment of nine ports but only four ports are actually available. If you use a multiport card, you should consider using a commercial communications add-on instead of MSCOMM.

Multiport cards are available in two basic types: so-called "dumb" boards and "intelligent" boards. Intelligent boards incorporate a dedicated microprocessor and additional memory for buffering. Dumb boards are simple extensions to standard serial hardware and offer few enhancements, except that they have independent IRQs.

Intelligent serial boards are not always superior to dumb multiport boards. They often have speed limitations that reduce their utility. The additional level of complexity adds one more place for software and hardware faults. The level of performance that intelligent boards promise may be better achieved by using multiple computers and dumb serial boards, all linked with a LAN. The redundancy that this offers is substantial and the cost may not exceed that of a single computer with intelligent boards.

Sometimes multiport serial cards are connected to external modems. An alternative to this is the use of an internal card with multiple modems on the card. There are some vendors of these cards listed in the Resources appendix. Cards with 2, 4 and 8 modems are available.

Another way to add multiple ports to post-Windows 95 computers is with USB serial port adapters. A number of manufacturers provide four or more serial ports in one USB-connected device.

1.3.6   Virtual Serial Ports

Virtual serial ports are software driver emulations of the hardware that comprises a standard serial port. This means that the UART that normally provides conversion of parallel data to an asynchronous serial data stream is replaced by a device driver that appears to provide Windows applications all of the facilities of a conventional UART. Windows treats the virtual serial port as a standard hardware-based port.

Virtual serial ports are a vital part of the USB serial port adapters and USB modems. The actual USB adapter provides connector that connects to the PC USB port, an RS-232C connector to connect to an external serial device, and hardware that provides the electrical interfaces required. The electrical interface covers both the USB connection, and the serial connection. Some USB serial adapters are available that support RS-422/485 in addition to RS-232. There is no physical UART; that part of the adapter is provided by the "virtual serial port" device driver. Likewise, a USB modem has no physical UART, only virtual. The actual comport number associated with the serial USB device is assigned when the device is installed. Since this device is "virtual," the assigned port may be changed to any other free port number using Control Panel. See the Debugging chapter for comments about problems that can occur when using virtual rather than physical serial ports.

Virtual serial ports also are an important part of many Ethernet-based serial servers. These servers can be used to extend serial connectivity range by using any TCP/IP network (potentially, world-wide using the Internet). A number of manufacturers and vendors of such systems will be found in the Resources Appendix. A device driver is installed on the PC that associates the IP address of the remotely connected port with a virtual serial port on the PC, so that conventional serial methods may be used to access that remote serial port. Many Ethernet serial servers support RS-232, RS-422, and RS-485 interfaces.

There also are virtual serial ports that **truly** are "virtual." That is, there may be no physical hardware associated with them. These may be configured to provide two or more comports on a PC that are "wired" together via the driver. Thus, a virtual serial port may be opened by one application and a second virtual serial port may be opened by a second application. Serial data sent by one application will then be received by the other application — all without either application becoming aware that no physical serial hardware is being used. It also is possible to use such a device driver to broadcast serial information to multiple receiving applications, all without additional hardware or cables. This sort of device driver can be a useful tool for debugging and testing, and may find application in other more complex designs. See the Resources Appendix for more information on this type of application.

Virtual serial ports also can be incorporated into your own programs and need not rely on an external application. An example of this is the **Eltima**Virtual Serial Port ActiveX control. It allows your application to create custom additional virtual serial port in system and fully control it in code. A demo version of the Eltima VSP ActiveX control is included on the CD ROM that accompanies this book. This ActiveX control provides the API to interface to the associated virtual serial port device driver. I have included a small example program that illustrates its application. See the ExtraExamples/VirtualSerialPort folder.

Franson Technology AB provides Virtual Serial Port support in their Serial Tools package. A demo version of this fine product is included on the CD ROM. I recommend this product highly.

1.4   Error-corrected File Transfer

When sending data from one computer to another, errors inevitably will happen. The probability that an error will happen largely depends on the signal-to-noise ratio of the communications link. Communications theory shows how likely are such errors. Even if two computers are connected with a short cable, there is a small but calculable probability of error. The probability of error increases as noise increases and the resistance to error decreases with an increase in speed of transfer. Errors become quite likely when you use high-speed modems.

The most important way to counter modem errors is to employ error correction in the modems. Modem error-correction is necessary but often is not sufficient.

One way to move a file from one computer to another over a serial link is to use an error-corrected file transfer protocol. There are many possible protocols. Some of the more popular are described in the next several sections.

It is possible to write software in Visual Basic that transfers files from one computer to another without using one of these popular protocols and without employing error-correction. The probability that an error will occur on a direct link between computers is small, and modem error-correction makes the probability of error small on most modem-connected links.

So, why use an error-corrected file transfer protocol?

There are errors that can happen that have nothing to do with a data error caused by noise. These errors are significant in a multitasking environment like Windows because multitasking increases the probability that they will happen. These are receive buffer overrun and UART overflow errors. An error-corrected file transfer protocol will recover from such an error that otherwise would require re-transmission of the entire file.

OK. Why use one of the standard protocols?

That's easy. All of the heavy lifting has been done for us. Others have worked out reliable methods that are used in the popular protocols. Those protocols have been encapsulated in DLLs, VBXs, and OCXs. That means that we do not have to debug the protocol itself but only the code that we design to use it. This work by others is worth its weight in gold.

I should make one point before proceeding. You cannot use (Kermit excepted) software (Xon/Xoff) flow control or handshaking when you use a file-transfer protocol. All other protocols would interpret the flow control characters as data and errors would be caused, perhaps even causing COMM.DRV to lock up.

The following sections describe the popular file-transfer protocols.

1.4.1   XMODEM

XMODEM (often spelled in all capitals) file transfer protocol, also called MODEM7 or XMODEM/Checksum, is an error-checking protocol originally developed in 1978 by Ward Christensen and made public domain. It is available on most BBSs and online systems that offer error-checking protocols. If a communications program offers more than one error-checking protocol, XMODEM is usually one of them.

XMODEM transfers files in 128-byte blocks. It adds an extra byte, called a checksum, to each block. The receiving system uses the checksum to calculate whether or not the block was correctly received. A fairly simple algorithm based on the contents of the block is used for this calculation. If the calculated block checksum does not agree with the checksum that accompanies the block, the receiving computer requests the sending computer to retransmit the packet by sending a single character called a NAK. A NAK is a Chr$(21). Otherwise, the receiver sends an ACK, Chr$(6). If the sending system receives an ACK, it transmits the next block. If the sending system receives neither a NAK nor an ACK within a specified time, it also retransmits the block.

This process is repeated for each block until the entire file is transferred or until the user aborts the transfer. It may also abort automatically if there are too many retries. A protocol that requires an acknowledgment before sending a subsequent block of data is known as half-duplex, even if it operates over a full-duplex communications link.

Chapter 4 includes code for a VB 5.0/VB 6.0 ActiveX custom control that implements XMODEM/checksum and XMODEM/CRC. The CD ROM that accompanies this book also includes a class module that implements XMODEM. You can incorporate this class module in your VB 4.0 or later programs. A VB2/3 example of the XMODEM protocol is in Appendix B.

The XMODEM protocol is the only file-transfer protocol that it is practical to implement in Visual Basic code. However, DLL and custom control implementations are easier.

XMODEM/Checksum offers approximately 96% reliability. That is, it detects and corrects approximately 96% of the errors that might be encountered. This is OK for a communications link that is relatively error free. It is not satisfactory for critical file transfers unless it is used over error-corrected links. Fortunately, all modems of recent vintage provide additional error-correction. This extra security makes XMODEM file transfers a viable alternative.

XMODEM CRC is an important variation. CRC (Cyclic Redundancy Check) replaces the single-byte checksum with a two-byte CRC. The calculation used to create the CRC is much more complex than that used in XMODEM/Checksum. This more complex calculation substantially improves error detection. XMODEM CRC offers approximately 99.9996% reliability. That is, approximately 99.9996% of all possible errors will be detected. This is satisfactory for most communications links.

XMODEM 1K is a variation on XMODEM CRC. The block length is increased from 128 bytes to 1024 bytes. This improves data throughput. XMODEM 1K is not a very popular protocol because other higher-performance protocols are available that are just as easy to implement.

1.4.2   YMODEM

Ymodem operates in a manner similar to that of XMODEM 1K. Ymodem pads a block with extra null (Chr$(0)) characters to make certain it is exactly 1024 bytes in size. The extra block size is not a consideration. However, the downloaded file usually is longer than the source file — this can be disconcerting and, in some cases, can cause trouble.

Ymodem/Batch provides a valuable feature. Multiple files can be selected for transfer. The name of the file to be downloaded need not be specified. It is included in the actual data that is transferred. Thus, the program on the receiving computer can be much more flexible. Like XMODEM, Ymodem/Batch is a half-duplex file transfer protocol.

Ymodem/G is a variation designed for error-free links. No acknowledgments are required, thus speeding transfer. This variation achieves speed similar to full-duplex protocols. Because acknowledgments are not required, problems with communications link latency (delays in transmission, such as satellite hops) are eliminated.

1.4.3   Kermit

Kermit, named after Kermit the Frog, is a popular file-transfer protocol on UNIX and some mainframe systems. It was created at Columbia University in 1981. There are many variations. This description will cover only the salient details.

Kermit is similar to XMODEM in that it transfers files in blocks — or packets, as they are referred to in Kermit documentation. Like all file transfer protocols, it uses a checksum for error detection.

Kermit differs from other protocols in several ways. For example, it can transfer files using 7 data bits. Where necessary, Kermit converts 8-bit characters in a file to 7-bit characters by stripping the 8th bit and sending it as a separate byte. Kermit also converts control characters into other ASCII characters so they can be transmitted. This means that software (Xon/Xoff) flow control can be used on the communications link without causing errors in the transferred file.

Another feature of Kermit is that its packet sizes can be changed to accommodate fixed packet sizes on a remote system or varying transmission conditions. A 96-byte packet size is standard. However, some implementations of Kermit do not allow variable-length packets.

Kermit programs can re-synchronize their transmissions if interrupted by noise or other faults, something that is not done well under XMODEM.

Kermit also allows "wild cards" as part of the batch transfer process

Some versions of Kermit implement rudimentary data compression, in which repeating characters in certain kinds of files are sent only once; this can result in a significant saving of time.

1.4.4   Zmodem

The Zmodem protocol was developed for the public domain by Omen Technology.

Zmodem operates much faster than most other protocols. It sends a continuous stream of data blocks without waiting for an ACK or NAK. Zmodem is a full-duplex protocol. That is, an ACK or NAK can be sent while the sending system is in the process of sending a subsequent block. ACKs and NAKs indicate a block number. If the sending system receives a NAK from the remote system, it re-sends that block and all subsequent blocks. If the sending system receives an ACK, it can discard all blocks up to and including the acknowledged block. This mechanism is known as a "sliding-window" algorithm. Because the ACK or NAK contains the block number that it references, it is several bytes long.

Zmodem blocks are 512 bytes in size. The protocol is relatively efficient because it is full duplex and because it operates well on links with significant latency (delays). Zmodem allows batch transfers, wild cards, and it includes the filename in the file protocol.

Zmodem is the most popular (non-Internet) file transfer protocol for PCs that is in use. It is the one that I prefer to use.

1.4.5   Other File Transfer Protocols

There are other file transfer protocols that I will not discuss in detail but which can be used on occasion.

One such protocol is **CompuServe-B**, often used to transfer files to and from the CompuServe ® online service. Visual Basic communications add-ons commonly support CompuServe-B. It is an efficient protocol, similar to Zmodem, but is optimized for packet switched networks. Its use is limited to communicating with CompuServe hosts.

Another protocol is **Sealink**, developed by System Enhancement Associates. Like Ymodem, it uses batch file transfers and includes the filename in the protocol. Like XMODEM, it uses a fixed-length 128-byte block. And, like Zmodem it uses a full-duplex sliding-window algorithm to overcome latency and to improve data throughput.

Protocols that are not mentioned in the preceding sections are either not implemented in any add-ons for Visual Basic or are designed for use over the Internet. Internet protocols are left for another time.

1.5   Terminal Emulation

Terminal emulation is software that allows your program to act like a standard terminal such as those that are used with minicomputer and mainframe systems. Applications that run on these hosts often interact with the terminal so that information can be displayed in appropriate areas onscreen with highlighting, color, or inverse video. These applications also can assign special meanings to the terminal function keys.

The host, using what is called an "escape sequence", controls character positioning and other screen attributes. That is because the Esc character (Chr$(27)) is used to signal the terminal that other characters that follow the Esc are used for screen attributes and are not to be displayed.

Popular emulations are ANSI, VT-52, VT-100, and VT-220 (Digital Equipment Corporation), and TV-910, TV- 912, and TV-925 (Televideo Corporation).

Most commercial communications add-on vendors provide several of these as built-in functions. This makes interfacing to big iron fairly simple.

You also can refer to the CD ROM for a VT-100 emulation written in VB6.

Chapter 2   Modems and Other Serial Devices

What is a modem?

The word modem stands for **mo**dulator **dem**odulator. Why it is called that and what it does is presented next.

The telephone network, referred to by the official acronym GSTN (General Switched Telephone Network) or sometimes PSTN (Public Switched Telephone Network) and often by the canard POTS (Plain Old Telephone System), is an analog system. I'll call it POTS from now on; I'm never one to stand on ceremony.

At its heart, in the telephone central offices and the trunks that interconnect them, a POTS often is digital. However, its extremities define its analog nature — the line that connects your modem to the telephone office. The actual telephone has gradually evolved from the device that Alexander Graham Bell invented to the one we use today. This evolution has not changed the basic nature of the device, except for ISDN (described later). Electrical signals are sent to and from the central office as alternating currents that represent the audio signal of a telephone conversation. Two wires are used for the circuit between you and the central office; these two wires carry the voice signal in both directions.

These analog signals intentionally are limited in their frequency range to between, approximately, 300 and 3300-3800 Hz. Historically, technological limitations of 100+ years ago, a desire to limit crosstalk, and later the need to reduce costs by limiting telephone bandwidth so that multiple signals could be multiplexed (combined), all have dictated this 3.5 KHz bandwidth limit.

Bandwidth is limited by low-pass filters ("brick wall" filters) at the point where the customer signal enters the telephone-switching network. A high-pass filter is also used to keep noise such as crosstalk from power lines (50 or 60 Hz) from interfering with the signals carried by the telephone system.

This bandwidth limitation is fine for voice signals but it is not very satisfactory for digital signals. Modems send data over POTS by converting the serial digital data from the computer to an analog signal to send over the telephone line. This is modulation. Analog signals from the telephone line are converted to serial digital data for input to a computer.

There are many different modulation techniques used by conventional modems. The actual details of these techniques are important to the designers of modems and telephone systems but an understanding of them is not needed to use them. An exception to this last statement is that it is often desirable to understand the theoretical and practical limitations of modems. If you misunderstand the limitations of the modems that you are using, you may expect more than is possible or you may misinterpret the source of some problem. So, I will discuss some of those limitations.

If you want to pursue the theoretical aspects of modems, I suggest that you get the most recent list of books from the IEEE Press. There are many good books there. One notable recent publication is mentioned in the Resources appendix.

I will limit my discussion of modem limitations to the most recent (as of this writing) standards. These limitations apply in greater or lesser degree to older designs and also will apply to future enhancements of current designs.

Claude Shannon and Warren Weaver published a book titled *The Mathematical Theory of Communication*. This book, and some earlier papers, developed what is now known as Shannon's theorem. My copy of the book was published in 1964 by Illini Books. I know there are some readers even older than I, so I do not feel too bad.

In so many words, Shannon's theorem tells us that the highest number of bits per second that a circuit can support is a function of the circuit's bandwidth and the signal-to-noise ratio. If we use Shannon's theorem to calculate the theoretical limit of common telephone lines, the highest data rate that can be approached is about 40k bps.

This limit is only approached with sophisticated trellis-code modulation schemes using multi-level phase and amplitude modulation. This allows each phase/amplitude combination of the signal to represent many bits. The most recent version of the ITU (International Telephone Union) V.34-1996 standard places 1664 points in the phase/amplitude envelope. If the bandwidth of the telephone line and the noise present allow it, this modulation scheme can encode 33.6k bps. This limit is seldom actually achieved and more complex schemes probably are not practical. The V.34-1996 standard also provides for a 200 bps auxiliary channel intended to convey modem control (such as flow control) data.

ITU standard V.90 modems use an asymmetric mode of operation when connected to the Internet that allows higher speed receipt of data from the Internet. Data transmitted still is sent at the raw connect negotiated connect speed. Often you will see rates in excess of 40 Kbps for receive data (53 Kbps is the "legal speed limit" for receive data but this speed is never seen in practice). These modems achieve their magic by replacing one of the analog segments that is a limiting factor in the normal modem communications with a purely digital link. This is done through the use of special modems in the telephone central office that do not perform one of the digital to analog conversions that would be done in conventional modem-to-modem communications.

V.90 (and higher numbered variants) modems can improve data throughput when downloading information from the Internet. However, they offer no advantages over V.34 modems when not used over the Internet.

What does all of this mean? The actual raw data rate that a modem can support depends on several things. The primary controlling element is the quality of the telephone lines that make up the circuit between the two ends of the system. The most recent modem standards allow the modems to test the maximum rate that the end-to-end connection will support and to set the raw speed to match that speed. An additional factor is the actual modem design. One modem manufacturer may simply do a better job than another. Like other areas, modem cost is not an exact guide to quality but there is a correlation.

In practical situations, high-speed (V.34 and later) modems can connect at raw data rates from 19.2k bps to 33.6k bps, all depending on modem compatibility and the capability of the telephone system to carry the data.

The sections on error-correction and data compression will discuss other factors that can affect modem performance.

There are several useful non-modem serial devices that also will be discussed in this chapter.

2.1   Smart Modems

About the time that microcomputers became popular, and before the advent of the first IBM PC, Dennis Hayes recognized that serial communications via modem was a niche that begged to be filled. His first product was called the Smart Modem 300. Its cost was low enough to be attractive to a large number of users. But its main claim to fame was the fact that it incorporated its own microprocessor that was programmed to respond to serial commands.

Modems had been developed before this that had built-in microprocessors that responded to serial commands. However, the comparatively low cost and the ease of use of the Smart Modem 300 quickly became a de facto industry standard. Even though this modem was limited to a maximum speed of 300 bps, it found lots of buyers.

Soon after the IBM PC was released, Hayes released the Smart Modem 1200. The Smart Modem 1200 used the same basic serial command set as the Smart Modem 300. And it also supported the much faster Bell 212 modem standard, allowing it to communicate at 1200 bps.

Manufacturing costs can be reduced by building internal modems instead of external modems that connect to the computer serial port. Internal modems plug into the computer bus meaning they do not need a separate power supply or case. They also do not need RS-232 interface circuitry or cables. The sales volume of internal modems is now more than three times that of external modems.

The mid-1980's brought the advent of 2400 bps modems, now using the international V.22bis standard. This higher speed increased the popularity of modems and the "Hayes" command set that was used in earlier modems was expanded to include new features demanded by these higher speed modems.

The step up in speed to 2400 bps made clear that one feature was needed in 2400 bps and higher speed modems. With the increase in speed, there was a parallel increase in the probability that data errors would be encountered. This could be a serious issue, even with error-corrected file transfers. If an error was encountered at a critical point at the start of a file transfer, before errors could be detected, the transfer might fail. Because errors were now an important issue, companies started work on modems with built-in error-correction. A leader in this field was Microcom Corporation. More details on modem-based error-correction are in the next section.

The CCITT (the Consultative Committee on International Telegraphy and Telephony), now called the ITU (the International Telephony Union) adopts international modem standards. These standards are intended to make modems and other data communications equipment inter-operable worldwide. The first international standard that saw wide use in the US was V.22bis. Although there have been occasional attempts to use proprietary non-standard modems since the mid-1980's, these have not been really successful.

The V.32 modem standard was adopted in mid-1980. V.32 modems supported communications at 9600 bps, using echo-cancellation and trellis-coded QAM modulation. These methods required substantial digital signal processing power and early units were very large and very expensive. However, by the end of the 1980's, the cost and size of V/32 modems were reduced enough to make them attractive for the common user.

By 1991, V.32bis modems (14.4k bps) became available. The next step up in power and speed came by 1994 with V.34 modems that supported 28.8k bps. We were near the point where V.34bis would be adopted, presumably to support 33.6k bps. With conventional, full-duplex schemes, this was probably the maximum practical modem speed for non-Internet applications. In 1988 the V.42 modem error-correction standard was adopted, and the V.42bis standard, which specifies data compression, was added shortly thereafter. More on these later.

In 1997 the V.90 modem standard was adopted. This allowed "56k bps" modems to communicate at somewhat higher speeds. See the description later in this chapter for limitations associated with this kind of modem.

None of these modem standards addressed the set of commands that are issued to modems over the serial link. One attempt at an international standard was X.25. It was awkward to use and the momentum of the de facto "Hayes standard" made X.25's widespread adoption impossible. So, all smart modems (modems with built-in intelligence) use a variation of the "standard" that Hayes developed.

The basic command set used by smart modems prefaces most modem commands with an "AT" and terminates a command with a carriage return (Chr$(13) or vbCr). Appendix F outlines this basic AT command set.

Each modem vendor added features that required new commands. With no standards-making body to guide these commands, each vendor was left with the task of creating them ad hoc. This has led to variations in commands from different vendors, and on occasion, even between models from a single vendor. Anarchy reins. What this means is that certain critical functions will be activated by using different commands for different modems.

Databases of recommended configurations for modems from various vendors are a selling point for some communications add-ons. I have compiled one that is on the CD ROM that accompanies this book.

Windows 95/98 and Windows NT 4.0 (and later versions of Windows) take a different tack. Microsoft has asked modem vendors to create files that specify configurations for their modems. Data in these files is then available through the TAPI (Telephony API) interface for applications. This is a step in the right direction but it is not perfect. There are still lots of things that can go wrong (now most frequently with a bug in the files provided by the modem vendors).

The microprocessor that makes a smart modem "smart" may be used for more than modem control. If the processor is sufficiently powerful, error control and data compression can be implemented in the modem

Some low-price modems reduce costs by moving error control and data compression from the modem microprocessor to a software driver that runs in the computer. One example of this is called RPI™ (Rockwell Proprietary Interface).

Most modem manufacturers now make similar so-called "WinModems." WinModems move even more processing to the computer software, including some digital signal processing. These approaches work but, in my opinion, they are rather "penny wise and pound foolish." Modern computers have enough demand on their processing capability without requiring that they also do the real-time error control and data compression and the even more significant demands of digital signal processing. My advice is to stay away from WinModem and other software-based techniques. However, the trend is for computer manufacturers to bundle these lower-cost modems with their systems. No doubt you will have to use them at some point.

2.1.1   Error Control

High-speed modems need built-in error detection and correction. This need was recognized in the early 1980's. Microcom Corporation was a leader in providing modems with built-in error control. They called their method MNP® (Microcom Networking Protocol).

Modem error control is very much like that implemented in error-correcting file transfer protocols. Serial data is buffered in memory and divided into packets or blocks. A CRC checksum is calculated for each block and is appended to each packet. If the receiving modem detects an errant packet, it requests that the sending modem resend that packet. If a packet of data is received correctly, that packet is acknowledged and the sending modem discards it. This form of error control protocol is called ARQ (Automatic ReQuest for retransmission).

These packets can contain more information than just data. Supervisory information is also included. This supervisory information consists of ACKs and NAKs, flow control, and modem performance information. The function of ACKs and NAKs has been described in the file transfers section.

Flow control is needed for a couple of reasons. The computers that are using the modem may need to temporarily halt the transfer of data in order to avoid a receive buffer overflow. The microprocessor in the modem converts the RTS from the computer to a software packet that is sent to the other modem so that the other modem can change the state of CTS to agree. Likewise, the modem may need extra time to re-transmit packets where an error has been detected. So, the modem itself can use the flow control packets.

Communications software that is used with modems using error-correction and data compression must enable hardware (RTS/CTS) flow control.

Modem performance information allows the modems to dynamically alter their operation to optimize their performance. One thing that they can do is reduce packet length. If there are large numbers of errors, a reduced packet size can improve data throughput by reducing the probability that any one packet has an error. If the error rate decreases, because of an improvement in the line conditions, the modems may negotiate a longer packet size, thus providing a corresponding improvement in throughput. Modem performance data can also be used for diagnostic purposes.

V.42 was the first popular international error control standard. It is based on LAPM (Link Access Protocol Modem) which is based on LAP, similar to part of the X.25 standard. MNP error control was incorporated as an optional part of V.42, thus providing compatibility with the earlier Microcom protocol. One enhancement of V.42 over MNP is the use of a 32-bit CRC instead of MNP's 16-bit CRC. This provides a much greater range of error detection. Now the probability of an undetected error is one in several billion.

How do modems establish an error-corrected (and compressed) link?

Error control and data compression are negotiated after the basic modem handshake. The initial modem handshake is an analog signal exchange. Once the modems have connected, data can be sent back and forth (between the modems). The error control and compression link is established by the exchange of a series of data frames that establish the capability of each modem. This exchange of data is not normally passed as data on the modem serial port. However, if a modem does not support this negotiation (some older modems do not), this attempt to establish error-correction will be passed to the serial port. On occasion this can cause a problem. One such problem area is discussed in the Alphanumeric Paging section.

Between the times when the modem begins its initial analog handshake and when the final error control and data compression negotiation is completed, the modem keeps CTS (Clear to Send) false. Usually, CD (Carrier Detect) is made true when the analog portion of the handshake is complete, followed by CTS when the other link negotiations are complete.

2.1.2   Data Compression

The rate at which data can be exchanged by modem is limited by the physical telephone connection. However there are improvements that will increase throughput. The primary improvement results from data compression.

The first kind of compression is a result of the actual protocol that is used to implement modem error control. This protocol is synchronous, so start and stop bits are stripped from each asynchronous character. Each actual byte online requires only 8-bits instead of the 10-bits required for each asynchronous character. Of course, there is some overhead required for framing, supervision, and CRC but the gain still is significant.

More important is intentional data compression using mathematical algorithms to analyze and eliminate redundancy in the data. Text and other data files contain significant redundancy where patterns of data and repeated characters can be represented with substantially fewer bytes than the uncompressed data.

Older compression methods include run length encoding (RLE), trailing blank suppression, and fixed word/phrase dictionary. These techniques result in modest improvements in throughput. Huffman and Lempel-Ziv algorithms are the basis for current modem compression methods. Huffman Code, a variable bit-length "minimum redundancy code" based on the known distribution of the frequency of occurrence of a particular character or code, is a starting point for a number of algorithms.

Variations on Huffman Code, such as used in MNP-5, use a dynamic dictionary. This is applied to an arbitrary sequence of characters and is not limited to the frequency of characters in English text or some other natural language. Starting from a small built-in dictionary, the dynamic dictionary learns frequently used words. Over a period of time, the dictionary develops an optimal coding for the data being transferred.

Lempel-Ziv is a generalization of the dynamic dictionary technique applied to arbitrary sequences of characters. The algorithm successively learns longer repeated sequences of characters and assigns codes to them. Unlike dictionary approaches, Lempel-Ziv learns partial strings as easily as entire words. Practical implementations require large dictionaries with built-in statistics to facilitate periodic pruning for optimization.

V.42bis data compression is a Lempel-Ziv implementation developed at British Telecom. It offers compression ratios as high as 4 to 1. Of course, this theoretical maximum is not often reached. However, compression ratios of 2 or 3 to 1 are commonly reached.

Compression works less effectively for some forms of binary data than it does for text data. However, do not assume that binary data cannot be compressed. Some forms of binary data have substantial redundancy. Executable files and bitmaps are examples of binary data that often compress well.

You cannot benefit from compression unless the serial data rate is substantially higher than the actual physical or "raw" connect rate. If you are using V.34/V.42bis modems that can connect as fast as 33.6k bps, you should use a serial speed high enough to achieve maximum benefit from compression. If your hardware will support it, you should use 115k bps. A speed of 57.6k bps often will provide almost as much throughput. Use 38.4k bps only if your hardware has trouble at the higher speeds.

One thing to note is that file transfers benefit from modem error-correction much more than interactive use. The data must flow almost continuously; otherwise much of the improvement in throughput due to compression will be masked by dead time.

2.2   40+ Popular Modem Questions and Answers

These are questions that often come up. They are not presented in any special order, so scan the questions to see if what you wanted to ask is there. Some of the answers are discussed elsewhere in the book, too.

How do AT commands work?

Modem commands begin with an '**AT**' prefix that gets the modem's **AT**tention. The modem uses the 'AT' characters to automatically determine the serial speed and character format. The 'AT' prefix is followed by the command(s). The modem executes the command line when it detects a carriage return (Chr$(13) or vbCr). If you enter commands from the keyboard, the Enter (or Return) key generates a carriage return.

For example, the command line ATV0 (remember the command terminates with a carriage return) gets the modem's attention and uses the V0 command option to tell the modem to display the responses to commands (result codes) as numbers rather than as text. V1 is the factory default, so this command changes the operation of the modem until the modem is reset or the VI command is issued.

Many commands are used to select between two or more options. Numeric parameters indicate the form of the command you want to be in effect. If you issue a command without specifying a numeric parameter, the modem assumes the 0 command option. For example, Q0 tells the modem to respond to commands with result codes. Issuing Q1 tells the modem not to issue responses (result codes).

There are two modem commands that do not begin with an 'AT'. These are 'A/', which instructs the modem to re-execute the previous command, and '+++' that is issued while the modem is online and instructs the modem to return to command mode (while remaining online).

See Appendix F for more details of modem commands.

My modem is online and I have issued a '+++' to return my modem to command mode. I then attempt to reconnect with the remote system by issuing the 'ATO' command but nothing happens. I have to hang up and re-dial. What happened?

The remote system is probably designed to echo your keystrokes to you. If the person who designed that system failed to disable the "Return to Command Mode" function, the echo of the '+++' caused the remote modem to also enter Command Mode. The remote modem is effectively "locked up." This poor design can be overcome by your changing your own "escape character" from '+' to something that is equally easy to remember. A common suggestion is to use the '=' character instead, because it is on the same key as the '+' character. To configure the modem to use '=' instead of '+', issue the modem command, ATS2=65.

My modem connects but I cannot receive any data. What's up?

Almost all modems require that RTS (Request To Send) is enabled, even if hardware flow control is not used. Make certain that your software asserts RTS. For example, using Visual Basic and MSCOMM set MSCOMM1.RTSEnable = True.

If you have a fax modem, make certain that it has not been left in fax mode. Also, you must initialize your modem so that it will be in data mode before it connects.

If you have a software-based modem, make certain that all drivers that were furnished with your modem are properly installed. Check with the modem vendor for any updates.

How do I dial a number and detect when a voice answers?

Unfortunately, this feature is not available in most modems. The digital signal processing capability is in the modem but it is not used for this purpose. Unless the hardware, usually a modem, provides this feature (look in your modem manual), there is nothing that you can do. I designed modems in the mid-1980's that had this feature so it certainly can be done. Perhaps modems will again provide this feature

Some Computer Telephony Interface (CTI) boards do provide this feature. See the Resources appendix for a listing of a few of the manufacturers of CTI products.

How do I re-dial when the phone is busy?

I have included a code example that shows how to do this using MSCOMM. You can use any communications add-on or the Windows communications API to do this. The critical thing that must be done is to enable tones detection and responses in the modem that you are using. The modem will then output the response "BUSY" when it detects a busy tone. Your code parses modem responses and acts appropriately. The most common modem initialization sequence requiring enabling responses and tones detection is "ATV1Q0X4". Check your modem manual to make sure.

There are examples of this in the code samples later in the book and on CD ROM.

What are "voice" modems?

Voice modems fall into one of three categories, although I expect these to evolve over time:

1. Modems that can function as an answering machine or voice-mail system. Some also function as a speakerphone.

2. Modems that can transmit data or voice over the same connection. Radish "Voice-View" is a current popular system. In the mid-1980's I was a design engineer for Prentice Corporation. We designed several modems that allowed alternate voice and data using the same telephone line, including two that were accessories to the Hewlett Packard Portable Plus computer (one of the first practical laptop computers).

3. Modems that can transmit data and voice simultaneously over the same connection. (DSVD). When voice is used, the speed of data transmission drops because it has to share the available bandwidth with digitized voice. Interoperability between modems of different vendors is limited because standards for voice are not yet universally adopted.

Voice modems are becoming more popular all the time. As standards are adopted that assure interoperability, the costs will fall and usage will increase, especially for interactive conferencing.

You cannot receive a voice call while your modem is in use. Call waiting is not supported. You cannot use these modems as Internet phones. Internet phones use a different technique for sending and receiving voice data.

How can I check the phone line to see if I can get a dial tone?

Before you dial, enable full modem tones detection. Initialize the modem with ATX4. If there is no dial tone, the modem will respond with either "NO DIALTONE" or "NO DIAL TONE" instead of dialing.

How do I determine my modem throughput?

File transfer protocols that are implemented using commercial add-ons usually provide a status window that displays throughput and that allows the transfer to be canceled. However, this does not help when you are not using a file transfer protocol. Often you must calculate this on your own by averaging data over time.

Some modems have a diagnostic query that will report the most recent session's average throughput from the time that carrier was established until disconnect. Look in the modem manual. Sometimes the ATI6 command will return this estimate.

There is some additional information on measuring throughput in the answer to the modem test modes question. So, look there too.

You may be able to use a feature of Windows 95/98 for TAPI enabled 32-bit applications such as HyperTerminal.

The System Monitor accessory that comes with Windows 95/98 has a number of features, among them the ability to display modem throughput in real time. To use it,

1. Open Control Panel/Modems. Select your modem. Click on Properties/Connection /Advanced. Check the 'Record a log file' and then click OK to close each dialog.

2. Make a connection with your modem using 32-bit software (e.g., HyperTerminal, which comes with Windows 95/98).

To start the System Monitor; choose Edit /Add Item. You should see your modem under Category. Select your modem. Under Item select "Bytes received/sec." and/or "Bytes sent/sec." Finally click Open.

Another more general-purpose method is to use a commercial product like TurboCom95 from Pacific CommWare. This comes with a utility that will display or log online activity including throughput.

If I set my serial port to 14.4k or 28.8k bps, why won't my modem work?

V.32bis modems connect at a maximum raw data rate of 14400 bps; V.34 modems connect at a maximum raw data rate of 28.8k bps (or 33.6k bps). This is the speed on the telephone line. However, most modems require a higher asynchronous serial speed, such as 19.2K, 38.4K, or 57.6k bps. These higher speeds allow the modems to use data compression to improve data throughput.

Why does my modem say CONNECT 14400 when the serial speed is higher? Or, why does my modem say CONNECT 38400 when I know that it has actually connected at a lower speed?

These are examples of several similar questions that have almost the same answer.

Modem manufacturers have their own notion of what information you would like to have. Often this is not what you actually want! Some modem manufacturers allow you to configure the modem so that it outputs a response that you find useful; others do not. So, once again you have to read the modem manual to understand the modem capability and what to expect.

After my modem connects, all that I receive is garbage data. Or, I get "Framing Errors" when my modem connects. What should I do?

Let's make the assumption that your modem is working. As a last resort, if none of these other things helps, try a new modem.

You must make certain that the modem keeps its serial port rate fixed at its initialized speed, regardless of the actual raw connection rate. Often the default modem configuration switches the serial rate to match the connection rate.

If data is output by the modem at a speed different from the comm port setting, you will get framing errors and the data will be meaningless.

You must use hardware flow control if the serial port speed is fixed, because it will always be equal to or higher than the actual connect rate. Therefore, transmit data can overflow the modem input buffer. To avoid data loss, the modem must have a way to signal "send no more data" until some of the pending data has been sent.

Both of these may require that you initialize the modem differently than its default configuration. Once again, you have to read your modem user manual.

Also, make certain that your serial port data length and parity settings match those used by the computer to which you are connected. Modems only support data that have 10-bit lengths. That means that you can have eight data bits, one start, and one stop bit, or you can have seven data bits, one parity bit, one start bit, and one stop bit. No other combination will work (except at speeds under 1200 bps).

Lastly, the telephone line can be too noisy to transmit data without error. If your modem supports error-correction, enable it. If not, try a lower speed, perhaps 2400 bps or even 1200 bps; lower speeds are able to pass error-free data at higher noise levels than higher speeds.

My modem does not answer automatically. What's wrong?

There are three things that could cause this.

First, the obvious — the modem may not be connected to the telephone line. There are two RJ modular connectors on a modem. One is used to connect the modem to the telephone line. This one is often marked "wall" or "line" or similar. The other is used to connect a local telephone set. This is often marked "phone." If you inadvertently connect the telephone line to the wrong connector, the modem will not answer.

Second, the modem must be initialized to auto-answer. The modem default is NOT auto-answer. The mode S0 register must be set to the number of rings on which to answer. For example, ATS0=2 will cause the modem to answer after it detects the second ring. Some modems have an option switch (often a dip switch) that can be used to select auto-answer.

Third, DTR (Data Terminal Ready) must be asserted, otherwise the modem will not answer. When using MSCOMM, make certain that the DTREnable property is set to True.

Last, make certain that the RS-232 cable used with an external modem is functional and wired correctly.

I have Call Waiting. If a call comes in while I am sending or receiving data, it can be interrupted. What can I do?

You can disable Call Waiting for the duration of the modem call by sending "ATDT*70" prior to dialing a number. If you are using pulse dialing, you can usually send "ATDP1170" prior to dialing. Check with your phone company to be sure. In either case, you must allow sufficient time between sending the command to disable Call Waiting and dialing the actual telephone number. One way to handle this is to include the disable command followed by a comma and the phone number The comma is the modem dial delay character

For example,

```
ATDT*70,555-1234
```

or

```
ATDP1170,555-1234
```

As soon as you disconnect, the Call Waiting feature will be restored by your telephone company.

If you are answering an incoming call, you cannot disable Call Waiting. It is best to not use Call Waiting on telephone lines that are often used for an answering modem.

I have only one telephone line. It has Call Waiting. If a call comes in while my modem is connected, I want to drop the modem connection so that I can answer the call. However, my modem does not disconnect and I get no indication that the phone has rung. What can I do?

See? No one is ever satisfied. First you want to disable Call Waiting so that your data is not interrupted. Now, you want to allow an interruption. What can I tell you?

There is not much that you can do. V.34 modems (or higher), those that support 28.8k bps connections and higher, are impervious to the sound that signals Call Waiting. They just treat it as noise like any other noise and they continue to work. Your only options are to set up Voice Mail with your telephone company or get a second phone line for your modem and fax. With Voice Mail, an incoming call will roll over to your mailbox and you can call back. With a second phone line you have multitasking, not task-switching. Sorry.

I send a command to my modem but I do not get a response. What should I try?

Well, first make certain that modem responses are enabled. The easiest way (this book is about software, after all) is to initialize the modem with the "ATQ0" command. This enables responses. If you want verbose (text, not numeric) responses, add a V1 to the command à la using MSCOMM, MSComm1.Output = "ATQ0V1" & VbCr. Sometimes there is an option switch (often a dip switch) that must be enabled.

However, a common error that programmers make is failing to understand that modems require some time to act on a command and to issue a response. Another thing that must be realized is that statements like Buffer = MSCOMM1.Input reads data that has already been received. The assignment statement does not wait until there are character(s) available; it executes immediately.

A code example of what you might do with MSCOMM is,

```
Do
    DoEvents
    Response$ = Response$ & Comm1.Input
    If InStr(Response$, "OK" & VbCrLf) Then
        'OK received
        Exit Do
    End If
Loop
```

Other ways to handle this situation are to wait for a specific number of characters to be received or a specific delay to allow the modem to respond. This second option is quite useful if you need to parse the modem response to determine what has happened. Another method is to search the response for a specific delimiter such as a carriage return and line feed. Each of these methods has uses and there are examples of them in the code chapters.

What if this does not work? I hate to say it but you may also have to check a few more things.

First, make certain your modem is installed correctly AND connected to the computer. Check the RS-232 cable, if your modem is external. I know, you did that. Check it again.

Second, make certain that there are no hardware conflicts. Your modem comm port must use an independent IRQ.

Third, did you open the port before you sent commands? If errors are not trapped and displayed, you may have a runtime error and not have noticed.

Fourth, is the modem in command mode? If, somehow, the modem is connected to another modem, it will not respond to commands. You have to return it to command mode by disconnecting or by issuing the escape sequence (usually "+++", with a slight delay between each "+", then wait for the "OK" response before issuing modem commands).

My modem is connected to a printer (or other device). I do not want modem responses printed. How do I tell the modem to output serial data only after connect?

Finally, an easy question. Just issue the command "ATQ1" to the modem or set the modem dip switches to disable modem responses.

I want to dial a series of numbers with delays between each set of numbers. My modem will not dial all of the digits that I send it. What do I do?

The modem command buffer is limited, often to 40 characters and on occasion even as few as 36. However, you can dial a series of digits that exceed the maximum length buffer length by breaking the series into subsets, each ending with a semicolon (";"). The exception to the semicolon is the final digits, if you want the modem to handshake and enter data mode. You must wait for the modem to issue an "OK" response before dialing subsequent digits. Here is an example in pseudo-code,

```
ATDT1-41-20-55-12345;
WaitFor "OK"
Delay 1                 'wait one second
ATDT999-8765;
WaitFor "OK"
Delay 2                 'wait for 2 seconds
ATDT#*6789;
WaitFor "OK"
Delay .5                  'wait for .5 second
ATDT777-654321          'now, connect
```

The semicolon forces the modem to return to command mode after dialing so that additional commands can be sent to it without causing it to disconnect.

See the "Why does my dialer fail?" question for notes about why this may not work as advertised.

I need to dial an access number, wait for a dial tone, and then dial another number. How do I do that?

Most modems support a dial modifier called "Wait For Second Dialtone" or similar. This usually, although not always, is the character "W". Here are two pseudo code examples,

```
ATDT9W555-9876
ATDT1-800-555-9876W555-1234
```

Well, that's not quite what I want to do. Actually I want to dial a number, wait until it is answered, and then dial some additional digits. Is there a way?

Probably. Most modems also have a dial modifier called "Wait For Quiet Answer". This also uses the tones detection capability of the modem to determine when ringing has stopped (the phone was answered) and, usually, for an additional five seconds of silence. Here is a pseudo-code example,

```
ATDT555-1234@986-3143;
WaitFor "OK"
ATH0
```

This sort of thing can be useful for numeric paging where you dial a service, wait for an answer and subsequent voice prompts, and then send the telephone number that you want to be displayed on the pager. Note that I have waited until the modem responded "OK" and then I sent the modem command to hang up. Omit the semicolon, the WaitFor, and the hang up command if you want to connect with another modem and to enter data mode.

I expand on this idea in the next question.

See the "Why does my dialer fail?" question for notes about why this may not work as advertised.

I want to use my modem as a dialer. I do not want it to connect and enter data mode. Can I use it to detect a busy tone and then re-dial if busy?

Perhaps, but it takes a little code and it is a little tricky (and it is not 100% certain to work but it is worth trying). You can use the "Wait For Quiet Answer" dial modifier to determine if a quiet answer has happened. If it has, the modem will return an "OK" response; otherwise, the return is "BUSY" or "NO CARRIER". Here is some code that assumes MSCOMM or the equivalent. It also assumes that you want to send tones to signal a number. You can uncomment the first MsgBox and replace, with some other appropriate code, the code that dials the second number after receiving "OK".

```
Dim Buffer As String
Comm1.Output = "ATX4V1Q0DT555-1234@;" & VbCr
Do
    DoEvents
    Buffer = Buffer & Comm1.Input
    If InStr(Buffer, "BUSY" & vbCrLF) Then
        DialerHangup
        MsgBox "Telephone Busy.  Try Again later"
        Exit Do
    ElseIf InStr(Buffer, "NO CARRIER" & vbCrLF) Then
        DialerHangup
        MsgBox "No answer.  Try Again later"
        Exit Do
    ElseIf InStr(Buffer, "OK" & vbCrLF") Then
        Comm1.Output = "ATDT986-2179;" & VbCr
        Buffer = ""
        Do Until InStr(Buffer, "OK" & vbCrLF)
            DoEvents
            Buffer = Buffer & Comm1.Input
        Loop
        DialerHangup
        Exit Do
    End If
Loop
Buffer = ""
'other stuff here
Sub DialerHangup ()
    Comm1.Output = "ATH0" & VbCr
    Do Until InStr(Buffer, "OK" & vbCrLF)
        DoEvents
        Buffer = Buffer & Comm1.Input
    Loop
End Sub
```

There are several tricks here. First, tone detection, text responses, and modem-responses were enabled as part of the dial string. Second, there is a small subroutine that handles disconnect for each of the states after dialing. Third, if the modem detects five seconds of quiet, it dials the additional number and then disconnects.

This code does not account for any problem that might cause the modem to fail to output one of the expected responses. Such a failure is unlikely but it is possible. So, you may want to add code that tests time to make certain that a loop will be exited if a timeout is exceeded. The anticipation of unexpected events is the essence of good, reliable design.

Note that the VB 5/6 constant vbCrLf is Chr$(13) & Chr$(10) in VB 4 or earlier.

See the "Why does my dialer fail?" question for notes about why this may not work as advertised.

Suppose that I want to create an auto-dialer program that dials a telephone number that I know has a PBX with a directory. I also know that the PBX accepts a party's name instead of an extension to route the call. Can I use my modem to do this sort of dialing?

Maybe. Some modems have a built-in conversion routine (dial modifier) that converts alphabetic characters to the equivalent tone digit. This pseudo code may work. If not, check your modem manual.

```
ATDT 555-1234 @ 986-3143;

WaitFor "OK"
other-code-here
```

You may want to look at the preceding example to enhance this so that busy is handled.

I have followed your suggestions for creating an auto-dialer. Why does my dialer fail?

Modem tones detection can be poorly implemented. Some noises after the remote system answers can be falsely detected as busy when these were not busy tones at all. Even voice can be mistaken as busy, so these routines can fail when, logically, they should work.

Your only option is to proceed with the understanding that you depend on the quality of the modem design. There is no amount of software that you can implement to work around a mediocre modem design.

What is Caller ID and how do I use it?

Caller ID is a technology that allows a called party to read the telephone number of the caller when the incoming call is ringing. Some modems are furnished with Caller ID functions. These modems will output a modem message with the telephone number of the caller. The format of this message has not been standardized so you will have to read your modem manual — and do the right thing.

Caller ID is an optional service that is being offered by most local telephone companies. In 1995, the FCC required nationwide availability. However, it may not be available yet, and it can be blocked by the caller anyway. So do not design your software with the assumption that all inbound calls will have CID information.

The telephone system portion of Caller ID has been standardized in the US. However, the methods used in North America are not the same as those available in the rest of the world. So, modems designed for the US probably will not provide reliable CID elsewhere.

There are standalone CID devices that can be connected to a computer serial port. Again, there is no standard for the format of the data that these devices provide, so you are on your own.

TAPI also includes methods for CID. This requires that your modem has an appropriate TAPI service provider (essentially a software driver furnished by modem manufacturer). See the TAPI chapter for more information.

My modem is connected to a remote system. The program code that I use to disconnect does not work reliably. Why?

The problem may come from using the modem escape sequence to return the modem to command mode and then issuing the hang-up command to disconnect. The timing between each character in the escape sequence may not meet the modem requirements, as in the pseudo code below.

```
wait '+' delay '+' delay '+' WaitForOK  'ATH0'
```

There is a minimum time before the first '+' with no other data being output and minimum delays between each '+'. These are called "guard time." In addition, you must wait until the modem responds 'OK' before sending the 'ATH0' command to hang-up.

A much more reliable way to disconnect is to initialize the modem to disconnect when DTR is lowered. Then (using MSCOMM), MSCOMM1.DTREnable = False is all that you need to force disconnect. To initialize the modem to disconnect when DTR is lowered, usually not the default, issue the AT&D2 command or include &D2 in the modem initialization sequence.

How does a modem maintain the maximum throughput that is possible? Or, what do the terms "fallback" and "fall-forward" mean?

When high-speed modems (modems that support speeds of 9600 bps or higher) connect, they negotiate a compatible raw serial speed. This raw speed can be less than the maximum speed that the modems support. It will be governed by the amount of noise that the modems encounter during connect as well as by the telephone line characteristics such as bandwidth, equalization, and amount of echo.

Higher speed modems like V.32bis (14.4k bps) modems can fallback to even lower speeds after connect if line conditions change. This can improve data throughput by reducing the number of errors that have to be corrected by the modem. However, once they are operating at this lower speed, they cannot "fall forward" even if conditions improve.

V.34 modems (28.8k bps and higher) can fallback or fall-forward, thus optimizing throughput as telephone line conditions change.

These speed changes are normally transparent to you as the user. But what they mean is that the actual throughput that you get over your modem may be less than you might have expected.

Can I use my modem with my cellular telephone?

Maybe. First, you may be able to purchase a modem designed to work with your cell phone. This is the best option because the cell phone manufacturer handles the mechanical and electrical interface issues. Motorola is a vendor of popular cell modems and cell phones.

There are some manufacturers who sell adapters that allow conventional modems to work with a variety of other vendor's cell phones. See the Resources appendix for some catalog sources.

How well these solutions work depends on several things. Luck is an important commodity, one that is needed here. Cell phones provide lower bandwidth and higher noise levels than conventional telephone systems. Also, a cellular connection can "drop out" for an extended period of time when the connection is handed off from one cell to another. Data throughput will be lower than you have come to expect using a conventional telephone line. In any case, many people are happy with the extra flexibility that they get with cellular modems.

I have a digital PBX. Can I use my modem with it?

Yes, but like many other things, there are some limitations.

Konnex is a manufacturer of a variety of adapters that allow you to connect your modem to digital telephone systems. These allow you to make outgoing calls fairly easily. However, inbound calls are not too practical. Also, these devices can reduce the maximum speed at which your modem can connect.

What modem should I buy?

That question's too hard. Try a different one. Modem manufactures are continuously bringing out new models with new features, and sometimes bugs. The modem business is extremely competitive. So, you get tremendous capability for the amount of money that you spend.

But, like other areas of the computer business, cost is a rough guide to quality. The best advice that I can give is to never purchase the lowest cost modem. Then, buy one with a money back guaranty. Finally, test it in a real life environment to make certain you want to keep it.

How do I get my modem to differentiate a fax from a data call, and then to do the right thing?

Some modems have "selective answer" and can identify the type of an incoming call. Besides TAPI, there are no standard methods for writing an application to use this feature. Most TAPI applications to date do not support this feature, even if the modem provides it.

What I do is to use two separate modems and/or fax machines, one for data and the other for fax, with a telephone line-sharing device. This device listens for fax CNG tones and rings the fax device if those are detected; otherwise it rings the data modem. Then the fax and modem may have individual applications that do not have to work in concert. In the short run, this is a satisfactory solution, and it may be my choice for the long term, too. Some fax machines have this feature built in.

What is Distinctive-Ring service and how might I use it?

Distinctive-Ring service is one way to solve the problem in the preceding question. The telephone company can provide an optional service that allows you to use two or more different telephone numbers on a single telephone line. Each telephone number will generate a distinctive ring pattern when it is called.

If you are going to use Distinctive Ring with modem or fax devices, you need to have hardware that identifies the ring pattern and connects the incoming call to the correct device. These are generically called "Distinctive-Ring call routers" and are available from a variety of sources.

Distinctive Ring is supported by some modems. However, as with "selective answer," TAPI applications would have to support this mode and they are few and far between. Actually, it is possible for a non-TAPI application to be written to use Distinctive Ring but writing such an application would be hard. The reason is that fax add-ons for VB are incompatible with other communications add-ons. This may change in the future.

Did I mention that I get overrun or CRC errors? How can I get my modem to transfer data faster but without these errors?

Probably this is not the fault of your modem. Overrun errors indicate that data is being received faster than Windows COMM.DRV can remove them from the UART.

If your modem is internal, it should have a 16550 AF UART as part of its design. If the modem is external, the serial port to which it connects should have a 16550 AF UART. If it does not then you will have to reduce the serial speed to avoid overrun errors.

If you have a 16550 AF UART and are still experiencing overrun errors, you need to look at your PC. Is it fast enough? Any other programs that are multitasking can slow operation sufficiently to cause this sort of error. Also, some programs under Windows 3.x or Windows 95/98 can access DOS routines that disable interrupts long enough to cause an overrun.

If you are downloading large files at high speed under Windows and are getting overrun or CRC errors, try minimizing the application. If your application is updating status or otherwise displaying data onscreen, you will avoid that overhead when the application is minimized. Repeatedly painting this window places a burden on the processor. So, you can improve reliability by minimizing the application after beginning the transfer, then restoring the application when you expect the transfer to have completed.

Avoid doing things that place an undue load on the computer. If you start a file transfer and then load Word for Windows or some other "heavy" program, you may see errors that you would not have seen had you loaded the "heavy" program before you started the file transfer.

Always make sure that you enable hardware flow control (CTS/RTS) in both your software and in all modems used in the system.

How do I protect my modem from telephone and power-line transients?

Transients or power surges can damage modems and the computers that are connected to the modems. A modem will have built-in protection from telephone-line transients. This protection is specified by the FCC and is designed to make the device safe to use. That is, if a transient, say caused by a lightning strike, happens, it should not be transferred to anyone who might be using the modem

However, this protection is designed for humans, not for the modem. It is acceptable, according to the FCC, for the modem or a connected computer to be damaged, sometimes irreversibly.

There are external devices, usually in the form of power strips, which are designed to protect computers and modems from power and telephone-line transients. These are well worth the extra investment. The more expensive and more effective, transient protection devices come with an insurance policy that will pay for the replacement of any equipment that is damaged by a transient when connected to them.

If I call into a system that has "Call Forwarding", are there any problems that I should anticipate?

Call Forwarding can be used to extend the distance of a local dialing area or for other valid reasons. However, this can cause extra telephone-line impairments and it can reduce the speed at which your modem can connect. Because of this, a V.34 modem might connect at 19.2k bps or lower speeds.

I have a number of phones, fax machines, and modems connected to my telephone lines. Can this cause a problem?

I guess that you know the answer to that question, since you asked it!

Yes. These other devices can cause extra noise that lowers connect speed. They can increase the ringer load, so that a modem may not detect ringing reliably.

Also, look for premises faulty wiring problems. These can range from loose connections to noise problems caused by crosstalk from other telephone lines or from the power system.

I have been told that I should get a telephone line that is conditioned for data to use my modem. What do you think?

Unless you own stock in the telephone company, you should not buy any extra-cost options. All that you need is a good voice-grade telephone line. However, if you hear noise, hum, an echo, or some other impairment on the line when you use it with a normal telephone set, you may have trouble with your modem.

The telephone company often will charge extra if they have to track down problems on the line, especially if the cause is at your location. But, it can be done.

Do not bother with noise reduction filters or other patent medicine. Anything added to reduce noise also reduces bandwidth, a quantity already in short supply. These are a loosing proposition.

What are "split speeds" and how might they improve my modem throughput?

The V.34 specification has an option that allows the modems to maintain different transmit and receive speeds. Thus, a modem can optimize throughput to the maximum that the telephone line will support in both directions.

For example, one thing that a modem must do is digitally cancel echo(s) of its transmitted signal so that the echo is not interpreted as receive data. If the echo(s) are more easily canceled in one direction than in the other, it may be possible to have a higher raw data rate in the direction where echo is less significant. Likewise, noise or bandwidth may not be the same in both directions.

Since this feature is optional, few modems have it. Look in the specifications or quiz the vendor if maximized performance is more important than cost.

I'm doing a file transfer. Sometimes the data flow seems to "stop" and then resume. What can cause this?

There are several things that might have happened.

The modem may have encountered errors that required that it retransmit data. If another error, or errors, happens during this retransmission, you may see a perceptible pause.

Also, the modems may encounter enough errors that they renegotiate the connect speed and the line equalization that they implement in the digital signal processor. This is called "re-training," and it can take a few seconds, sometimes more than just a few.

If you are using a cell modem, there may have been a hand-off from one cell to another. Such a hand-off can interrupt data flow for a few seconds.

The remote computer may have had its processing load increased because of multitasking or other program demands, so it may simply have slowed down and caused a pause in the flow of data.

How do I use the modem test modes?

Check your modem manual. Often modems support testing via the AT&T_ commands.

For example, AT&T1 can be used to place a modem into Local Analog Loopback (LAL). This is a single-modem test that connects the transmitted analog signal to the receive, internal to the modem. Any data that is sent will also be received. This is useful to test some of the analog portions of the modem but it does not test the telephone interface itself.

AT&T3 can be used to place the modem into Local Digital Loopback (LDL). This loops serial data back to the serial port. This is useful for testing the serial port and simple tests of your software.

AT&T4 can be used to grant Remote Digital Loopback (RDL), while AT&T5 can prohibit RDL. AT&T6 can be used to initiate RDL for two modems that are already connected. RDL is an excellent way to test the performance of the complete communications path. Any data that you send is returned to you. You can measure the number of errors that are made, giving you a good idea of how well the modems are performing over actual telephone lines.

To facilitate RDL testing, the modems may allow you to specify whether or not modem error-correction is enabled. If error-correction is not enabled then you can measure the error rate. If error-correction is enabled (perhaps using the AT&T7 command) then you can measure maximum throughput. You should recognize that real-life throughput is determined by more than the modem throughput. However, this will give you a maximum number that you know cannot be exceeded, regardless of other factors.

The AT&T0 command will terminate any of these modem tests.

What's this stuff about V.80 and H.324, etc.?

The H.324 specification of the ITU (International Telecommunications Union) defines video conferencing. This specifies the software protocol or stack that is used to send and receive interactive video and audio data. More alphabet soup defines how audio data, G.723, and video data, H.263, are compressed. More fun, H.223 specifies how audio and video data is multiplexed into a single synchronized data stream.

The V.80 specification enables video conferencing over standard telephone lines (POTS).

All of this is beyond practical application in Visual Basic until someone enables all of these low-level operations with one or more OCXs. As of this writing, that's all in the future.

What are V.90/92/94 modems?

There are many modem manufacturers who are producing 56k bps modems. Some of the manufacturers who are working in this arena are U.S. Robotics, Motorola, Lucent, Rockwell, and others. The V.90 (etc.) standard governing this type of modem recently has been adopted.

These modems offer a maximum 56k bps transfer speed for data from the telephone central office (actually, less than 50k bps is the result) but only 28.8K to 33.6k bps maximum from your computer to the telephone central office. This system can work because it relies on error-free, almost unlimited bandwidth of the digital portion of the telephone system.

The telephone central office will provide access numbers for the complementary V.9x modems that must be used and located in the central office. This means that these modems will be useful for high-volume users like Internet Service Providers (ISPs).

Applications other than the Internet normally will not benefit from these attempts to circumvent Shannon's law.

What are Password Security and/or Callback?

Password Security and callback are methods for implementing a secure modem network. It can be implemented as an option in some modems, or with an add-on device. This secure system allows the answering modem to connect but requires that a password is received from the calling system before data can be transferred. For additional security, the modems may disconnect after the password is received. If the password is valid, the callback device calls the originating modem so that data can be transferred.

How do I send or receive a WAV or other audio file using my modem?

I assume that you want to play or record a WAV file using your "voice" modem. So, either you dial and wait for a human to answer or you use your modem to answer a voice call and play or record an audio file after going offhook.

This can be done with any voice modem. This kind of modem has features of both a sound card and a conventional modem. The documentation that comes with the modem will provide the "AT" commands that you use to play and/or record audio files. These commands are not standardized. So, your best bet may be to rely on TAPI to perform these functions. See the TAPI section for more information.

What is Remote Modem Configuration?

Some modems have a password-protected mode that allows them to be configured from a remote location. This means that a network administrator or other person responsible for their operation can call them and reconfigure them from a distant location. This option will not be used very often but, when it is needed, it can save a lot of time and money.

Tell me about non-standard modems.

If that's a question then I'll refer you to the next sections for information on less conventional modems.

2.3 Network Modems

Network modems allow LAN-based users to utilize the modems connected to the remote access server or communications server for dialing-out to bulletin boards and other on-line services (see *Figure 2.1*). Utilizing the modem(s) that are attached to this server eliminates the need to have a dedicated modem connected to each PC, thus lowering hardware costs and support requirements.

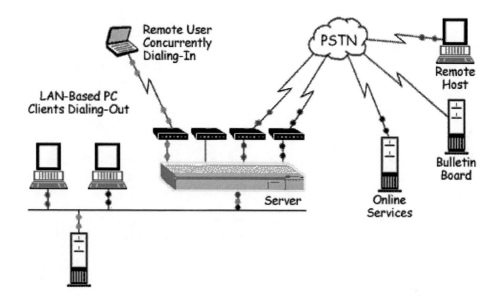

Figure 2.1 Network Modems

If there is more than one modem connected to the server, it is usually called a modem pool. Some systems are designed for Netware ™ and others are designed for mixed-protocol networks.

For example, some Telnet software supplied by the TCP/IP stack vendor has an INT 14 interface. This redirects the INT 14 traffic over the Telnet session established between the PC and the communications server.

Incoming calls will be handled by the server software which often has been designed to allow remote access to the network.

Outgoing calls can be handled by NASI (Netware Asynchronous Serial Interface) or INT14 (a special DOS interrupt) services. These services were originally designed for DOS-based applications but they have been extended to Windows by various server communications software and modem-pool vendors.

Using NASI or INT14 services for dial-out connection is simple (make certain that your communications server supports either NASI or INT14). Instead of selecting a comm port such as Com1, select NASI as the interface. After that, treat the shared modem the same way that you would treat a local modem. There must be a NASI or INT14 service, often called a "redirector", running on your PC.

The only Visual Basic add-ons that I know about that provide INT14/NASI support are Crescent's PDQComm (discontinued), Magna Carta Software's CommTools, and Greenleaf's PowerComm. How well this may work seems to depend largely on the INT14 implementation. There will be more information on CommTools in a later section of the book.

A much more popular way to implement modem sharing via a network connection is to use an Ethernet serial port server or hub. These are less expensive than network modems, though you must add the actual external modems in addition to the server, and are better supported by a variety of manufacturers. See section 1.3.6 and the Resources Appendix for more information.

2.4   Other Modems

There are a wide variety of modems that do not work over the telephone system, our old friend POTS. Many of these support asynchronous communications and a few even have Hayes "AT" command sets to make them quite easy to use. I'll discuss some of these here.

There are some non-modem devices that may be similar to some of these modems. These non-modem devices are called line drivers and they will be discussed in the RS-232, RS-422, and RS-485 sections.

Manufactures and vendors of these modems will be listed in the Resources appendix.

2.4.1   Leased-Line Modems

Modems in this category are designed to operate over telephone circuits. These telephone circuits are not switched like a conventional telephone. Rather, they are dedicated signal paths that connect two or more locations. The circuit is continuously open and is always available for use. It is a telephone circuit because the signal characteristics are the same as those of a conventional dial telephone system. These circuits may be furnished by Ma Bell (or these days, Ma's offspring) or they may be part of a private network. The telephone circuit is often characterized as unconditioned 3002B, voice-grade. Distances over which these modems can work, technically, are unlimited. The signal levels are maintained within specific ranges by the telephone system.

Common use of these modems is to connect large installations over significant distances. The economy of having a dedicated circuit can be four-fold.

First, it avoids long-distance toll charges — in exchange for the cost of the circuit (often hundreds or thousands of dollars per month).

Second, there is no delay required to dial and establish communications. The circuit is always ready to send or receive data. This can be crucial if real-time data is to be transferred.

Third, an organization that needs this form of communication often needs many circuits for different applications, both voice and data. It may have the infrastructure to provide the circuit already in place, thus incurring small incremental cost. Prime examples of this type of organization are banks and public utilities.

Fourth, it is fairly common that this form of communication is used to connect locations where conventional dial circuits do not exist or are expensive to add. They are often used in remote areas via microwave systems.

There are two leased-line architectures that are used. One is called point-to-point and the other is called multidrop or multipoint.

Point-to-point is self-explanatory. Two locations are connected together with a dedicated circuit.

Multipoint or multidrop is less obvious, perhaps. With this architecture, several locations are interconnected. Here is how this works. One location is designated the "Master." All other locations on the network are called "Slaves." This is not a peer-to-peer network where any location can send data to and receive data from any other location. In a master/slave network, the master controls access to the network by sending "polls" to slaves. When a slave receives a poll, it is allowed to send data. When that session is complete, the slave must release the network so that a different slave can be polled and, thus, receive control. This system usually has a way for the master to send data to multiple slaves or to broadcast to all slaves simultaneously.

The protocol used to communicate on the network determines how polls are sent, that is, how individual slaves are addressed. These protocols are not standardized, although there are a few commonly used protocols that are used by networks of synchronous modems. However, there are no such protocols defined for networks of asynchronous modems. Some asynchronous protocols are modeled on synchronous SDLC or HDLC protocols, while others use the more familiar file transfer protocols that are used with dial modems, enhanced to support the master/slave addressing scheme. I think that this second method has significant advantages when used in a Visual Basic environment.

I will not go into detail on the protocol or its implementation. This is quite application-specific. If you do not have in-house expertise in this area, consultation is available.

There are two physical types of point-to-point and multidrop networks. One is called two-wire. Two-wire systems use the same wire pair to send and to receive data. Thus, the master modem must release the circuit after it sends a poll and wait for a response from the address slave. The other is a four-wire system. Four wires allow the master and addressed slave to send and receive data simultaneously. The four-wire system is the only practical physical connection if you are to use conventional error-checked file transfers.

Leased-line modems may use several modulation schemes that are standardized. Examples are Bell 202, Bell 206, V.27 and V.29. V.29-based modems communicate at the highest speed, usually 9600 bps. However, variations on the implementations of V.29 modems make interoperability from manufacturer to manufacturer problematic at best. V.33 modems are available that raise the maximum speed to 14400 bps but these usually are point-to-point only.

Leased-line modems usually are manufactured by large companies. Some of these are Codex (Motorola), Penril, and Gandalf.

2.4.2   Limited-Distance Modems

Limited-distance modems, also called short-haul or short range, are similar in operation and application to leased-line modems. The differences are that they do not use telephone lines but instead use dedicated wire circuits. Also, the distance over which they can operate is limited by the signal loss in the circuit and by the speed at which they operate.

Limited-distance modems can operate at higher (sometimes much higher) speeds than leased-line modems. The maximum speed is a function of the modem design and the wire size used. For example, a typical limited-distance modem might be able to be used at 19200 bps over a distance of only 1.5 miles using 26AWG wire. The distance could increase to three miles using 19AWG wire.

Communications protocols that can be used with limited-distance modems are the same as can be used with leased-line modems. Both point-to-point and multipoint operation are possible.

Modems used in the network must all come from the same manufacturer. Typically, they must all be the same model.

Care should be used when designing the wire routing for a network of multipoint, limited-distance modems. The circuit should go from modem to modem and the last modem should be terminated with whatever resistance is specified by the manufacturer. A "star architecture," with the master at the center of the network and slaves at the points, usually is not suitable and will not work well.

2.4.3   Fiber Optic Modems

Fiber optic modems often are misnamed. The most common are not modems but are line drivers. Line drivers encode and decode light signals in baseband. The light signals are not modulated in a baseband system. Rather, a baseband system may represent a 1 (one) by the light on and a 0 (zero) by the light off. The light used might be generated by LED or Laser.

There are high-performance fiber optic modems that are true modems. These high-performance devices use much more of the available bandwidth and are correspondingly more expensive than the baseband versions.

However, I'll also discuss both baseband and true fiber optic modems here. Baseband fiber optic modems are more common than true fiber optic modems and are almost universally misnamed modems. For our purposes, the main difference between them is cost and performance.

Fiber optic modems come in two flavors. One is multi-mode; the other is single-mode. These terms refer to the type of fiber optic cable that is used to convey the signals. Multi-mode fibers have a larger diameter (some are as large as 125 mils) while single-mode fiber diameter is typically 8 mils. Multi-mode fiber has more loss than single-mode but its size lends itself to single and double-fiber runs. Single-mode fiber has less loss but it needs to run in a bundle with other fibers. It is not strong enough, unless heavily jacketed, to run by itself.

A modem designed for multi-mode fiber will not work with single-mode fiber. The reverse is also true.

The loss-budget of fiber systems governs the distance over which they will operate. Typical, inexpensive modems will operate over distances of two miles at 19200 bps. More expensive versions might be capable of working over three miles at up to 115k bps.

Fiber optic modems often are used in point-to-point networks but multipoint networks are possible with some designs. A multipoint fiber optic network would be connected as a physical loop or a physical star. Software would have to be developed that accounts for the unusual nature of a physical loop. A physical star network would function like a multipoint leased-line or limited-distance modem network.

Why use fiber optic modems?

These devices are almost completely immune to interference from outside signals. EMI and RFI (electromagnetic and radio frequency interference) are eliminated. They use no electrical conductors so there is no chance of short circuits or hazards from electrical devices or power systems. Ground loops cannot be a problem. Fiber optic modems are ideal in harsh environments. Transients from lightning cannot cause data errors. The data transferred is as error-free as if the computers had been directly connected.

2.4.4   Wireless Modems

Suppose that you need to communicate over a wide area but do not want to use dial (conventional or cellular), leased-line, short-haul, or fiber optic modems. One option might be to use wireless modems. Examples of these sorts of applications are many. They might be PC to mobile units on a factory floor (inventory control, for example), SCADA (Supervisory Control And Data Acquisition) in an industrial area, scoreboard control in an athletic arena, diagnostics or data acquisition from moving vehicles, medical data from mobile patients or animals, robotics, or office applications like a wireless link to a remote printer.

These modems are based on radio transceivers. The radio signal can be modulated in a similar way to that of telephone-line modems. But, often, more complex methods are used to modulate the radio signal, to share radio bandwidth with a variety of users, and to avoid interference.

Wireless modems operate on a number of different frequencies and use a variety of methods to assure that the data link between units is maintained.

Short-range units, with a range of a few hundred feet, might use a simple conventional RF-modulation scheme. Typical units operate in the 902-928 MHz or 2.4 GHz bands and require no FCC license. Data rates up to 115200 bps are possible. Bluetooth and WiFi (802.11b) based modems have become a popular and lower-priced variant in this category.

Longer-range units may use spread spectrum access and a packetized point-to-point or multipoint protocol. These packet protocols usually provide error-free communications. Power output and operating frequency determine whether or not these units require an FCC license. Often they do not. Range of these units can be 300 meters or more. Data rates up to 19200 bps are usual.

The longest-range wireless modems use packetized protocols and sometimes spread spectrum. More often, they use packet burst communications or simpler encoding schemes. Line-of-sight communications can provide ranges up to 15-20 miles without repeaters and much further with repeaters. Point-to-point or multipoint operation is possible. Some units have Hayes AT command sets for easy use. Data rates of 9600 bps are common, although lower speed units may cost slightly less than higher speed modems. Modems with the longest ranges usually require FCC licenses.

The most sophisticated modems in this category are satellite telephone modems. There are several manufacturers of portable satellite telephones. These devices are notebook-computer-sized units with a detachable top that functions as an antenna. They can be used in a way similar to cellular telephones to place telephone calls from many locations on the globe. Data and fax options are available. Cost to use this system is about $5 per minute at this writing.

Current satellite telephones use the Inmarsat satellite system. Future systems that use dedicated satellites should reduce the cost and increase the flexibility and speed of these telephone and data systems.

There is one group of devices that are neither wireless nor directly wired. So, I'm including them here. They are called Power Line Carrier Systems (PLCS). PLCS encode data in one of several formats, from FM to spread spectrum, and place the signals on the power line. Any device on the same power system branch can then communicate using the power system as the wiring. This has the obvious advantage that no extra wiring is necessary.

PLCS are limited in distance, usually to a single branch or circuit. The signals are blocked by power line transformers so common application limits the communications link to a single building. Even modestly sized buildings have multiple circuits that are fed from separate transformer taps. Each power circuit, unless it has special bypass hardware, is isolated from any other circuit.

Power utilities have used PLCS for many years over long-distance power transmission systems for protection signaling and coordination (operation of circuit breakers, etc.). Lower power versions are now used for residential and industrial automation such as lighting, environmental control, alarm systems, and other useful applications.

2.5 ISDN

The Integrated Services Digital Network has the potential to provide data rates that exceed those possible with analog modems such as V.34 with a raw speed up to 33.6k bps and V.90 (or similar) mixed analog/digital modems with raw speeds up to 56k bps.

ISDN is an international standard for digital communications so it has the potential to be inter-operable across national boundaries. This potential for inter-operability is somewhat unfulfilled at this writing.

ISDN covers more than just local connections such as found in a home office. It also is used in some digital PBXs and multiple channels. Less than T1 bandwidth is also available. But, my discussion will be limited to the sort of ISDN connection that might be used with a PC.

ISDN is a completely digital solution. In place of a conventional telephone connection, the telephone company furnishes what is called a Basic Rate Interface, or BRI. The BRI provides three digital channels. Two of these are 64k bps and one is 16k bps. The two 64k bps channels, designated "B", can be used individually for voice or data and they can be combined or "bonded" to furnish one 128k bps data channel. Bonding the two B channels commonly uses the Multilink PPP protocol, a TCP/IP protocol often used for Internet access.

There are two basic ways that ISDN can be used with a PC. You may have an internal ISDN adapter that has a TAPI compliant service provider (this is the driver software that Windows uses to identify and use the adapter). Or, the ISDN connection may be a unit that connects to the PC using a conventional serial port.

ISDN units that use a standard serial port are called Terminal Adapters (TA's) or ISDN modems. Of course, they are not true modems because ISDN is digital all the way. However, the word modem explains their usage more clearly than does the TA acronym.

ISDN modems usually use the Hayes AT command set to configure, dial, and answer. Unlike an analog modem, the time required to connect after dialing is very short; by comparison it seems almost instantaneous. ISDN modems usually offer Lempel-Ziv data compression so data throughput as high as 460k bps may be possible. Throughput this high requires a correspondingly high serial port speed. A replacement communications driver and high-speed serial port will be required to use them at any speed in excess of 115k bps.

A number of ISDN manufacturers are offering TA's with analog telephone support including built-in fax and V.34 modems. These, of course, use the same software that is used by conventional analog "smart modems."

For our purposes, ISDN modems will be treated just like conventional smart modems, whether communications uses built-in analog modems or purely digital channels.

For several reasons, ISDN is not used as commonly in North America as it is in Europe. ISDN, even 10 years after its introduction, is not widely available in North America. Where ISDN is available, the costs can exceed analog telephone lines, sometimes substantially. Analog modems with data compression offer performance that approaches ISDN without compression, so the cost-to-performance ratio may not justify the use of ISDN. Finally, ISDN is not well understood by telephone companies and others. Actually getting it to work correctly in all variations is much harder than it should be.

2.6   RS-232-C

There are many communications standards that have been promulgated by the EIA (Electronic Industries Association). These standards are prefaced by the two letters RS, probably derived from Radio Standard. The EIA is the successor to the Radio Manufacturers Association and the Radio-Electronics-Television Association. Many contributions to the ITU (formerly CCITT) are made through the auspices of the EIA.

The RS-232-C (also called EIA-232) standard, hereafter referred to as RS-232, specifies the control functions and signal paths of 25 signal lines interconnecting modems and terminals or computers. The RS-232 connector contains 25 pins, 22 of which are defined and three are unassigned. RS-232 is very similar to the international standards V.24 and V.28 but it's not identical. For practical implementations, they can be thought of as the same.

Two related specifications, RS-334 and RS-366, describe the electrical signal parameters of RS-232 signals. These specifications are often simply described as part of RS-232 even though, technically, they are separate.

RS-232 signals are bipolar, i.e., the signal excursions are between ± V. The voltage, V can be as little as ± 3V to as much as ± 18V. Actually, most RS-232 receivers work satisfactorily for unipolar signals, say 0V to + V. A bipolar signal will improve noise immunity, however. The signal need not be symmetrical about ground; +12 to -9V excursions are fairly common.

RS-232 signals are also single-ended. That means that only one signal line is used for data transmission. A ground is required as a signal return path. The ground is shared by all signals in the circuit so each connected device must share the same ground and any substantial potential difference on ground can cause faulty operation.

Some of the limitations of RS-232 include its cable length limit of 50 feet and its bit-rate limit of 20k bps. These limitations are observed by ignoring them. The distance limitation is a more obvious physical limitation and should not be challenged (too much). However, improvements in serial port hardware have increased the maximum bit-rate, especially for short cables, to in excess of 500k bps.

Maximum cable length and bit-rate are inversely related. You might be able to get satisfactory operation at 19.2k bps and 100 feet while 115k bps may be satisfactory at a distance of 25 feet using a low-capacitance cable.

The RS-232 standard specifies a cable and connectors with 25 lines. With the advent of the IBM PC, and limiting data to asynchronous only, it is common to see nine-pin connectors. These are called RS-232 also. They are not, of course, but who am I to argue with history?

The most confusing aspect of RS-232 is how it is wired. The reason for this is that the transmit signal on one computer must, somehow, be converted to the receive signal on the computer with which it is linked. Likewise, certain handshake signals are output from one computer and become, again somehow, inputs to the other computer. How does this happen?

Modems or equivalent devices perform these conversions for you. A modem is known as a DCE (Data Communications Device). A computer or terminal is a DTE (Data Terminal Device). A DCE is linked to another DCE while each DCE connects to a DTE using RS-232.

The second most confusing aspect of RS-232 is that the connectors consist of two types. These are called male and female. The male connectors have pins that extend from the base of the connector while female connectors consist of receptacles that mate with the corresponding male pins. I could easily joke about this terminology but my mind does not work that way. Which gender connector is used for a specific application depends on the choice made by the manufacturer of the DTE or DCE. This anarchic approach makes things hard. Fortunately, most manufacturers of common devices have stuck to some basic arrangements — 25-pin RS-232 ports on PCs are usually female while connectors on modems usually are female, too. This means that the cable used to connect a PC to a modem, often termed a "modem cable" for want of a better term, has male connectors on each end. Non-modem DCE manufacturers are less consistent. Naturally, 9-pin RS-232 ports on PCs are usually male! Thus, a 9-pin "modem cable" will have male connectors on the PC end with female connectors on the modem end.

The next section details null modems, cables, and adapters. This should allow you to configure any RS-232 connection that you need

2.7   Null Modems, Cables, and Adapters

Null modems are RS-232 cables or adapters that allow you to connect two DTE devices together. A null modem can take the place of a conventional modem that might have been used. It is a rather simple device with some slight complications, depending on how the flow control lines are handled. The figures that follow will cover every variation that I know. The actual null modem might be a cable with wires cross connected at the connectors to perform the function. Or, it may be a small modular adapter with cross connects done internally. Regardless, the electrical function is the same.

The simplest null modem consists of three connections. Transmit and receive data is crossed over and signal ground connects straight through. The 3-wire null modem has no connections for handshaking (flow control). This kind of connection is suitable for low data rates or transferring small amounts of data. It can also be used if software flow control is used. *Figure 2.7.1* shows this connection. The pin numbers for a 25-pin connector are shown with the corresponding 9-pin connector pin numbers in parenthesis ().

The connector body usually has the pin number imprinted or embossed next to the pin. Make certain that you observe the numbering on the connector body. This will assure that you use the correct pin regardless of whether the connector is male or female.

Connections that cross are not connected to each other unless there is a "dot" or small circle to indicate a connection.

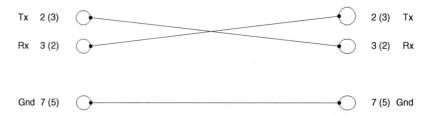

Figure 2.7.1 3-wire Null Modem

An equivalent connection is found in many null modem adapters (or cables). In the simplest null modem adapters, the handshaking (flow control) pins are connected back (loopback handshaking) to the corresponding pins. You should use care when using a commercial null modem. If the adapter or cable uses this kind of connection, it is limited to low data rates, small amounts of data, or when software flow control is used. An example of this is shown in *Figure 2.7.2*.

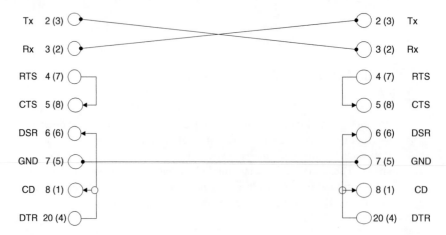

Figure 2.7.2 *3-wire Null Modem with loopback handshaking*

In *Figure 2.7.2* DSR, CD, and DTR are connected together, as are CTS and RTS.

The most useful null modem is one that implements full handshaking. *Figure 2.7.3* shows the most common variation of this null modem. It can be used with virtually any pair of DTE devices, regardless of whether or not the DTE employs flow control.

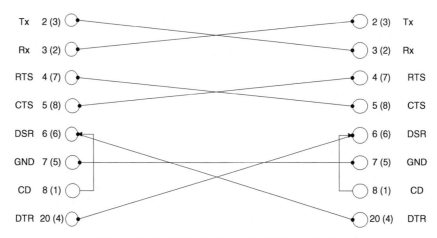

Figure 2.7.3 *Null Modem with full handshaking*

In *Figure 2.7.3*, DSR and CD are connected together, while CTS is cross connected to RTS and DTR is cross connected to DSR.

There is one other serial adapter that is a type of null modem. That is a serial printer adapter. This is, usually, a cable that connects your computer to a serial printer. Most serial printers are a DTE so the transmit data must connect to receive data on the printer. Also, most serial printers use DTR for flow control instead of RTS. So, the printer null modem cable shown in *Figure 2.7.4* is often required. Check your printer manual to make sure, though.

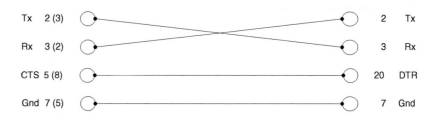

Figure 2.7.4 *Serial Printer Cable*

The serial printer cable shown here may be slightly different than one that might be used if the printer were to be used with a Windows printer driver rather than with serial data directly from a Visual Basic program. Some printer drivers may assume that the printer DTR is connected to the serial port DSR. However, Visual Basic programs that use the serial port instead of the printer object will use the Windows communications API in one form or another. In that case, CTS, not DSR, should be used for handshaking (hardware flow control). Of course, either cable would work if software (Xon/Xoff) flow control is used.

You can use other adapters to convert a connector gender from male to female and/or to convert 9-pin connectors to 25-pin. These adapters are available at reasonable cost from a variety of sources. Your local Radio Shack™ store will often have them in stock.

How can you find out what adapter or null modem you need? Well, ideally, the documentation for the device that you need to connect to will tell you. But, if it does not, there are tools that will help you figure it out.

Many manufacturers build RS-232 "breakout boxes," test, and wiring adapters. With these, you can use standard, straight through cables and a breakout box to create any combination of connections that you might need. Fortunately, RS-232 is pretty well connected. If you make a wiring error with a breakout box, you are not likely to cause any damage. If the breakout box or RS-232 tester has LED indicators, you will know which lines are active and can easily avoid inadvertently connecting one active line to another.

Using a breakout box, you can test one of the connection schemes shown previously. A breakout box can be a real time saver. See Chapter 9 and the Resources appendix for more information.

2.8   RS-232 Switches

RS-232 switches can be used to connect a single serial port to multiple serial devices. These can take one of two forms: manual or code-operated switches. Some switches, called serial crosspoint switches, allow you to connect any port to any other.

A manual switch is just what its name suggests. It is, usually, a box with a rotary switch that allows you to connect one serial port to another. These boxes may have mixed gender connectors. Some may offer built-in null modems.

Code-operated switches allow you to send a port select command from your computer to the switch box. A microprocessor detects the codes that you send and electronically selects the port specified by your command. Code-operated switches are quite useful for use with polled data acquisition systems to select one of several printers or other serial devices.

Some code-operated switches, often designed for remote industrial data acquisition or control systems, have built-in modems. These remote code-operated switches often may be remotely programmed and have built-in security systems.

The only caution is that the code used to select ports usually cannot be duplicated in the actual data that is transferred, unless the manufacturer of the switch advertises "transparent operation." In that case, some extra method, such as the state of a hardware handshake line, sequence after modem connect, or some special method is required to isolate the select command from data.

2.9 RS-422 and Line Drivers

RS-422 specifies the electrical characteristics of a serial data interface that uses balanced voltage signals. Each signal path consists of a pair of wires. The voltage difference imposed between each wire in the pair represents a digital signal. RS-422 provides point-to-point or point-to-multipoint communications (up to 10 receivers can be connected to one transmitter).

Unlike RS-232, where the signal is with reference to ground, RS-422 does not use ground as a signal path. That means that RS-422 signals are much less subject to noise and are, basically, immune to problems caused by differences in the ground potential between systems. The nominal voltage difference between the two wires of a pair is 0-5V. The RS-422 pair is terminated at the receiver with a resistor that matches the characteristic impedance of the wire pair. This reduces reflection of signal back to the transmitter that might otherwise cause data errors.

The differential receivers in an RS-422 link will reject any noise that is imposed on the pair. An RS-422 system often will allow common mode input ranges of ±7V or more while they are sensitive to differential signals as low as 200mV.

These characteristics allow RS-422 signals to be transmitted at data rates up to 10M bits per second and up to 4000 feet in distance. Of course, just like RS-232, there is a trade-off in speed and distance. Communication at 10M bits per second is limited to 40 feet but the specification allows 100K bits per second at 4000 feet. The RS-422 specification provides much higher performance than RS-232.

There are add-in boards for PCs that provide RS-422 interfaces. These use conventional UARTs and are used like a conventional serial port. There also are several manufacturers of RS-232 to RS-422 adapters.

I have included line drivers in this section but they are not exactly the same as RS-422. There is enough similarity that it is convenient to discuss them together, however.

Line drivers are DCE devices (as are modems) used to extend the operational distance of RS-232. These line drivers often use RS-422 signaling between each line driver. They are not modems because the signals are baseband, not modulated.

Some line drivers offer additional features beyond the ability to extend the operational distance of RS-232. Useful features are transformer coupling or optical isolation. This allows the line drivers to electrically isolate the connected systems which can be very valuable in industrial or other harsh environments. Line drivers with transformer or optical isolation often offer 500V or higher isolation between connected devices.

2.10   RS-485

RS-485 is similar to RS-422. It, too, is a differential system with similar speed and range specifications. The primary difference is that it supports true multipoint communications. Up to 32 drivers (transmitters) and receivers can be connected to the same bus, i.e., the same pair of wires.

An RS-485 driver that is not transmitting must be disabled. This disabling is called "TRI-STATE." This allows another driver to send data on the shared wire pair. RS-485 drivers can withstand bus contention, i.e., if two or more drivers try to send data at the same time. However, they do not detect such contention and, if it happens, all data that is transmitted during contention is lost. Eliminating bus contention so that data is sent and received without error can be done in one of two ways. First, the software that sends data must make certain that no other point on the network is transmitting. This is usually done by implementing a half-duplex, Master/Slave architecture. This means that one computer is the Master and it controls access to the network by sending commands to Slaves. When a Slave receives a command (each Slave must have a unique address and the address is included in the command so that the Slave can identify a command to it), it can then send data. The Master waits until a complete message is received from a Slave and then it sends a message to another Slave to allow it to access the network. The ability to broadcast messages from Master to all Slaves is often implemented. Depending on the design of the communications protocol, a Slave may or may not be permitted to send messages to another Slave without the intervention of the Master.

The device that controls each RS-485 driver is responsible for disabling the driver when no data is to be sent and enabling the driver when data is to be sent. There is no standard method for doing this; how it is handled is determined by the design of the RS-485 interface that you are using. So make certain that you read the manufacturer's instructions. There are three methods that are commonly used by RS-485 interfaces.

The software that sends data can enable the RS-485 driver by asserting RTS. RTS is often connected to the transmit enable of the RS-485 driver. This is the approach that is used by a number of different commercial RS-485 adapters. The advantage is that it requires very little extra hardware. The disadvantage is that RTS is handled by COMM.DRV under Windows. COMM.DRV often implements a 200 mS delay when RTS changes state. This delay can slow data transfer substantially and it must be accounted for by the software that sends data. Alternatively, a replacement for COMM.DRV can be purchased that eliminates this built-in delay. Here is example code, using MSCOMM, which you might employ:

```
MSComm1.RTSEnable = True
Delay 200                         'delay 200 mS
MSComm1.Output = SomeData
Do Until MSComm1.OutBufferCount < 1
    DoEvents                      'wait for all data to be sent
Loop
MSComm1.RTSEnable = False
```

If the version of COMM.DRV that you are using has no built-in RTS delay, you can eliminate the first delay but you may need to add a short delay (similar to the one shown in the next code fragment) before RTS is lowered to assure that all data is transmitted.

Another approach that is used in some RS-485 add-in cards (boards that are used instead of a conventional RS-232 serial port) is to use an I/O port that is not used by the standard serial port to control the driver. The problem with this is that Visual Basic provides no Inp or Out functions to access I/O ports. You have to use an add-on DLL or VxD to control this I/O port. Examples that use such DLLs and VxD are included in the sample code in the book, although not for this specific application, so you will have a little homework to do if you use such a RS-485 card. Here is example code, using MSCOMM and an add-on DLL that has the required IO port function, which might be used:

```
WritePort &H3FF, 1      'enable RS-485 driver
MSComm1.Output = SomeData
Do Until MSComm1.OutBufferCount < 1
    DoEvents             'wait for all data to be sent
Loop
Delay 16/PortSpeed       'PortSpeed in bytes/S
WritePort &H3FF, 0       'disable RS-486 driver
```

I should comment that the delay after OutBufferCount < 1 is to allow all data in a 16550 AF UART to be sent before the driver is disabled. At 19200 bps, this delay is about 9 mS. The assumption is that the transmit FIFO is full. However, there is no way to be sure so this is a way to assure that all data will be transmitted.

One approach that is used by a number of add-in card manufactures, and by RS-232 to RS-485 adapter manufacturers, is to have special hardware on the card or adapter that enables the driver when data is present in the UART transmit buffer and disables the driver when no data is in the UART buffer. This method is the simplest because it requires no special program code to enable and disable the RS-485 driver. Your code can send data as fast as possible with no extra delays as are required for other RS-485 cards or adapters. These cards and adapters tend to be slightly more expensive than other, simpler cards. However, their ease of use may justify the extra cost.

B&B Electronics has a more complete description of the workings of both RS-422 and RS-485. You can download it from their web site. See Appendix A for their URL.

2.11   Current Loop

Current loop perhaps is the oldest way to connect serial devices. Teletype machines predating WWII used current loop signaling. Actually, the original telegraph system designed by Samuel Morse used current loop! Typically, a 20 mA loop current indicates a Marking condition and 0 mA indicates a Space. There is no standard for current loop signaling so their design was usually ad hoc. A current loop system offers high noise immunity and good isolation. However, speed and signaling distance are somewhat limited. Several manufacturers offer hardware converters to interface between current loop and RS-232. If you need to interface to vintage equipment, it can be done. Using current loop introduces no software issues except those caused by slow, sometimes mechanical, systems.

In some industrial control environments where high-noise immunity is a priority, 4-20 mA current loop systems are still commonly used. One popular communications protocol that can be used in this environment is called the Hart protocol.

2.12   Checksums and CRCs in Visual Basic

Problem: Please show me some Visual Basic code that I can use to validate serial communications. When digital messages are transmitted and received, some errors can be expected to appear. Errors occur because of interference between channels, fading of signals, atmospheric conditions, and various sources of noise. You have spoken of checksums and CRCs. How about it?

Solution: First, take a look at the code in Chapter 4 of this book. The XMCommCRC ActiveX control source contains both checksum and CRC calculations. Commonly used methods for detecting errors include checksums, longitudinal redundancy code, and cyclic redundancy checks. I will break out a couple of other variations here.

Checksums

A checksum calculation often is the simplest way to validate data. The actual algorithms for calculating checksums encompass a variety of arithmetic operations. The checksum used by XMODEM uses these rules: "Add each byte in the packet (including packet header) one byte at a time, modulus &HFF." Here is a code fragment that calculates an eight-bit XMODEM checksum for a string buffer. The code in XMCommCRC does the equivalent calculation using an array of type Byte.

```
Private Function XMODEMchecksum (Buffer As String) As String
Dim I As Integer
Dim Checksum As Byte
     For I = 1 to Len(Buffer)
            Checksum = (Checksum +  Asc(Mid$(Buffer, I, 1))) And &HFF
     Next I
     XMODEMchecksum = Chr$(Checksum)
End Function
0
```

A similar checksum calculation used in some other communications protocols replaces the addition operator in the checksum with a bitwise logical Xor operator. For example,

```
Private Function Xorchecksum (Buffer As String) As String
Dim I As Integer
Dim Checksum As Byte
     For I = 1 to Len(Buffer)
            Checksum = Checksum Xor (Asc(Mid$(Buffer, I, 1)))
     Next I
     Xorchecksum = Chr$(Checksum)
End Function
```

You easily can imagine other similar calculations. This type of single-byte checksum calculation can detect about 95% of all possible errors. Thus, they are good – but not great!

CRCs

Let us improve on the error detection capability provided by a simple checksum calculation. One way would be to extend the checksum by simply making it larger. A two-byte arithmetic checksum accumulates a 16-bit result and would improve error detection by about 2%. Thus, about 97% of all possible errors would be detected. This scheme is used every day. TCP/IP (the protocol stack used for Internet communications) uses a simple two-byte checksum for error detection. This is satisfactory for communications where the error rate is low. Internet communications are fairly immune to errors because other error-correction methods are used. Often connections to the Internet use modems that implement much more secure error detection and correction using CRCs.

Common CRC (cyclic redundancy check) calculations result in 16 or 32-bit (uncommonly, even larger CRCs are possible) results. Much less common CRC calculations result in 8-bit or 24-bit results. I will present code examples here that do the more common 16-bit and 32-bit calculations. CRC calculations improve the probability that an error will be detected. CRC-16 will detect 99.99% of all possible errors. A CRC-32 lowers the probability of an undetected error substantially. Only one in several billion possible errors will go undetected when using a CRC-32

The actual algorithm used for CRC calculations varies. Each uses a polynomial that identifies the actual algorithm. The following table lists a few of the polynomials that may be used. I will not discuss 8-bit CRC calculation. However, I have included an example on the accompanying CD ROM that illustrates the 8-bit calculation used by various Dallas Semiconductor products such as the DS-1820 digital thermometer.

The following table illustrates some possible CRC calculations and the associated polynomial.

CRC		Polynomial
CRC-16	16-bit	$X^{16} + X^{15} + X^2 + 1$
SDLC (CCITT)	16-bit	$X^{16} + X^{12} + X^5 + 1$
CRC-32	32-bit	$X^{32} + X^{26} + X^{23} + X^{22} + X^{16} + X^{12} + X^{11} + X^{10} + X^8$ $+ X^7 + X^5 + X^4 + X^2 + X + 1$

Table 2.1 CRC polynomials

The following code is found in the clsCRC16.cls module. Refer also to the CRC calculation in the XMCommCRC section that follows this section.

```
Public Function CRC16A(Buffer() As Byte) As Long
Dim I As Long
Dim Temp As Long
Dim CRC As Long
Dim J As Integer
    For I = 0 To UBound(Buffer) - 1
        Temp = Buffer(I) * &H100&
        CRC = CRC Xor Temp
            For J = 0 To 7
                If (CRC And &H8000&) Then
                    CRC = ((CRC * 2) Xor &H1021&) _
                                    And &HFFFF&
                Else
                    CRC = (CRC * 2) And &HFFFF&
                End If
            Next J
    Next I
    CRC16A = CRC And &HFFFF
End Function
```

This code and the code in the XMCommCRC ActiveX control in the next section illustrate straightforward implementations of the calculations required for the polynomial used.

The nested loops that are used in these calculations are fairly expensive. That is, they can take a significant amount of computational resource. Calculating a 16-bit CRC using this sort of technique executes about 2000 statements for a 128-byte packet. The execution time might be important on a small or low-power system.

Calculation of a 32-bit CRC is much more complex and time consuming. Implementation of a 32-bit CRC calculation using a technique similar to the above would require execution of perhaps as many as 64000 statements for the same packet size! The execution time might prove to be prohibitive. However, one can reduce this by using a table-lookup method. When the object is instantiated (the class Initialize event is generated), a table is created in memory that represents intermediate steps in the calculation of the polynomial. This table data can be "reused" for each CRC calculation. The code in the next example illustrates a method that improves CRC calculation performance by table lookup. The lookup method reduces by four- or five-fold the number of statements that might need to be executed.

CRC32 Class Module (clsCRC32.cls)

```
Private CRCTable(0 To 255) As Long
```

CRCTable is an array of Long Integers (32-bits) that will be used to store a set of values that will be used for each CRC calculation.

The actual values in CRCTable are calculated when the object is created via the Initialize event.

```
Private Sub Class_Initialize()
Dim I As Integer
Dim J As Integer
Dim Limit As Long
Dim CRC As Long
Dim Temp1 As Long
  Limit = &HEDB88320
  For I = 0 To 255
    CRC = I
    For J = 8 To 1 Step -1
      If CRC < 0 Then
        Temp1 = CRC And &H7FFFFFFF
        Temp1 = Temp1 \ 2
        Temp1 = Temp1 Or &H40000000
      Else
        Temp1 = CRC \ 2
      End If
      If CRC And 1 Then
        CRC = Temp1 Xor Limit
      Else
        CRC = Temp1
      End If
    Next J
    CRCTable(I) = CRC
```

```
    Next I
End Sub
```

OK. Now we can call the CalcCRC32 function. It loops through an array of data, usually a serial communications data packet. The values previously calculated and stored in the CRCTable are used for one of the vital Xor operations needed to calculate the CRC. The table-lookup substantially reduces the required number of statements needed.

```
Public Function CalcCRC32(ByteArray() As Byte) As Long
Dim Limit As Long
Dim CRC As Long
Dim Temp1 As Long
Dim Temp2 As Long
Dim I As Long
   Limit = UBound(ByteArray)
   CRC = -1
   For I = 0 To Limit
     If CRC < 0 Then
       Temp1 = CRC And &H7FFFFFFF
       Temp1 = Temp1 \ 256
       Temp1 = (Temp1 Or &H800000) And &HFFFFFF
     Else
       Temp1 = (CRC \ 256) And &HFFFFFF
     End If
     Temp2 = ByteArray(I)     ' get the byte
     Temp2 = CRCTable((CRC Xor Temp2) And &HFF)
     CRC = Temp1 Xor Temp2
   Next I
   CRC = CRC Xor &HFFFFFFFF
   CalcCRC32 = CRC
End Function
```

How do you decide what form of checksum or CRC calculation should be used? That is a system-design problem. This critical design decision should be made early on. Fortunately, often this decision has already been made for us. We are told what algorithm to use based on the requirement of the equipment with which we are working.

If left on our own, we should base the decision on the error rate that we expect to encounter. If we expect to see an error only occasionally (perhaps one error in 10^6 characters), then some simple form of error detection (checksum or CRC-16) should be satisfactory. Examples of systems where low error rates would be expected are a direct-wired connection or a connection over modems that implement their own error-correction methods. A fiber-optic system should have a **very** low probability of error.

However, if we anticipate that the error rate may be significant (perhaps one error in 10^5 characters, or worse) then a more resilient error detection scheme such as CRC-32 may be justified. An example of a system where high error rates might be observed is a non-fixed point radio system. Such a radio system would experience multiple sources of interference and potential loss of signal integrity.

Chapter 3   Windows

Windows is a truly powerful multitasking environment. With this power comes
responsibility. Windows provides two methods, called APIs, for programs to use serial
communications. One, called, for want of a better term, the Windows API, is built into the
Windows kernel. The other, designed for use with Windows 95/98 and Windows NT 4.0,
or later, is called TAPI (Telephony API). TAPI goes far beyond serial communications but
that is the only aspect that we will examine here.

3.1   The Windows API

The Windows API provides a number of functions that you use for serial communications.
It is the interface between a program and the Windows communications driver. The
Windows communications driver virtualizes the UART. It has the interrupt service routines
that manipulate serial data, buffers transmit and receive data, and it does UART error
handling. All of this is done 'behind the scenes." An understanding of the basics can be
useful, however.

The MSCOMM custom control comes with Visual Basic Professional Edition and higher. It
also is included with Visual Studio.NET products. MSCOMM encapsulates the Windows
API, making it very easy to use. All commercial communications add-ons, either DLLs or
custom controls, also encapsulate the Windows API while adding features beyond basic
serial I/O. MSCOMM and various commercial products will be discussed in detail in later
chapters.

MSCOMM, or a commercial communications add-on, is better than using the Windows
API. The add-on's functions are easier to use. And, you usually get better performance
because things that might have to be done in Visual Basic code instead are done in compiled
C or assembly language. The C or assembly language routines will be faster than the
comparable Visual Basic code. Features like error-corrected file transfers demand the
performance that an add-on will provide.

You may want to use the Windows API anyway. You may be using Visual Basic Standard
Edition. You may not want to spend the money to upgrade to the Professional Edition or to
purchase a commercial add-on. Or, you may just want to learn how to use the Windows
Communications API. Whatever the reason, here is what you need to know.

There are several basic steps to using the Windows communications API.

1. Use the **OpenComm** (16-bit Windows) or the CreateFile (32-bit Windows) functions to open a serial port. This notifies Windows that your application is using the port and keeps other applications from accessing the port.

2. Configure the serial port using the **SetCommState** function.

3. Use the **SetCommEventMask** function to enable detection of any UART hardware events that interest you.

4. Write and read serial data. To write data you will use the **WriteComm** function if you are using the 16-bit API. You use the **WriteFile** function if you use the 32-bit API. To read data you will use the **ReadComm** function for the 16-bit API and the **ReadFile** function for the 32-bit API.

5. Use the **GetCommError**, **GetCommEventMask** and **GetCommEvent** functions to get information on errors and hardware events. **GetCommEvent** resets the event state so that new errors and events will be detected.

6. Use the **EscapeCommFunction** to set the state of DTR, RTS, and for other housekeeping tasks.

7. Use the **CloseComm** function to release the port so that other applications can use it.

The Device Control Block (DCB) is a data structure that the API functions use to specify the serial port configuration. The 16-bit DCB and the 32-bit DCB differ in detail. Each of those, along with the 16-bit and 32-bit declarations, will be further described in the chapters that illustrate the Windows API in code.

There are some other functions in the API that can be used. *Table 3.1.1* shows all communications API functions and gives a description of each.

I should issue one caution when using the Windows communications API. A serial port that is opened must be closed before it again can be opened, even by the same application. Visual Basic allows you to stop the execution of a program within the design environment by clicking on the "End" button or by selecting "End" from the VB/IDE (Integrated Design Environment) Run menu. These actions do not close an open serial port. If you design a program that uses the OpenComm API, you will inevitably run it in the IDE to debug. Equally inevitably, you will end the program without calling CloseComm. When you do this, you will not be able to again execute OpenComm for that serial port. You will have to exit Visual Basic. At that point, Windows closes the opened port. You can then restart VB and rerun your program.

Function	Description
BuildCommDCB	This builds a Device Control Block (DCB) to be used with OpenComm, etc.
ClearCommBreak	Clears the UART break state.
CloseComm	Close the serial port.
EnableCommNotification	Specifies which window receives communications notification messages. VB 5 is the first version that allows callbacks via pointers. Earlier versions need an add-on custom control to use this function.
EscapeCommFunction	Performs serial port housekeeping tasks.
FlushComm	Clears the serial port transmit and receive buffers.
GetCommError	Return the serial port status.
GetCommEventMask	Return the serial port event mask.
GetCommState	Return the current Device Control Block (DCB).
OpenComm CreateFile	Open the serial port.
ReadComm ReadFile	Read data from an open serial port.
SetCommBreak	Send break.
SetCommEventMask	Specify which serial port events to detect.
SetCommState	Configure a serial port using the specified DCB
TransmitCommChar	Transmit a single character, bypassing the transmit buffer.
UngetCommChar	Place a character into the receive buffer so that it will be read next. It need not be a character that had been read previously.
WriteComm WriteFile	Write data to an open serial port.

Table 3.1.1 Communications API Functions

3.2   TAPI

The Windows Telephony API is part of the Microsoft Windows Open Services Architecture (WOSA) that provides a single set of interfaces to enterprise computing services. WOSA encompasses a number of APIs, providing applications and corporate developers with an open set of interfaces to which applications can be written.

Windows Telephony API consists of two interfaces. Developers write to an applications programming interface (API). The other interface, referred to as the service provider interface (SPI), is used to establish the connection to the specific telephone network. This model is similar to the computer industry model whereby printer manufacturers provide printer drivers for Windows-based applications. Figure 3.2.1 shows the relationship between the "front-end" Windows Telephony API and the "back-end" Windows Telephony SPI.

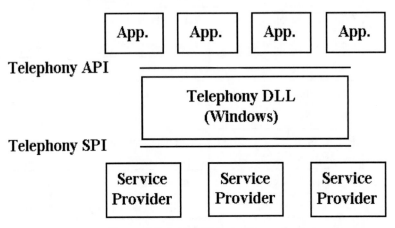

Figure 3.2.1 TAPI Block Diagram

TAPI allows you to build applications that can place, receive, and manage calls from your computer. It supports both voice and data calls. Your application can place calls using a variety of hardware devices on telephone networks that offer complex services, like voice mail, e-mail, conferencing, and call waiting. All these services use different technologies to manage calls and transmit voice and data; however, TAPI hides these service specific details from your application.

The only TAPI applications that I will examine in later sections of the book are those that use a modem for data communications and simple telephone dialing.

What does TAPI give us that we might not have otherwise? In simple terms, it allows us to write applications that are hardware independent. One of the biggest headaches when using modems in a serial communications application is the fact that there are few standards that define what commands a modem manufacturer uses to activate a specific mode of operation in a modem.

For example, we might want to make certain that a modem connects in error-correcting mode, uses modem data compression, keeps the serial port speed fixed (regardless of the actual raw modem speed) and uses hardware (RTS/CTS) flow control. There are large numbers of different modem initialization strings used for different modems that would be used to do this. The modem manufacturer creates a file that defines all of his modem characteristics and the modem string used to activate different modes. This is done in a standard way, the TSP interface in Figure 3.2.1, because a modem is a Telephony Service Provider.

What this means is that your TAPI application need not know what modem is being used. Even more importantly, users of your program have a standardized way to install modems or other telephony devices. Users do not need to know what initialization string is used for any specific option. If users install a new modem, an application that you write using TAPI will not have to change; it should work transparently.

These benefits are tremendous. They are also, at this writing, not always fulfilled. Like every aspect of computer development, TAPI will become more reliable and flexible as it matures.

TAPI is only available on Windows 95/98 and Windows NT 4.0 platforms. So, any application that must use Windows 3.x or Windows 3.51 must rely on non-TAPI methods of modem control and access.

There are two forms of TAPI that we might use.

Simple TAPI, also called Assisted Telephony, uses a modem to dial a telephone number. It does not do anything else. Simple TAPI is a useful, if limited, function. There is an example in a later section that illustrates how it might be used.

Full TAPI is much more powerful than Simple TAPI. It also is much harder to use. In fact, Full TAPI requires callback procedures. Visual Basic through version 4 does not provide callbacks via pointer. Visual Basic 5 does support procedure pointers via the new AddressOf operator.

Use of full TAPI for serial data will often involve a commercial communications add-on. One that is currently available, PDQComm from Progress Software, Crescent Division, comes with the PDQTapi custom control. This makes use of TAPI for serial communications fairly easy. There is an example later in the book that shows the use of PDQTapi. It also mentions some of the limitations that may be of some concern.

One feature of TAPI, including Simple TAPI, is the use of a standardized method for dialing regardless of location (country, area code, etc.). To dial a number, a dialable address must be formed from the canonical address. A dialable address contains the necessary dialing and routing information to place a call on a specific line at a specific location. Here is a description of the Canonical Address Format:

+ CountryCode Delimiter [AreaCodeDelimiter] SubscriberNumber [| Subaddress] [^ Name] [CRLF ...]

Where fields between [ ] are optional

+ is ASCII Hex (2B). It indicates that the number that follows uses the canonical format.

CountryCode is a variable length string containing only the digits 0–9. It identifies the country in which the address is located.

Delimiter is any character except 0–9 + | ^ CRLF. It is used to delimit the end of the CountryCode part and the beginning of the Area Code or SubscriberNumber, and the Area Code and SubscriberNumber parts of an address. A hyphen (-) is often used as a delimiter.

Area Code is a variable length string containing only the digits 0–9. It is optional. Area Code is the area code part of the address. If present, it must be followed by a delimiter.

SubscriberNumber is a variable length string containing only the digits 0–9 as well as delimiters but excluding + $ | ^ CRLF. Delimiters embedded in the SubscriberNumber are ignored.

| is ASCII Hex (7C) and is optional. If present, the information following it up to the next + | ^ CRLF, or the end of the canonical address string, is treated as sub-address information, as for an ISDN sub-address.

Subaddress is a variable length string containing a sub-address. The string is delimited by + | ^ CRLF or the end of the address string. During dialing, sub-address information is passed to the remote party. It can be such things as an ISDN sub-address or an email address.

^ is ASCII Hex (5E) and is optional. If present, the information following it up to the next CRLF or the end of the canonical address string is treated as an ISDN name.

Name is a variable sized string treated as name information. Name is delimited by CRLF or the end of the canonical address string and may contain other delimiters. During dialing, name information is passed to the remote party.

CRLF is ASCII Hex (0D) followed by ASCII Hex (0A). It is optional. If present, it indicates that another canonical number is following this one. It is used to separate multiple canonical addresses as part of a single address string for use with inverse multiplexing. Inverse multiplexing is typically used to allow multiple channels to be combined into a single call to increase the call's bandwidth. This technique is infrequently used, so you can disregard it for now.

Here is a description of a dialable number:

DialableNumber [| Subaddress] [^ Name][CRLF ...]

Where:

DialableNumber is digits and modifiers 0–9 A–D * # , ! W w P p T t @ $? ; delimited by | ^ CRLF or the end of the dialable address string.

Within the DialableNumber are:

0–9 A–D * # are ASCII characters corresponding to the DTMF and/or pulse digits. Although most telephones support only 12 DTMF tones in a 3x4 array, the DTMF standard supports a 4x4 array with 16 tones. The characters A–D correspond to the extra 4 DTMF tones.

! is ASCII Hex (21). It indicates that a flash is to be inserted in the dial string.

P p either ASCII Hex (50) or Hex (70). It indicates that pulse dialing is to be used for the digits following it.

T t either ASCII Hex (54) or Hex (74). It indicates that tone (DTMF) dialing is to be used for the digits following it.

, is ASCII Hex (27). It indicates that dialing is to be paused. The duration of a pause is device-specific and can be retrieved from the line's device capabilities. Multiple commas can be used to provide longer pauses.

W w is ASCII Hex (57) or Hex (77). It is an uppercase or lowercase W indicates that dialing should proceed only after a dial tone has been detected.

@ is ASCII Hex (40). It indicates that dialing is to "wait for quiet answer" before dialing the remainder of the dialable address. This means to wait for at least one ringback (AKA remote ring) tone followed by several seconds of silence.

$ is ASCII Hex (24). It indicates that dialing the billing information is to wait for a "billing signal" (such as a credit card prompt tone).

? is ASCII Hex (3F). It indicates that the user is to be prompted before continuing with dialing. The provider does not actually do the prompting but the presence of the ? forces the provider to reject the string as invalid. It alerts the app to the need to break the string into pieces and prompt the user in-between.

; is ASCII Hex (3B). If placed at the end of a partially specified dialable address string, it indicates that the dialable number information is incomplete and more address information will be provided later. ; is only allowed in the DialableNumber portion of an address.

| is ASCII Hex (7C). It is optional. If present, the information following it up to the next | ^ CRLF, or the end of the dialable address string, is treated as sub-address information (as for an ISDN sub-address).

Subaddress is a variable length string containing a sub-address. The string is delimited by the next + | ^ CRLF or the end of the address string. When dialing, sub-address information is passed to the remote party. It can be an ISDN sub-address, an email address, and so on.

^ is ASCII Hex (5E). It is optional. If present, the information following it up to the next CRLF or the end of the dialable address string is treated as an ISDN name.

Name is variable length string treated as name information. Name is delimited by CRLF or the end of the disable address string. When dialing, name information is passed to the remote party.

CRLF is ASCII Hex (0D) followed by ASCII Hex (0A). It is optional. If present, this indicates that another dialable number is following this one. It is used to separate multiple canonical addresses as part of a single address string (inverse multiplexing).

TAPI supplies the function lineTranslateAddress to convert a canonical address to a dialable address. The function uses a canonical address, a line identifier, and information about the caller's current location to synthesize a dialable address. Running the Telephony control panel applet typically specifies information about the caller's current geographic location.

For example, a valid canonical address for Denver telephone information would be "+1-303-555-1212" where the first 1 is the country code for the USA and 303 is the Denver area code. If this number is passed to lineTranslateAddress and if you are calling from a business phone in Denver, the function would return a number similar to "t9w5551212" (depending on the features of the line you're using). The "t" in the dial string says to use DTMF (Dual-Tone Multi-Frequency signals or Touch Tone) dialing. It first dials a 9 and waits for a dial tone, then dials 555-1212. Notice that the 1 and 303 area code were removed by TAPI since you're calling from within the 303 calling area.

TAPI provides much more than dialing, of course. There are dozens of functions that support data, fax, PBX, and voice communications. These are beyond the scope of this book. However, a number of the non-data functions will be found in ExceleTel's TeleTools TAPI OCX. Refer to Appendix A for contact information for this company. In addition, I have included a description of X/Page OCX (provides a TAPI interface for Alphanumeric paging) from Logisoft, Inc in the chapter on Alphanumeric paging. More and more tools will become available to aid VB programmers.

3.3 General Programming Techniques

This section will address some programming issues that are common to any Visual Basic program. I'll present code examples that use MSCOMM. However the discussion will apply to any other custom control and, in general, to API or DLL-based code.

You often will have the option to read data from a serial port one character at a time (perhaps looking for a specific terminating character), appending each character to a string buffer, or to append all available data to a string buffer and to parse data out of that string using the VB InStr function or perhaps the StringVariableName.Index method in VB .NET.

You will **always** get better performance with the second option. That is the method that I will show in all of my example code. This method has the additional advantage that parsing receive data from a string buffer is necessary if the data that you are looking for consists of more than a single character. Using MSCOMM or custom controls that have equivalent functionality, set the property MSCOMM1.InputLen = 0.

How do you decide what buffer size to use when you configure a comm port? You can select the same or different sizes for the transmit and receive buffers that are managed by the Windows communications driver. There is an optimum size for the transmit buffer that is governed by the device that you most immediately connect to. If this is a modem, there will usually be an input buffer of 1K to 2K bytes. If this buffer is filled, the modem must use flow control to make certain that the buffer does not overflow, resulting in loss of data. Therefore, a 1K transmit buffer size is most often the best size to choose when using a modem.

Less guidance is available for determining the receive buffer size. A good rule-of-thumb is to base your receive buffer size on the maximum data rate. You should allow a minimum of two seconds of receive buffer depth. This will allow a reasonable margin for multitasking, network activity, and any other processes that require significant processor time. If your maximum data rate is 57.6k bps, one byte might be received every 173 microseconds, 5760 bytes per second. So, a good choice might be 12K bytes. Your receive buffer probably should not exceed 16K bytes, though. Any computer that is used for very high-speed data should be sufficiently fast and it must use appropriate hardware such as 16550 AF UARTs. Higher-speed computers can allow you to use smaller receive buffers. Faster is better!

As discussed in an earlier section in the book, if you have the option, use hardware flow control. Flow control is required for most applications at any speed in excess of 9600 bps. Hardware (RTS/CTS) flow control is more reliable than software (Xon/Xoff) flow control.

3.3.1   Procedural and Event Driven Code

Visual Basic programs epitomize the event driven methodology. When a VB form loads, a series of events is generated automatically. The form's load, activate, resize, paint, and focus events may all fire. A VB program can be written to take advantage of these automatic events to stage, or force execution of, a series of code statements. The code that executes in these events can be logically linked but the order of execution is based on the design of Visual Basic and, to an extent, the function of the code itself.

Command buttons, Timers, Text boxes, and communications custom controls are a few examples of the many controls that can be added to a form. Each of these controls adds its own set of events that can be used to sequence code in response to user action or other computer activity.

You have the option to write code in Visual Basic that does not rely on any events at all. If your code starts in a basic module named Sub Main, it is possible to write an entire program that has no event driven code at all. However, this is contrary to the spirit of Windows programming and it is not often done.

However, it is practical to write code that is not logically event driven, although the code itself executes as a result of Visual Basic events. This sounds contradictory but it is not in reality. The next section will discuss state machines. They are critical to the design of all communications programs.

Code that executes "in-line" or that proceeds from state to state as a result of program operation — not Visual Basic events — is what I term procedural code. A simple example of procedural code is,

```
Sub TestDial_Click ()
    Dim Buffer As String

    TestDial.Enabled = False
    Comm1.Rthreshold = 0
    Comm1.InputLen = 0
    Comm1.Output = "ATZ" & VbCr
    Do
        DoEvents
        Buffer = Buffer & Comm1.Input
        If InStr(Buffer, "OK" & vbCrLF) Then
            Exit Do
        End If
    Loop
    Buffer = ""
    Comm1.Output = "ATDT555-1234" & VbCr
    Do
        DoEvents
        Buffer = Buffer & Comm1.Input
        If InStr (Buffer, "CONNECT") Then
            Online
            Exit Do
    Elseif InStr (Buffer, "NO CARRIER") Then
            Offline
            Exit Do
    End If
    Loop
    Buffer = ""
    'etc.
    TestDial.Enabled = True
End Sub
```

This code fragment shows a Visual Basic event (TestDial_Click) that activates a series of "in-line" procedures. The important things that this fragment does is to send an ATZ command to a modem, wait for an OK response, dial a telephone number and wait for a modem response that indicates whether or not the modem connected. If it connects, the Online procedure is called; otherwise, the Offline procedure is called. No Visual Basic events were involved in this process except the Click event that started it.

A number of experts in the Visual Basic developer community have observed that DoEvents loops, like the ones that I have used to wait for modem responses, are dangerous. Some have gone so far as to say that they are an anathema and should be avoided completely. The reason for using DoEvents in the loops is to allow multitasking in 16-bit versions of the program and to allow Visual Basic to process messages (events) for other objects in the program.

There are two problems with DoEvents loops. First, each DoEvents call has processor overhead. This overhead is small under Windows 3.x, medium under Windows 95/98, and more significant under Windows NT. Each call to DoEvents invokes three or more processor task switches. Each task switch involves saving the entire state of the machine (all registers, address space, and links to the previous task), then loading a new execution state. The microprocessor has built-in support for some of this effort but nonetheless several microseconds are required (from 17 to 100 on a 100 MHz Pentium depending on OS and 3 to 4 on a Pentium III running at 800 MHz under Windows 2K).

The second issue, and one that can be more important than the first, is that one of the reasons to use DoEvents is to allow Visual Basic to process messages or events. However, this allows unintended reentrancy. Suppose the user clicks on the TestDial command button while one of the DoEvents loops is processing a previous click. This would cause a serious problem. However, I have avoided that possibility by disabling the command button on entry to the Click event and re-enabling it on exit. This prevents reentry. Any code that uses a DoEvents loop to wait may require equivalent technique, using either a control Enabled property or by using some sort of flag.

Well, cannot we avoid the "DoEvents issues" by a different technique? The obvious answer is, "Yes." Two things might be done. You could convert the step-by-step process, essentially a synchronous system, to an event driven process. There are two ways that this might be done.

First, you could use a Timer control to implement a Timer event driven process. Second, you might use the OnComm receive event processing that is available by all communications custom controls.

Here is an example of the first idea. Add a Timer named DialTimer to your form.

```
Sub TestDial_Click ()
    TestDial.Enabled = False
    Comm1.Rthreshold = 0
    Comm1.InputLen = 0
    Comm1.Output = "ATZ" & VbCr
    DialTimer.Enabled = True
End Sub

Sub DialTimer_Timer ()
    Static State As String
    Static Buffer As String

    If State = "" Then
```

```
            State = "INIT"
            Buffer = ""
        End If
        Select Case State
            Case "INIT"
                Buffer = Buffer & Comm1.Input
                If InStr(Buffer, "OK" & vbCrLF) Then
                    State = "DIALING"
                    Comm1.Output = "ATDT555-1234" & VbCr
                    Buffer = ""
                End If
            Case "DIALING"
                Buffer = Buffer & Comm1.Input
                If InStr (Buffer, "CONNECT") Then
                    Online
                    DialTimer.Enabled = False
                    TestDial.Enabled = True
                    State = ""
                Elseif InStr (Buffer, "NO CARRIER") Then
                    Offline
                    DialTimer.Enabled = False
                    TestDial.Enabled = True
                    State = ""
                End If
        End Select
    End Sub
```

Well, this takes a little more code than the previous example but it uses no DoEvents. So, it must be preferred, right?

You might have guessed that the answer is, "Maybe."

Timers are a limited resource. It is always possible to run out of them on heavily loaded systems or in complex programs where quite a few timers may be needed.

The smallest possible Timer interval is 55 mS (exception: Windows 2K and later reduce the minimum Timer interval to 1 mS). This means that, on average, there usually will be a minimum of 55 mS between the time a message is received and when your code will respond to it. A DoEvents loop will respond as quickly as it is possible to respond. This delay usually is not an issue. However, it is there, and it should be recognized.

Like the previous example, where a DoEvents loop was used, the command button event that initiated this operation must not be reentered.

Let's look at the other way that the problem might be tackled, using OnComm receive processing.

```
Sub TestDial_Click ()
    TestDial.Enabled = False
    Comm1.Rthreshold = 1
    Comm1.InputLen = 0
    Comm1.Output = "ATZ" & VbCr
End Sub

Sub Comm1_OnComm ()
    Static State As String
    Static Buffer As String

    If State = "" Then
        State = "INIT"
        Buffer = ""
    End If
    Select Case State
        Case "INIT"
            Buffer = Buffer & Comm1.Input
            If InStr(Buffer, "OK" & vbCrLF) Then
                State = "DIALING"
                Comm1.Output = "ATDT555-1234" & VbCr
                Buffer = ""
            End If
        Case "DIALING"
            Buffer = Buffer & Comm1.Input
            If InStr (Buffer, "CONNECT") Then
                Online
                TestDial.Enabled = True
                State = ""
            Elseif InStr (Buffer, "NO CARRIER") Then
                Offline
                TestDial.Enabled = True
                State = ""
            End If
    End Select
End Sub
```

This looks very much like the code used in the Timer example. But, since OnComm is a built-in procedure in all communications controls, it is better, is not it?

Yes, and maybe. There is no timer required, so this type of code will not use that limited resource. However, under Windows 3.x, OnComm receive events **do** use a timer. It turns out that communications event notification at speeds in excess of 9600 bps is not reliable under Windows 3.x. So, MSCOMM and other VBXs are built using a Windows timer to poll for receive data and other communications events. Windows 95/98 and Windows NT do not have this problem, so OCXs built for 32-bit versions of Visual Basic do not use a Windows timer.

However, the biggest problem here is complexity. Your communications processes seldom will be as simple as this example. The more states that have to be accounted for in OnComm, the more complex and difficult debugging becomes.

The second most significant problem is that it is hard, though not impossible, to add a timeout function to the receive process. If you must perform some remedial action if no response is received within a specified time after a message or command is sent, the timeout can be added to the inline code.

Here is my advice.

- Use "in-line" processing of serial receive data **if the receive message is a direct response to a transmitted command or request**. This means that you parse responses that are appended to a string buffer inside a DoEvents loop if you expect a response that is directly related to the transmitted message. Add a check for timeout, if needed.

- Use OnComm processing **if receive messages are asynchronous**, i.e., if your code must respond to receive messages or commands that are issued by the other system but that are not directly connected to messages that are sent by your program. An example of this might be a host computer that is waiting for its modem to connect. After connect, the host may send a message requesting a log on. It can then process a log on and password and respond to other asynchronously generated requests from the remote system.

There is an exception to this guideline. Polled systems sometimes can be thought of as asynchronous even though responses to polls are a direct result of the poll message or command. There are two examples of polling in the sample code later in the book. Each example uses a timer to poll the connected instrument.

One might choose to process receive data in OnComm. In that case, we could profitably decouple the receive code from the transmit code. If there is no response to a poll, no error occurs.

Another example, Simple Logger, polls the connected instrument in a timer event. If there is no response, an error message is displayed. The second case uses inline, rather than OnComm, event processing of receive data.

Why would I make a distinction for polled systems? Polling is a regular, repeated process. Frequently, polling is done at timed intervals. Most importantly, when a poll is sent, often the response need not be received before the next poll.

Do not confuse the terminology that I have used here. Asynchronous messages refer to ones that may come at any time and are not connected to another message or event. Synchronous replies are a direct result of a transmitted message.

Of course, practical systems often will mix synchronous and asynchronous messaging. Several of the examples in this book show such a mix. For example, a host computer will send an initialization string to its modem. It will wait for the modem to respond to the initialization command with an "OK" reply and then will proceed to another state. This will be done using inline code. However, it may make sense to wait for remote systems to connect and to issue commands to the host. This is most easily done by using OnComm receive processing.

We have not abandoned the idea of using a Timer, though. OnComm is a procedure that is provided by communications VBXs and OCXs. Event driven communications, with the same result as using OnComm, is possible using DLLs if you use a Timer to poll functions in the DLL. There are several commercial communications DLLs that are popular products. These DLLs can be lower overhead solutions, creating smaller programs that are easier to install and that may be preferred because the DLLs can be more easily used by non-Visual Basic programs.

If you choose to use a DLL and have need for asynchronous receive processing (asynchronous as described above), then it is a reasonable approach to use a Timer to simulate OnComm.

Some communications DLLs (and the Windows API) allow you to use callback functions to create an event driven program. VB 5.0 and later versions of VB support callbacks using the AddressOf operator, so they can be a valid alternative in that environment.

None of the code fragments that I have shown here is truly complete. A working system will require error and exception handling. Errors can happen and they have to be anticipated. For example, suppose that your code attempts to open a comm port that is already open or that does not exist. You must trap the error and report it to the user — your program cannot crash. Exceptions are things that happen that are out of the ordinary but that logically may be expected to occur. An exception that should be handled by dialing code is a "BUSY" response from the modem. Another exception that may happen is a failed file transfer after connect. I'll show examples of error and exception handling later in the book. However, I cannot anticipate all that may happen in a real-world application, so be ready to anticipate these.

3.3.2   State Machines

Have not we already discussed state machines? Yes, all computer programs are state machines and state machines are an even more important concept in communications systems. For example, a modem will have two states: offline, and online. Likewise, a program that uses a modem will have these same two states and perhaps several more — all related to the logic of the overall design.

Let me be a little clearer. What we are talking about is finite-state software design. The fundamental concept is that communications programs can be viewed from the perspective of finite-state machines (FSM) and can therefore be implemented by methods adapted from the hardware design of FSMs.

An FSM is characterized by a fairly complex logic structure that has a relatively large number of flow-control statements (loops, If/Then/Else, Case Select statements, etc.) compared to the number of input/output statements.

The first step in designing a FSM is to completely describe its behavior by listing each possible output along with the sequence of inputs that produce it.

Let's look at a simple dialer as an example. The goal would be to use a modem to dial a telephone number. In this simple example, we would send a command to the modem to lookup the telephone number in a database. At that point, we are in the READY state. To dial the number, we would send the dial command to the modem using the phone number from the database. We are now in the WAITING state. If the called party answers, we enter the TALKING state. If, on the other hand, the phone number is busy, we enter the BUSY state. To exit the BUSY state, the user can click a button that returns us to the READY state, or we may implement a timer that, after an appropriate delay, dials again and takes us back to the WAITING state.

Input	State: Ready	State: Waiting	State: Talking	State: BusyWait
Select Number	(no output) KReady	(no output) KWaiting	(no output) KTalking	(no output) KBusyWait
Dial_Click	ATDT555-1234; KWaiting	(no output) KWaiting	(no output) KTalking	ATDT555-1234; disable Busy timer KWaiting
Cancel_Click	(no output) KReady	ATH0 KReady	ATH0 KReady	(no output) KReady
OK	(no output) KReady	(no output) KTalking	(no output) KTalking	(no output) KBusyWait
BUSY	(no output) KReady	enable Busy timer KBusyWait	(no output) KReady	(no output) KBusyWait
Busy Timeout	(no output) KReady	(no output) KWaiting	(no output) KTalking	ATDT555-1234; disable Busy timer KWaiting

Table 3.3.2.1 *Finite-State Machine Transition Table*

A convenient way to specify an FSM is to use a state-transition table. Each row in this table corresponds to a different input and each column represents a distinct state. Each box at the intersection of a state column and an input row contains two fields, the next output and the next state. I have used the symbol K to indicate the next state in **Table 3.3.2.1**. We can predict how the state machine will react to any valid input by inspecting the intersection of the column associated with that state and the appropriate input row. The table tells us the next output and state. This process can be repeated for each new state, thus defining the FSM action for any arbitrary input sequence

Any table entry where there is no output and state change for a specific input is either a "do not care" or an input that is impossible in that state. In this example, some inputs are user commands, others are modem responses, and one is a timer timeout that results from a busy signal.

This example is comparatively trivial. However, it represents the approach that you might use to design an FSM. There may be more than one state transition table that will solve any specific problem. If you lay them out, side-by-side, the most efficient one will (probably) be obvious by inspection. In general, the smallest table that is complete is the optimal design. But, you must be certain that the table is sufficiently large to cover all possible inputs, including exceptions, and that all possible states have been identified.

Design work that you do in advance of coding can help you make intelligent decisions and they can make the actual coding much easier. A design technique like state transition tables is also good for documentation and code maintenance.

One powerful way to implement an FSM that permits modification of the FSM without recompilation of the VB source code is through the use of a database to store the transition state table. As long as no new states are added, any required changes could be implemented by simply modifying records in the database.

3.3.3   Delays

All communications programs need to implement delays in code. You often will want to perform some action, wait a specified interval, and then look for a response. Or, you may need to accommodate some physical limitation, or just a "fact of life," of the system with which you are communicating. The next section discusses some of these issues.

The simplest form of delay uses the Visual Basic Timer function. The Timer function returns the number of seconds since midnight. The resolution of Timer is approximately 55mS, often satisfactory for non-critical applications. This technique is used in many of the examples later in the book. Here is the basic form of the code,

```
Sub Delay (DelayTime As Single)
    Dim Timeout As Single
    Timeout = Timer + DelayTime
    Do Until Timer >= Timeout
        DoEvents
    Loop
```

```
    End Sub
```

The routine will enter the Do/Loop until the number of seconds or fractions of a second in DelayTime have elapsed. The DoEvents call allows the VB program to process other messages and allows multitasking in 16-bit versions of the program.

However, there is a problem with this routine. Remember that I said that the Timer function returns the number of seconds since midnight? That means that one-second before midnight, the value that is returned is 86399. Exactly at midnight, the value that is returned is 0. If the DelayTime plus the current value of Timer is greater than 86399, then the loop will execute **forever**. Forever is probably too long.

How do we fix this? One suggestion is the following code fragment, which you will often see in code. **Do not use it. It does not always work.**

```
Sub Pause (Secs As Single)
    Dim Start As Single
    Dim EndTime As Single

    Start = Timer
    EndTime = Start + Secs
    Do
        DoEvents
        If Start > Timer Then
        '-- Adjust for Midnight  ???
            EndTime = EndTime - 24 * 60 * 60
        End If
    Loop Until Timer >= EndTime
End Sub
```

This code adjusts for midnight but it assumes that the entry time is within one-timer-tick of midnight. Surely this is not correct.

Here is a delay that spans midnight and makes no assumption about how close to midnight we are at entry. I suggest this routine to avoid trouble.

```
Sub Delay (DelayTime As Single)
    Dim Timeout As Single
    Timeout = Timer + DelayTime
    If Timeout <= 86399 Then
        Do Until Timer >= Timeout
            DoEvents
        Loop
    Else
        Timeout = Timeout - 86399
        Do Until Timer < 1
            DoEvents
        Loop
        Do Until Timer >= Timeout
```

```
            DoEvents
        Loop
    End If
End Sub
```

Is there another approach that we should consider to implement programmatic delays? Yes, Windows provides several API functions that deal with time. The Windows Kernel API functions are GetCurrentTime, GetMessageTime, GetTickCount (the same as GetCurrentTime), and GetTimerResolution. However, these provide no advantage over the VB Timer function (Timer encapsulates the GetCurrentTime API, in fact). However, the Windows multimedia subsystem also has a couple of timer functions. The one that is most useful, in my opinion, is the TimeGetTime function.

The TimeGetTime (timeGetTime in 32-bit versions of Windows) function has several advantages over the VB Timer function. One advantage is that the resolution of TimeGetTime is 1mS, compared to the 55mS resolution of Timer. That means that much smaller timing increments are possible. Another advantage is that TimeGetTime returns a Long (four bytes) that represents milliseconds. Like the Timer function, this function rolls over. Unlike Timer, it rolls over every 49.7 days (every 4,294,967,296 milliseconds). For all practical purposes, this rollover is not a concern and it requires no compensation. Although the probability of error because of rollover is small, it is still possible — so, a very critical application would modify the following routine in a similar fashion to the Delay routine that I presented to compensate for midnight rollover.

```
Declare Function TimeGetTime& Lib "MMSYSTEM.DLL" ()
'or, for 32-bits
'Declare Function timeGetTime& Lib "WINMM.DLL" ()

Sub Wait (Timeout As Single)
    Dim WaitTime As Long
    WaitTime = Int(Timeout * 1000) + timeGetTime()
    Do Until timeGetTime() >= WaitTime
        DoEvents
    Loop
End Sub
```

The Timeout argument is in seconds or fractions of a second greater than 1mS. NOTE: Capitalization in the names of functions and subroutines under 32-bit versions of Windows is important. Make certain to use timeGetTime (not TimeGetTime).

This is a very good way to implement millisecond-resolution delays. One of the sample programs that I present later in the book (the Flashlite control program) uses this routine to do what I call "character pacing."

Some systems with which we must communicate implement their serial receive functions in a way that, charitably, might be termed, "sub-optimal." What can happen is that the system designer may have assumed that the system is connected to a terminal, not a computer. If the designer makes this assumption, the serial receive routines may be polled, not interrupt driven. Or, even if interrupt driven, the routine may be too slow to handle back-to-back data. If data arrive too quickly, some characters may be lost. To overcome this limitation, it may be necessary to place a short delay between each character that is sent from your program.

If you need to implement character pacing, you would like to avoid adding more delay between characters than is needed. If you had to rely on the VB Timer for these delays, your program might appear to be sluggish. Of course, you know that the fault lies with the connected system but try selling that to your boss or to a customer.

Before we leave this subject, we should look at how to use the Visual Basic Timer control for delays.

The Visual Basic Timer controls are not subject to the vagaries of rollovers. However, they are somewhat limited. The maximum Timer.Interval is 65535 (a little more than one minute). A VB Timer uses a Windows timer and Windows timers are a limited resource, so it is possible for the mix of software that is running on the computer to use them up. The Visual Basic Timer control has a resolution of 55 mS (1 mS under Windows 2K and later). Lastly, Windows will queue only a single timer message to a timer control. If the preceding message has not been processed, because of multitasking or something else that slows the program execution, the new timer message is not added to the queue. This means that Timer controls often run "slow." Nonetheless, they are useful. Here is example code that uses a Timer control for a delay.

```
DelayTimer.Interval = DelayRequiredInMilliseconds
DelayTimer.Enabled = True
Do Until DelayTimer.Enabled = False
    DoEvents
Loop
Sub DelayTimer_Timer ()
    DelayTimer.Enabled = False
End Sub
```

One common fault that all of these delay routines have is that they use a DoEvents loop. DoEvents is used to allow multitasking under Win16 and to allow your Visual Basic program to process other messages. The unintended consequence is that a VB message can cause an event that you did not anticipate. There is a way to avoid this — under Win32, and using VB 4.0/32 or VB 5/6. This method uses the Sleep API. Here is the declaration for the Sleep API.

```
Declare Sub Sleep Lib "Kernel32" (ByVal _
    dwMilliseconds As Long)
```

You should use the Sleep API if your 32-bit application needs a delay and if no other parts of your application need to process events (VB messages) while the delay is executing. The main advantage to Sleep over other delay methods is that your application uses no processing time while "asleep." This can improve the overall multitasking performance of your computer.

If you anticipate that your application needs a delay but that it needs to process occasional messages, you can combine Sleep with a counter to have the best of all worlds. Here is an example of this idea.

```
Sub SleepDelay (Timeout As Single)   'Timeout in mS
    Dim I As Integer
    For I = 1 to Timeout \ 50
        Sleep (20)
        DoEvents
    Next I
End Sub
```

This SleepDelay routine has a resolution of 20 mS. A one-second delay will execute 50 calls to DoEvents which will keep your application fairly "snappy" while returning the bulk of the processing time to the system. If your application processes messages during the SleepDelay process, it will run "long". So be aware of how it works.

Note for Visual Studio .NET programmers: Use the System.Threading.Thread. Sleep method.

3.3.4 Latency

Communications systems are comprised of many elements. These might be a computer at one location, a modem, telephone system, another modem, and computer. Of course, practical systems can be even more complex than this. Each of the elements in this link requires a finite amount of time to process data. The total time that is required is known as the link latency or, simply latency.

Why is this important? It can be a critical factor in getting a system to work reliably. You cannot send a command from one computer to another and expect a response in less time than is governed by the latency in both directions. Suppose that you send a command to a modem and expect a response back. The time required is the length of time required to send the data serially, the time required for the modem to decode the command and execute it, the time required for the modem to issue a reply, and the time required for your program to receive and interpret the reply. This may seem trivial. However, unless your program anticipates the time required plus some extra for safety, your code may cause a fault.

The simple example in the preceding paragraph is one of the areas where latency is often overlooked. If you send a command to a modem and then send another command to the modem without giving the modem sufficient time to respond to the previous command, two things may happen — both bad. The modem will not respond to the second command while it is executing another command or issuing a reply. Worse though, some modems may simply "lock-up" if this happens. This is a design flaw in the modem and it does happen. It is not pleasant to have to reset a modem or to have to reboot a computer to return a modem to life.

Do not forget that a modem response consists not only of the message, such as "OK", but it also includes carriage return and linefeed characters after the "OK". If you send a new command before **all** of these characters have been sent, the modem will not respond to the new command.

Latency can effect other operations, too. Some file transfers work better if the downloading (receiving) system activates its download code after the uploading (transmitting) system activates its upload code.

Some file transfer protocols have an "auto-download" feature. This means that the custom control is looking for the specific data pattern that indicates that an upload from the other computer has been started. This feature is not as reliable as I would like it to be and, often, uploads and downloads are specific states in a state machine that are implemented in code. State machine code and auto-download are not compatible — you do not want a transfer starting behind your back, with no notification.

So, in the case of a file transfer, it is a good idea to estimate the latency of the entire system and to account for it with a delay that allows the two computers to "synchronize." You will see examples of this in the Remote and BBS examples later in the book.

I offer one caution. Do not underestimate the potential delays in the telephone system. Terrestrial delays seldom exceed 100 mS but satellite delays can add 400-500 mS, sometimes more. If you anticipate that one or more satellite hops can be used in the link then add a corresponding software delay.

If might seem that these delays, especially if they are much longer than might be needed, are a waste of time (so to speak). If the extra delay allows you to avoid an error, even occasionally, it is time well spent.

It can be hard to estimate some of the latency that you encounter. Suppose that your code sends a request to the connected computer that causes a database lookup by that computer. The time to do the lookup can vary over a large range.

One way to remove most of the guesswork about latency is to build a communications protocol that is state and event driven . Pseudo-code (MSCOMM or other custom control) for a remote might look something like this,

```
Dim Ret As Integer
Dim ResponseString As String
Comm1.Output = "PASSWORD " & MyPassword$ & VbCr
```

```
Ret = WaitForResponse ("PROCEED" & VbCr, 60,_
    ResponseString)
If Ret = 1 Then
    MsgBox "Invalid response: " & ResponseString
    Disconnect
    Exit Sub
ElseIf Ret = 0
    MsgBox "Timeout.  Host did not respond."
    Disconnect
    Exit Sub
End If
'PROCEED was received, enter the next state.
```

The WaitForResponse routine would be similar to ones that are in the examples later in the book. I will not repeat it here. If the string, "PROCEED" is received before 60 seconds expires, then everything is OK and we can execute the next state. If some other string is received, perhaps " Invalid Password ", this will be displayed. If no response is received from the other computer within 60 seconds, then Timeout is displayed.

Complementary code on the other computer looks for various requests and parses them. For example, this fragment,

```
Comm1_OnComm ()
    Static Buffer As String
    Dim Position As Integer
    Dim Password
    Dim Ret As Integer
    If Comm1.CommEvent = 2 Then
        Buffer = Buffer & Comm1.Input
        Position = InStr(Buffer, "PASSWORD")
        If Position > 0 And InStr(Buffer, VbCr) Then
            Password = Mid$(Buffer, Position +1)
            Ret = Lookup(Password)
            If Ret = True Then
                Comm1.Output = "PROCEED" & VbCr
                Buffer = ""
            Else
                Comm1.Output = "Invalid Password" & vbCr
                Disconnect
                Buffer = ""
            End If
        End If
    'parse other requests here
    End If
End Sub
```

This OnComm routine receives and parses various requests; the password request is shown here. When the string "PASSWORD" is received, the routine looks for a carriage return. The data between "PASSWORD" and the carriage return is passed to the Lookup function for validation. If Lookup returns False, the routine sends "Invalid Password" and then disconnects. This disconnect may be important, if you want a secure system!

The time required to do the password lookup can be affected by the time it takes to do password decryption and the time required to actually lookup the password and to set any required security level. As long as the complete process takes no longer than 60 seconds, everything will work as quickly as possible.

3.3.5 Problems with VB Modal Dialogs

Sixteen-bit versions of Visual Basic do not process messages such as OnComm events while VB modal dialogs are displayed. The VB MsgBox and CommonDialog control are examples of VB modal dialogs. Do not use these dialogs if OnComm messages must be generated while a dialog is displayed. An alternative is to use the Windows API MessageBox function that is modal but does not interrupt messages. Or build your own dialogs using a Form and adding other components to create the functions that you need. Of course, these forms can be displayed modally.

You also may notice that event (message) processing is slightly different when executing your program inside the VB IDE (Integrated Design Environment) than when executing compiled code. Messages can be blocked when a Message Box or other modal dialog is displayed from 32-bit versions so VB in the IDE but not blocked when compiled. This can cause headaches when debugging applications, so be aware of this potential problem.

Chapter 4   MSCOMM Custom Controls

Visual Basic Professional versions 2 and later have been furnished with custom controls for serial communications. MSCOMM was furnished as MSCOMM.VBX for VB 2.0 and VB 3.0 and as MSCOMM16.OCX and MSCOMM32.OCX for VB 4. VB 5.0 and later are 32-bit only, so they only come with MSCOMM32.OCX.

Each of the VBX and OCX versions has slightly different features. Those features will be covered in this chapter.

MSCOMM encapsulates the Windows communications API and makes access to communications functions as simple as reading or setting properties. The Windows API allows communications event notification. However, Visual Basic before version 5 did not support procedure pointers for callbacks, so there was no way to implement event driven communications directly in VB code. However, MSCOMM, which is written in C and is compiled to machine code, uses function pointers and allows the API to perform event notification.

MSCOMM is hidden at runtime. When executing, MSCOMM has no visual interface; all input and output must be done in code.

I will not attempt to duplicate the information that Microsoft provides in the manuals or the Help files for MSCOMM. However, I will try to tell you what is wrong in those references. I will also try to explain (in English) what are the important properties and events and how they function. The best way to do this is by annotated examples. Preliminary to that, I'll discuss some of the salient features of each version of MSCOMM.

4.1   VB 2.0

The version of MSCOMM that was furnished with VB 2.0 was the first to provide event driven communications. There are situations where you want to be notified when a communications event happens. I have discussed situations where you might use OnComm events to receive serial data. Your program may also benefit from events caused by communications errors and/or changes in hardware state.

The serial port UART has hardware inputs for CD (Carrier Detect), RI (Ring Indicate), CTS (Clear To Send), and DSR (Data Set Ready). There are OnComm events that are generated whenever these inputs change state. The Help file refers to RTS. However, this is an output and there is no event associated with it.

The CommEvent property can be checked to identify what event triggered OnComm.

Each communications control you use corresponds to one serial port. If you need to access more than one serial port in your application, you must use more than one communications control.

VB 2.0 is compatible with Windows 3.0, so MSCOMM encapsulates the Windows 3.0 communications API. The Windows 3.0 API does not support event notification so all communications events are generated by MSCOMM via a polling mechanism. The VB programmer does not need to understand any of these hidden operations.

One caution that also applies to VB 3.0 is that a VB 2.0 or VB 3.0 program can run under Windows in Standard mode. Standard mode will run on 80286 microprocessors. Standard mode is a task switcher. It does not provide a good multitasking environment. If your program is run in Standard mode Windows, it can experience data loss if some other program runs.

Standard mode Windows is not a suitable environment for any communications program if multiple programs are run.

4.2   VB 3.0

All of the comments that I made for VB 2.0 and MSCOMM apply to VB 3.0. In addition, the MSCOMM that was furnished with VB 3.0 encapsulated the Windows 3.1 communications API. This allowed for a potential improvement in VB 3.0 because MSCOMM could use the Windows API event notification method. The MSCOMM that was furnished with VB 3.0 did, in fact, use the API notification method.

Unfortunately, the Windows 3.1 API event notification method simply did not work well at speeds in excess of 9600 bps. That meant that an updated version of MSCOMM was released almost immediately. The update, dated May 12, 1993, is version 2.1.0.1.

With the release of this updated version of MSCOMM, two properties were added. One property, named Notification, is hidden. It is not reflected in the properties Window at design- time. The default for the Notification property is 0. If Notification is 0, the Windows 3.0 API methods are used. If Notification is 1, Windows 3.1 event notification methods are used.

The other property that was added when MSCOMM was updated is the Interval property. The Interval property specifies how frequently, in milliseconds, MSCOMM polls the API to generate OnComm events (Notification property is set to default 0). The designer of MSCOMM chose an unfortunate default for the Interval property of 1000. This results in very sluggish performance.

My suggestion for these properties is to leave the Notification property at the default 0. If you are certain that your code will never require speeds in excess of 9600 bps then you might use Notification = 1. However, there is no real performance gain at low serial speeds so I would not take the chance. Set the Interval property to 55. That provides the maximum OnComm rate (18.2 times per second) and results in the most reliable implementation.

4.3   VB 4.0

VB 4.0 is the first version that creates applications in both 32-bits and 16-bits. It is also the last version to support 16-bit development. All later versions of Visual Basic will be 32 bits, at least until Windows itself becomes a 64-bit operating system.

The MSCOMM that is furnished with VB 4.0 comes in two flavors, MSCOMM16.OCX and MSCOMM32.OCX. Note that these controls are now OCXs instead of VBXs. MSCOMM16 is used with VB 4.0/16 and MSCOMM32 is used with VB 4.0/32. However, there is no additional functionality with these controls. So, every comment that I made about MSCOMM for VB 2.0 and VB 3.0 applies — except Notification and, peripherally, Interval.

The Notification property is still hidden. **However, it should not be changed from the default of 0.** If you set Notification to 1, you can cause unpredictable operation and possibly a GPF.

The Interval property does the same thing with MSCOMM16.OCX that it did with MSCOMM.VBX. However, when using MSCOMM32.OCX, Interval does nothing. Fortunately, the Windows 95/98 and Windows NT communications API event notification method works correctly so Interval is not needed. There were updates to both MSCOMM16.OCX and MSCOMM32.OCX dated January 12, 1996. These updated versions are 1.0.018 and 1.0.2805, respectively.

Under the 32-bit versions of Windows, you must make certain that you do not close a serial port when there is data in the output buffer, the buffer maintained by the serial driver. Here is the code that I use to make certain that the buffer is empty before the port is closed.

```
Sub PortClose (Comm1 As Control)
    Comm1.Handshaking = 0
    Do Until Comm1.OutBufferCount = 0
        DoEvents
    Loop
    Sleep (200)          'wait 200 mS
    Comm1.PortOpen = False
End Sub
```

I call this code anytime the port is to be closed, including the form Unload event for the form that contains the MSCOMM control.

VB 4.0 uses Unicode to store and manipulate strings. Unicode is a character set where 2 bytes are used to represent each character. Some other programs, such as the Windows 95/98 API, use ANSI (American National Standards Institute) or DBCS (Double-Byte Character Set) to store and manipulate strings. When you move strings outside of Visual Basic, you may encounter differences between Unicode and ANSI/DBCS.

ANSI is the most popular character standard used by personal computers. Because the ANSI standard uses only a single byte to represent each character, it is limited to a maximum of 256 character and punctuation codes. Although this is adequate for English, it does not fully support many other languages.

DBCS is used in Microsoft Windows systems that are distributed in most parts of Asia. It provides support for many different non-European-based language alphabets, including Arabic, Hebrew, Chinese, Japanese, and Korean (just to mention a few). DBCS uses the numbers 0 – 128 to represent the ASCII character set. Some numbers greater than 128 function as lead-byte characters that are not really characters but simply indicators that the next value is a character from a non-Latin character set. In DBCS, ASCII characters are only 1 byte in length whereas Japanese, Korean, and other East Asian characters are 2 bytes in length.

Unicode is a character-encoding scheme that uses 2 bytes for every character. The International Standards Organization (ISO) defines a number in the range of 0 to 65,535 ($2^{16} - 1$) for just about every character and symbol in every language (plus some empty spaces for future growth). On all 32-bit versions of Windows, the Component Object Model (COM), the basis for OLE and ActiveX technologies, uses Unicode. Unicode is fully supported by Windows NT. Although both Unicode and DBCS have double-byte characters, the encoding schemes are completely different.

Why is DBCS an issue? Under VB 4.0 and earlier versions of Visual Basic, strings are used to transmit and receive binary data. For example, if you were to create a string with all binary characters from 0 to &HFF, you could send it using the MSCOMM Output property.

```
Dim I As Integer
Dim Buffer As String
For I = 0 To &HFF
    Buffer = Buffer & Chr$(I)
Next I
Comm1.Output = Buffer
```

Under normal circumstances, this would not cause a problem. The Unicode to ANSI conversion that MSCOMM performs sends the correct data if the OS is not a DBCS version. So, you may not get what you expect with most Asian versions of Windows.

"DBCS is a different character set from Unicode. Because Visual Basic represents all strings internally in Unicode format, both ANSI characters and DBCS characters are converted to Unicode and Unicode characters are converted to ANSI characters or DBCS characters automatically whenever the conversion is needed." This is a quote from Microsoft documentation on DBCS. It also illustrates the problem with MSCOMM — the conversion is automatic but it may not be correct. Yikes!

What's the solution? Use MSCOMM from VB 5.0. It should also be possible to write a 16-bit OLE Server to wrap the 16-bit comm API. The 16-bit applications would not be affected because strings there are a single-byte per character. However, I have not been able to get satisfactory performance from my attempt to use this method.

4.4   VB 5.0/VB 6.0

VB 5.0 and VB 6.0 have several features that can be important in communications applications.

You may now build an application that compiles to native code. VB is still dependent on a runtime DLL for its internal functions. However, if you are creating your own special communications protocol, native code may offer a speed advantage. You can profile native code using native code compiler options and debug native code using the Visual C++ environment.

Visual Basic itself renders forms and graphical components more quickly. This speedup makes your applications look more professional. It includes native-code compilation, although a runtime DLL is still required. The native-code portions of an application can have various optimizations applied when compiled.

You may combine one or more existing controls or create your own from scratch. ActiveX controls created with Visual Basic have events, data binding support, licensing support, property pages, Internet features, and more.

There are many features that improve multi-programmer development and client server (Enterprise) development. The one feature that I want to discuss is, naturally, the current version of MSCOMM32.OCX.

The MSCOMM that is furnished with VB 5.0/VB 6.0 has one important property. The designers recognized the need to handle binary data in all environments differently than ASCII (or ANSI/DBCS) data is handled. This property is called InputMode. If InputMode = 0 - comInputModeText, data is Text (or ASCII/ANSI), while InputMode = 1-comInputModeBinary, data is binary. The Help file states that this property applies to input. However, it also specifies what may be assigned to output.

Text data is handled just as with previous versions of MSCOMM. String buffers are used for Input and Output data. However, binary data now uses a byte array as a buffer instead of a string. This eliminates the issue of Unicode conversion in DBCS systems. For example,

```
Private SendArray () As Byte
Private Sub SendData ()    'transmit binary data
    ReDim SendArray (&HFF) As Byte
    Dim I As Integer
    For I = 0 to &HFF
        SendArray(I) = I   'assign data to byte array
    Next I
    Comm1.Output = SendArray ()
End Sub
Private Sub Comm1_OnComm ()    'receive binary data
    Dim ReceiveArray () As Byte
    If Comm1.CommEvent = EV_comEV_RECEIVE Then   '2
```

```
        ReceiveArray() = Comm1.Input
        'do something useful with ReceiveArray
    End If
End Sub
```

If you are working with binary data, you should not use a string buffer. Rather, you should use an array of type Byte and set the .InputMode property to comInputModeBinary (or 1). This adds complication but improves reliability across all OS environments. What about backward compatibility? Earlier versions of MSCOMM did not have this property. Will they still work? Yes, as long as your program does not have to execute under a DBCS enabled OS. Microsoft released an update to MSCOMM32.OCX with VB5 Service Pack 2 that allows you to use string buffers for binary data. The version that was delivered with VB5 did not allow strings to be used to buffer binary data. The versions of MSCOMM that come with VB6 and VB.NET continue to allow you to use strings.

If you are writing communications code that uses DBCS, you must set the InputMode property to 1- InputModeBinary. Then you must convert each DBCS character to two bytes for transmission and receive two bytes for each DBCS character that is received. I will not go into any details on this process because I have never done it in a practical application.

The MSCOMM32.OCX that comes with VB5 has a couple of bugs that were fixed in the Service Pack 2 version (this version is also included in Service Pack 3). It is important that you use this updated version or the version that comes with VB6.

However, there is one important bug (not a feature!) in the MSCOMM32.OCX that affects transmission of data. The 32-bit API uses several timeout variables that are represented by the COMMTIMEOUTS structure. One of these, WriteTotalTimeoutConstant, is set to 5000 (five seconds) internally. This is not represented by an MSCOMM property. This constant determines how long the communications driver can spend transmitting a buffer before it halts transmission. Five seconds represents only 150 bytes at 300 bps and only 600 bytes at 1200 bps. It is quite possible, and likely, that one will want to send more data than that in one buffer.

This bug also can cause headaches for systems that use flow control, either software (Xon/Xoff) or hardware (CTS/RTS), even at high serial speeds. If flow control halts transmission while data is still in the transmit buffer, that data can be lost if the time halted exceeds five seconds. Five seconds can be quite a short time in some environments.

Not to worry, though. This version of MSCOMM has a vital property called CommID. The CommID is the handle returned by the API when the serial port is opened. This means that you can use the appropriate API functions to change the value of this constant after the port

has been opened. Here are the API declarations that you need.

```
Type COMMTIMEOUTS
    ReadIntervalTimeout As Long
    ReadTotalTimeoutMultiplier As Long
    ReadTotalTimeoutConstant As Long
    WriteTotalTimeoutMultiplier As Long
```

```
        WriteTotalTimeoutConstant As Long
    End Type
    Declare Function SetCommTimeouts Lib "Kernel32" (ByVal _
        hFile As Long, lpCommTimeouts As COMMTIMEOUTS) As Long
    Declare Function GetCommTimeouts Lib "Kernel32" (ByVal _
        hFile As Long, lpCommTimeouts As COMMTIMEOUTS) As Long
```

Then, in your code:

```
    Dim timeouts As COMMTIMEOUTS
    Dim Ret As Long
        If Comm1.PortOpen = False Then
            Comm1.PortOpen = True
        End If
        Ret = GetCommTimeouts(Comm1.CommID, timeouts)
        ' Set some default timeouts
        timeouts.ReadIntervalTimeout = 1
        timeouts.ReadTotalTimeoutMultiplier = 1
        timeouts.ReadTotalTimeoutConstant = 1
        timeouts.WriteTotalTimeoutMultiplier = 1
        timeouts.WriteTotalTimeoutConstant = _
            (Comm1.OutBufferSize \ Val(Comm1.Settings)) _
            * 10000 + 1000
        Ret = SetCommTimeouts(Comm1.CommID, timeouts)
```

This code will set the WriteTotalTimeoutConstant based on the OutBufferSize and the bit rate. Make an appropriate change to this code to allow extended timeouts to handle the potential problem with flow control. For example, you might want to change the code fragment to:

```
    Dim timeouts As COMMTIMEOUTS
    Dim Ret As Long
        If Comm1.PortOpen = False Then
            Comm1.PortOpen = True
        End If
        Ret = GetCommTimeouts(Comm1.CommID, timeouts)
        ' Set some default timeouts
        timeouts.ReadIntervalTimeout = 1
        timeouts.ReadTotalTimeoutMultiplier = 1
        timeouts.ReadTotalTimeoutConstant = 1
        timeouts.WriteTotalTimeoutMultiplier = 1000
        timeouts.WriteTotalTimeoutConstant = 3600_
        Ret = SetCommTimeouts(Comm1.CommID, timeouts)
```

This fragment specifies a WriteTotalTimeout of 3600 seconds (one hour). This should be sufficient for almost any laggardly flow control situation.

Earlier editions of this book failed to make an important point. So, I have added a short section here on the issue of the End Of File (EOF) character: Chr$(26) an ASCII character with the value of decimal 26 or hexadecimal 1A. EOF also is designated as the ASCII control character control-Z, sometimes shown as ^Z. MSComm provides a CommEvent that is designed to be generated when an EOF character is detected in the receive data stream (comEvEOF). **However, the conEvEOF event is not generated reliably**. So, we have to do this work on our own. Fortunately, the job is easy. Here is a simple function that may be called.

```
Private Function HasEOF(SerialData As String) As Boolean
Const EOF As Integer = 26
    If InStr(SerialData, Chr$(EOF)) Then HasEOF = True
End Function
```

This function is shown in its simplest form, but extending it is straight forward.

Another very important feature of the MSCOMM32.OCX that is furnished with VB 5.0 and later versions is that it supports speeds as high as 230k bps, although most common serial ports will support speeds up to (only) 115k bps. This is a limitation of the serial port hardware, not the software.

See the XMCommCRC ActiveX control project later in this chapter for examples of VB 5/VB 6 coding. Several additional examples are provided on the CD ROM that accompanies this book.

See Chapter 11 for example applets that use VB 3.0 and one that uses VB 5/VB 6 with the new MSCOMM32.OCX. The SerialMonitor presented benefits from these new versions of MSCOMM.

4.5   MSCOMM Example Code

The MSCOMM code in the following sections illustrates a variety of ways that the simplest communications custom control may be used and provides examples of a number of practical applications. Of course, there are many more things that might be done for which I have not included examples. However, most of what you might want to do is covered, if not in detail, then by intelligent extension.

All of the MSCOMM examples could be implemented using any of the commercial custom controls. All custom controls support the basic properties of MSCOMM that are shown here.

See Chapter 11 for applets that use MSCOMM to aid serial communications debugging

4.5.1 Dialer

Problem: Use MSCOMM to create a phone dialer.

Solution: There is a sample application furnished with VB 2.0-4, and a new version for VB 6. However, the VB 2.0-4 version has a couple of problems that this code fixes. The problem with the original code was that it would not disconnect the modem properly. All of the changes that I made are highlighted in the Dial subroutine and the About subroutine.

```
Sub Dial (Number$)
    Dim DialString$, FromModem$
    Dim StopString As String
    Dim Delay As Double
    '--- AT is the Hayes compatible ATTENTION command an
    '    is required to send commands to the modem.
    '--- DT means "Dial Tone" - The Dial command, using
    '    touch tones as opposed to pulse (DP = Pulse)
    '--- PhoneNumbers$(Index) is the phone number of the
    '    person you're dialing
    '--- A semicolon tells the modem to return to
    '    command mode after dialing (important)
    '--- A Carriage return, Chr$(13), is required when
    '    sending commands to the modem.
    DialString$ = "ATX4&D2DT" + Number$ + ";" + VbCr
    StopString = "ATH0" +  Chr$(13)
    '-- Comm port settings
    Comm1.Settings = "300,N,8,1"
    '-- Open the comm port
    On Error Resume Next
    Comm1.PortOpen = True
    If Err Then
        MsgBox "COM1: not available. Change the " _
            & "CommPort property to another port."
        Exit Sub
```

```
      End If
      '-- Flush the input buffer
      Comm1.InBufferCount = 0
      '-- Dial the number
      Comm1.Output = DialString$
      '-- Wait for "OK" to come back from the modem
      Do
          DoEvents
          '-- If there is data in the buffer, then read
          If Comm1.InBufferCount Then
              FromModem$ = FromModem$ + Comm1.Input
              '-- Check for "OK"
              If InStr(FromModem$, "OK") Then
                  '- Notify the user to pick up the 'phone
                  FromModem = ""
                  Beep
                  MsgBox "Please pick up the phone and" _
                      & " either press Enter, or click OK"
                  Exit Do
              ElseIf InStr(FromModem$, "NO DIAL TONE") Or _
                      InStr(FromModem$, "NO DIALTONE") Then
                  FromModem = ""
                  Beep
                  MsgBox "Dialtone not detected.  Please" _
                      & " check the telephone connection."
                  CancelFlag = True
                  Exit Do
              End If
          End If
          '-- Was Cancel pressed?
          If CancelFlag Then
              CancelFlag = False
              Exit Do
          End If
      Loop
      '-- Disconnect the modem
      Comm1.Output = StopString
      '-- Close the port
      Delay = Timer
      Do While Delay + 1 > Timer
          DoEvents
          '-- If there is data in the buffer, then read
          If Comm1.InBufferCount Then
              FromModem$ = FromModem$ + Comm1.Input
          End If
      Loop
      Comm1.DTREnable = False      'belts and suspenders
      Delay = Timer                'disconnect.
```

```
                                    'If StopString
      Do While Delay + 1 > Timer 'worked, this is not   needed.
          DoEvents
      Loop
      Comm1.DTREnable = True
      Comm1.PortOpen = False
End Sub
```

The changes that are highlighted in this code are small. However, they are important to assure that it is reliable. Of course, a "real application" would lookup phone numbers in a database rather than present hard-coded ones. That is a subject for another time.

There must be a delay after sending the "ATH" command to the modem to place the modem back on hook. The delay is required to allow the modem time to respond with an "OK" response. If the serial port is closed immediately after the "ATH" command, the port will close before the characters are sent and the modem will not disconnect.

Alternately, you may lower DTR. This will force the modem to go on hook. Lowering DTR is superfluous if the "ATH" command resulted in a hang up. However, I have included it as a valid alternate method.

4.5.2   VBTERM

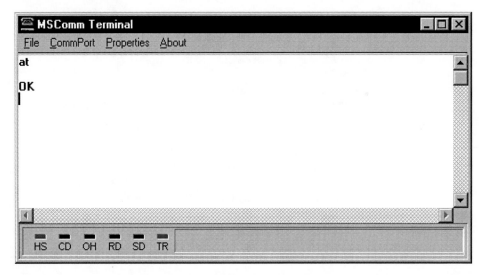

Problem: How do I create a simple terminal emulator using MSCOMM? I want to display modem status. Can I do that?

Solution: Visual Basic Professional Edition comes with example code for a simple terminal emulator. It has a couple of errors and omissions. It also lacks a way to show modem status. I have added code that monitors the status of transmit and receive data, and serial port status lines. This applet has been updated for VB6.

```
Private Sub CommOutput (OutputString As String)
    On Error Resume Next
    If Comm1.PortOpen = True Then
        Comm1.Output = OutputString
        If Err.Number > 0 Then
            MsgBox Err.Description, vbExclamation
        Else
            LED2(4) = True
        End If
    End If
End Sub
```

Simulating the LEDs of an external modem is an oft-asked-for feature. The CommOutput
routine is added so that the Tx data LED will flash on and off as data is sent.

```
Private Sub MDial_Click ()
    On Local Error Resume Next
    Static Num As String
    Dim SaveThreshold As Integer
    Dim Buffer As String
    '--- Get a number from the user.
    Num = InputBox$("Enter Phone Number:", _
        "Dial Number", Num)
    If Num = "" Then Exit Sub
    '--- Open the port if it is not already
    If Not MScomm1.PortOpen Then
        MOpen_Click
        If Err Then Exit Sub
    End If
    '--- Dial the number
    OH True                    'use Modem Lights
    MSComm1.Output = "AT&D2DT" + Num + VbCr
    Status.Caption = "Dialing... " & Num
    SaveThreshold = MSComm1.RThreshold
    MSComm1.RThreshold = 0     'keep all processing here
    MHangup.Enabled = True     'enable Hangup
    Do Until MSComm1.DTREnable = False
        DoEvents
        Buffer = Buffer & MSComm1.Input
        If InStr(Buffer, "CONNECT") Then
            Wait .2            'allow OnComm to reflect CD
            Exit Do
        ElseIf InStr(Buffer, "NO CARRIER" & VbCrLF _
                ) Then
            MsgBox "Lost Carrier!"
            OH False
            Status.Caption = ""
            Exit Do
        ElseIf InStr(Buffer, "ERROR" & VbCrLf) Then
```

```
              MsgBox "Check Number!"
              OH False
              Status.Caption = ""
              Exit Do
          End If
      Loop
      MSComm1.RThreshold = SaveThreshold
  End Sub
```

Changes were made to the Mdial routine for the modem status LEDs.

```
  '--- Toggle DTREnable to hang up the line
  Private Sub MHangup_Click ()
  Dim Ret As Integer
      Ret = MSComm1.DTREnable      'Save current setting
      MSComm1.DTREnable = True      'Turn DTR on
      Wait .2
      MSComm1.DTREnable = False      'Turn DTR off
      OH False                  'use Modem Lights
      Wait .2
      MSComm1.DTREnable = Ret       'Restore old setting
  End Sub
```

The changes made were to assure that DTR is held off for 200 milliseconds and to reset the OH modem status LED.

```
  Private Sub MOpen_Click ()
      On Error Resume Next
      MSComm1.PortOpen = Not MSComm1.PortOpen
      If Err.Number > 0 Then MsgBox Err.Description,
  vbExclamation
      MOpen.Checked = MSComm1.PortOpen
      MSendText.Enabled = MSComm1.PortOpen
      MHangup.Enabled = MSComm1.PortOpen
  End Sub
```

The MOpen_Click routine was changed to avoid the errors that would happen if a file were sent when the serial port is closed.

```
  Private Static Sub MSComm1_OnComm ()
      Dim EVMsg As String
      Dim ERMsg As String
      Dim Ret As Integer
      '--- Branch according to the CommEvent Prop..
      Select Case MSComm1.CommEvent
          '--- Event messages
          Case comEvReceive
              ShowData (MSComm1.Input)

          Case comEvCTS
```

```
            EVMsg = "Change in CTS Detected"
        Case comEvDSR
            EVMsg = "Change in DSR Detected"
        Case comEvCD
            EVMsg = "Change in CD Detected.  CD = "
            If MSComm1.CDHolding Then
                EVMsg = EVMsg & "True."
                OH True          'use Modem Lights
            Else
                OH False            'use Modem Lights
                EVMsg = EVMsg$ & "False."
            End If
        Case comEvRing
            EVMsg = "The Phone is Ringing"
        Case comEvEOF
            EVMsg = "End of File Detected"
        '--- Error messages
        Case comEventBreak
            EVMsg = "Break Received"
        Case comEventCTSTO
            ERMsg = "CTS Timeout"
        Case comEventDSRTO
            ERMsg = "DSR Timeout"
        Case comEventFrame
            EVMsg = "Framing Error"
        Case comEventOverrun
            ERMsg = "Overrun Error"
        Case comEventCDTO
            ERMsg = "Carrier Detect Timeout"
        Case comEventRxOver
            ERMsg = "Receive Buffer Overflow"
        Case comEventRxParity
            EVMsg = "Parity Error"
        Case comEventTxFull
            ERMsg = "Transmit Buffer Full"
        Case Else
            ERMsg = "Unknown error or event"
    End Select
    If Len(EVMsg) Then
        '--- Display event messages in label
        Status.Caption = EVMsg
        EVMsg = ""
    ElseIf Len(ERMsg) Then
        '--- Display error messages in an alert
        '    message box.
        Beep
        Ret = MsgBox(ERMsg, vbOKCancel, _
            "Press Cancel to Quit, Ok to ignore.")
```

```
                    ERMsg = ""
                    '--- If Cancel (2) was pressed
                    If Ret = vbCancel Then
                        MSComm1.PortOpen = 0    'Close the port and quit
                    End If
                End If
            End Sub
```

The OnComm routine does two things. First, it handles and displays any errors that are detected by the Windows communications driver. These errors cause MSCOMM events. It also displays changes of serial port status such as Ring Detected and Carrier Detected or Lost.

The second essential feature of the OnComm routine is to receive serial data. The ShowData routine is called to display receive data.

This code controls the OH LED. It assumes that the modem cannot be off hook if Carrier Detect is False. Since we have no way of knowing if the modem is offhook otherwise (except when we use the dial menu), this LED is just an approximation of the actual modem hook switch state.

```
    Private Sub MSendText_Click ()
        On Error Resume Next
        Dim hSend As Integer, BSize As Integer
        Dim LF As Long, Ret As Integer
        If MSComm1.CDHolding = False Then
            Ret = MsgBox("Carrier has not been " & _
                "established.  Select OK to continue to " & _
                "send the file.", vbOKCancel)
            If Ret = vbCancel Then Exit Sub
        End If
        MSendText.Enabled = False
        '--- Get Text File name from the user
        OpenLog.DialogTitle = "Send Text File"
        OpenLog.Filter = _
            "Text Files (*.TXT)|*.txt|All Files (*.*)|*.*"
        Do
            OpenLog.Filename = ""
            OpenLog.Action = 1
            If Err.Number <> 0Then Exit Sub
            Temp = OpenLog.Filename
            '--- If file does not exist, go back
            Ret = Len(Dir$(Temp))
            If Err.Number > 0 Then
                MsgBox Err.Description, vbExclamation
                MSendText.Enabled = True
                Exit Sub
            End If
            If Ret Then
```

```
                    Exit Do
            Else
                MsgBox Temp + " not found!", vbExclamation
            End If
        Loop
        '--- Open the log file
        hSend = FreeFile
        Open Temp For Binary Access Read As hSend
        If Err Then
            MsgBox Err.Description, vbExclamation
        Else
            '--- Display the Cancel dialog box
            CancelSend = False
            frmCanSend.Label1.Caption = "Sending Text File - " _
                & Temp
            frmCanSend.Show
            '--- Read the file in blocks the size of our
            '    transmit buffer.
            BSize = MSComm1.OutBufferSize
            LF = LOF(hSend)
            Do Until EOF(hSend) Or CancelSend
                '--- Do not read too much at the end
                If LF - Loc(hSend) <= BSize Then
                    BSize = LF - Loc(hSend) + 1
                End If
                '--- Read a block of data
                Temp = Space$(BSize)
                Get hSend, , Temp
                '--- Transmit the block
                CommOutput Temp     'use Modem Lights
                If Err.Number <> 0 Then
                    MsgBox Err.Description, vbExclamation
                    Exit Do
                End If
                '--- Wait for all the data to be sent
                Do
                    DoEvents
                Loop Until MSComm1.OutBufferCount = 0 _
                    Or CancelSend
            Loop
        End If
        Close hSend
        MSendText.Enabled = True
        CancelSend = True
        frmCanSend.Hide
    End Sub
```

The MSendText routine is used to send a text file. It has been modified to use the CommOutput routine so that the modem TX LED will be lit when data is sent.

```
Private Sub OH (OffHook As Integer)
    If OffHook = True Then
        LED2(2).BackColor = &HFF&
    Else
        LED2(2).BackColor = &H0&
        MHangup.Enabled = False
    End If
End Sub
```

The OH routine was added to display the assumed status of the modem hook switch.

```
Private Sub ShowData (Dta)
    On Error Resume Next
    Dim Nd As Integer, I As Integer
    LED2(3).BackColor = &HFF&           'use Modem Lights
    '--- Make sure the existing text does not get too large.
    Nd = Len(Term.Text)
    If Nd >= 16384 Then
        Term.Text = Mid$(Term.Text, 4097)
        Nd = Len(Term.Text)
    End If
    '--- Point to the end of Term's data
    Term.SelStart = Nd
    '--- Filter/handle Back Space characters
    Do
        I = InStr(Dta, Chr$(8))
        If I Then
            If I = 1 Then
                Term.SelStart = Nd - 1
                Term.SelLength = 1
                Dta = Mid$(Dta, I + 1)
            Else
                Dta = Left$(Dta, I - 2) + Mid$(Dta, I + 1)
            End If
        End If
    Loop While I
    '--- Eliminate Line Feeds (put back below)
    Do
        I = InStr(Dta, VbLf)
        If I Then
            Dta = Left$(Dta, I - 1) + Mid$(Dta, I + 1)
        End If
    Loop While I
    '--- Make sure all Carriage Returns have a Line Feed
    I = 1
    Do
```

```
                I = InStr(I, Dta, VbCr)
                If I Then
                    Dta = Left$(Dta, I) + VbLf + _
                        Mid$(Dta, I + 1)
                    I = I + 1
                End If
            Loop While I
            '--- Add the filtered data to .Text
            Term.SelText = Dta
            '--- Log data to file if requested
            If hLogFile Then
                I = 2
                Do
                    Err = 0
                    Put hLogFile, , Dta
                    If Err.Number <> 0 Then
                        I = MsgBox(Err.Description, vbOKCancel)
                        If I = vbCancel Then
                            MCloseLog_Click
                        End If
                    End If
                Loop While I <> vbCancel
            End If
        End Sub
```

The ShowData routine is vital. It displays receive data in the textbox that comprises the "terminal display" and it changes the RX data LED to indicate that data is being received.

A standard textbox cannot hold more than 32K characters so the ShowData routine arbitrarily limits the number of characters in the text box to approximately 16K.

Some systems with which you might use this terminal emulator may send only a carriage return to terminate a line instead of the carriage return and line feed that VB needs to properly display in a textbox, or they may reverse the order so that a line feed precedes the associated carriage return. If receive data was not filtered, the textbox would display funky little rectangles instead of a new line. In either case, ShowData replaces all carriage returns in the receive data stream with a carriage return/line feed pair. It also deletes any received line feed characters. This process assures that a new line will always be displayed as you expect.

ShowData also records all receive data to a log file if logging is enabled.

```
        Private Sub StatusTimer_Timer ()
            Static ClearCount As Integer
            Static StatusMessage As String
            If Status.Caption <> "" Then
                If Status.Caption = StatusMessage Then
                    ClearCount = ClearCount + 1
                    If ClearCount >= 100 Then
```

```
                            'Clear Status in 15 sec
                            ClearCount = 0
                            Status.Caption = ""
                  End If
            Else
                StatusMessage = Status.Caption
                ClearCount = 0
            End If
        End If
        If Val(MSComm1.Settings) > 1200 Then
            LED2(0).BackColor = &HFF&
        Else
            LED2(0).BackColor = &H0&
        End If
        If MSComm1.DTREnable = True Then
            LED2(5).BackColor = &HFF&
        Else
            LED2(5).BackColor = &H0&
        End If
        If MSComm1.OutBufferCount > 0 Then
            LED2(4).BackColor = &HFF&
        Else
            LED2(4).BackColor = &H0&
        End If
        If MSComm1.InBufferCount > 0 Then
            LED2(3).BackColor = &HFF&
        Else
            LED2(3).BackColor = &H0&
        End If
        If MSComm1.CDHolding = True Then
            LED2(1).BackColor = &HFF&
        Else
            LED2(1).BackColor = &H0&
        End If
    End Sub
```

I added the StatusTimer to update the modem LED's. It also clears the Status caption (set by events in the OnComm routine) at regular intervals.

```
Private Sub Term_KeyPress (KeyAscii As Integer)
    '--- If the port is opened,
    If MSComm1.PortOpen Then
        '--- Send the key stroke to the port
        CommOutput Chr$(KeyAscii) 'use Modem Lights
        '--- Unless Echo is on, there is no need to
        '   let the Text control display the key.
        If Not Echo Then KeyAscii = 0
    Else          'do not display, if port not open
        KeyAscii = 0
    End If
End Sub
```

This code causes the modem status TX LED to change state as typed characters are sent. Also, a minor bug in the original code was changed so that typed characters are displayed only when the comm port is open.

```
Sub Timer1_Timer ()
    Timer1.Enabled = False
End Sub
Sub Wait (WaitTime As Single)
    Timer1.Interval = WaitTime * 1000
    Timer1.Enabled = True
    Do Until Timer1.Enabled = False
        DoEvents
    Loop
End Sub
```

These last routines use a Timer control to implement a delay. WaitTime may be specified in millisecond increments. The minimum possible delay is about 55 mS.

I also made some housekeeping changes that are not reflected in the code. I changed the menu so that DTREnable was correctly checked when the program starts. I changed the MSComm1.RThreshold property to 1 instead of the default 0, and the MSComm1.RTSEnable to True instead of the default False.

4.5.3   PortFind

This code is presented as a standalone applet. However, I have also included it in several subsequent examples.

Problem: Locate the modem and return the serial port number and speed.

Solution: This applet finds only the first modem. If the system has more than one modem, the code must be modified to continue after the first one is found. Modem speeds up to 115200 bps are detected. The VB6 version of this code has been modified to use Enums for FindModem results. This makes the code cleaner and more reusable.

```
Private Function FindModem(BitRate As Long) As Integer
'this routine leaves the CommPort open, if found
Dim I As Integer
Dim PortNum As Integer
Dim ErrCode
    With Comm1
        If .PortOpen Then .PortOpen = False
        BitRate = 1200
        .RTSEnable = True
        .Settings = Format$(BitRate) & ",n,8,1"
            'lowest common denominator
        On Error Resume Next
         For I = 1 To 16
            '-- Check ports 1 through 16
            .CommPort = I
            '-- Try to open the port
            .PortOpen = True
            If .PortOpen Then
  '-- Port opened OK. Now test for modem by sending "AT"
                ErrCode = ModemCommand("ATV1Q0", 1)
                If ErrCode = OK Then
                    PortNum = I
                    Exit For
                Else
                '-- OK was not received. Close the port
                    Comm1.PortOpen = False
                End If
            End If
         Next I
         If PortNum > 0 Then
            'modem found?
             BitRate = 115200
             Do
             Comm1.Settings = Format$(BitRate) & ",n,8,1"
                'start at the highest speed
                '-- Port opened OK. Now test for modem
                'speed by sending "AT"
                ErrCode = ModemCommand("AT", 1)
                If ErrCode = OK Then Exit Do
                BitRate = BitRate \ 2
                If BitRate = 28800 Then BitRate = 38400
             Loop Until BitRate = 1200
             FindModem = PortNum
         End If
    End With
End Function
```

The FindModem routine opens comm ports in sequence. It starts at Com1 and proceeds
through each port in sequence to Com16. If the port can be opened, the serial port exists
and it is not in use by some other application. Then an "ATV1Q0" command is sent. If a
modem is attached to that port, it will respond with an "OK."

If "OK" is received, the next thing that FindModem does is to attempt to find the maximum serial speed that the modem will use. It starts at 115200 bps, sends an "AT" and waits for an OK response. If OK is not received, the next speed that is tried is 57600 bps, and so forth.

FindModem locates only the first available modem. It could be modified to return a list of all available modems, though. I'll leave that enhancement to you.

FindModem is used in several other examples later in the book. The code for it is not repeated there.

```
Sub Form_Load ()
    Dim BitRate As Long
    Dim PortNum As Integer
    PortNum = FindModem(BitRate)
    If PortNum = 0 Then
        MsgBox "No modem detected."
    Else
        MsgBox "Modem found.  Port=" & Format$(PortNum) & _
            ".  Supports at least " & Format$(BitRate) & _
            " bps."
    End If
    Unload Me
End Sub
```

Call FindModem when the PortFind applet starts.

```
    Private Function ModemCommand(ATCommand As String, ByVal _
                        WaitTime As Single) As ModemResponses
Dim Timeout As Long
Dim Buffer As String
Dim RThreshold As Integer
    RThreshold = Comm1.RThreshold
    Comm1.RThreshold = 0
    Comm1.Output = ATCommand & vbCr
    ModemCommand = Timeout
    'assume a timeout error
    Timeout = timeGetTime + WaitTime * 1000
    Do Until timeGetTime > Timeout
        DoEvents
        Buffer = Buffer & Comm1.Input
        If InStr(Buffer, "OK" & vbCrLf) Then
            ModemCommand = OK
            Exit Do
        ElseIf InStr(Buffer, "ERROR" & vbCrLf) Then
            ModemCommand = Error
            Exit Do
        ElseIf InStr(Buffer, "DIAL") Then
            ModemCommand = NoDialtone
            Exit Do
        ElseIf InStr(Buffer, "CONNECT") Then
            ModemCommand = Connect
            Exit Do
        ElseIf InStr(Buffer, "NO CARRIER" & vbCrLf) Then
            ModemCommand = NoCarrier
            Exit Do
        ElseIf InStr(Buffer, "BUSY" & vbCrLf) Then
            ModemCommand = Busy
            Exit Do
        End If
    Loop
    Comm1.RThreshold = RThreshold
End Function
```

The ModemCommand sends the string that is specified in the ATCommand argument out the serial port. The following Enum specifies the values returned from the ModemCommand function.

```
Private Enum ModemResponses
    Timeout = 0
    Busy = 1
    NoDialtone = 2
    NoCarrier = 3
    Connect = 4
    Error = 5
    OK = 6
End Enum
```

ModemCommand is used in a number of other examples in the book. Code for it is not repeated there. The PortFind routine is a good addition to any non-TAPI program that uses modems. You may want to add it to your toolbox.

There is an alternate approach that you may wish to investigate. Windows maintains a list of installed devices in the Registry. You can use code that searches through the Registry for these devices. Registry information for serial ports looks like this:

.ClassKey = HKEY_LOCAL_MACHINE
.SectionKey = "Hardware\DeviceMap\SerialComm"

If you enumerate all of the keys in this section, you will have all of the comm ports available to Windows. The only problem with this approach is that it does not identify ports that are not functioning and/or that are in use by other applications.

See the Additional Examples folder on the CD-ROM for an example that uses Registry information to enumerate installed ports.

4.5.4 MODEMCFG

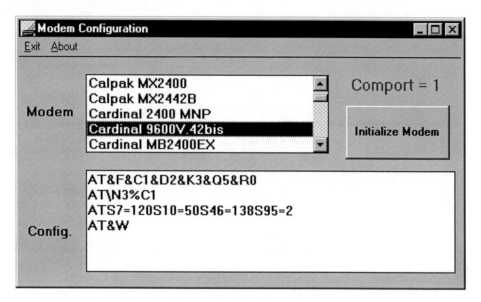

Problem: There are lots of modems available. How do I offer the correct configuration for them all? I want to use MSComm, so I cannot use TAPI.

Solution: You cannot do it. The best that you can do is to provide a database of configurations for various modems and hope that the one that you need is in the database. Here is an applet that includes a modem configuration database. It is not up-to-date but it may provide a starting point. It is a "flat file" so it can be edited using Notepad or some other text editor. It would be a comparatively simple process to export the configuration data to an Access database and to use ADO (Microsoft Active Data Objects) or equivalent for this. I will leave that as a future project.

```
Dim Config(200) As String

Sub Form_Load ()
    Dim Modem As String
    Dim I As Integer
    Dim PortNumber As Integer
    Dim BitRate As Long
    Dim Filenum As Integer
    Dim Ret As Integer
    Filenum = FreeFile
    Open App.Path & "\Modem.txt" For Input As #Filenum
    Do While Not EOF(1)
        Input #Filenum, Modem
        If Len(Modem) < 1 Then Exit Do
        Input #Filenum, Config(I)
```

```
            ModemList.AddItem Modem
                I = I + 1
        Loop
        Close
        Modmcfg.Show
        Modmcfg.Refresh
        ModemList.ListIndex = 0
        PortNumber = FindModem(BitRate)
        If PortNumber = 0 Then
            MsgBox "No modem detected"
        Else
            MsgBox "Modem found at port " & Str$(PortNumber)
            Label3.Caption = "Comm port =" & Str$(PortNumber)
            Label3.Visible = True
        End If
        Ret = WaitForResp("AT" & VbCr, "OK" & _
            VbCrLf, .5)
        If Ret <> 0 Then
            MsgBox "Response Error " & Str$(Ret%)
        End If
    End Sub
```

The modem configuration database that is furnished on the CD ROM (MODEM.TXT) is
read and modems are added to the ModemList list box.

Call FindModem to locate the first available modem.

```
    Sub InitModem_Click ()
    Dim I As Integer
    Dim ConfigString As String
    Dim Ret As Integer
    Dim ErrorDet As Integer
        For I = 1 To Len(Config(ModemList.ListIndex))
            If Mid$(Config(ModemList.ListIndex), I, 2) = _
                                                "^^" Then
                Ret = ModemCommand(ConfigString)
                If Ret = 1 Then
                    MsgBox ("An error was detected on command:" _
                        & ConfigString)
                    'needed because a subsequent string may be OK
                    ErrorDet = True
                End If
                ConfigString = ""
                I = I + 1
            Else
                ConfigString = ConfigString & _
                    Mid$(Config(ModemList.ListIndex), I, 1)
            End If
        Next I
```

```
            'pick up the last command
            Ret = ModemCommand(ConfigString)
            If Ret = 1 Then
                MsgBox ("An error was detected on command: " _
                    & ConfigString)
                ErrorDet = True
            End If
            If ErrorDet = False Then
                MsgBox ("The modem initialized successfully")
            End If
    End Sub
```

The InitModem procedure sends modem initialization strings to a modem. If an "OK" response is received, the modem string is valid. If an "ERROR" is received, the modem string is invalid.

The modem initialization strings are selected from the Config string-array. The Config array is populated when the form is loaded. The actual modem selected by the user is the index into the Config array.

Each entry in the Config array consists of one or more modem initialization strings. Each such string is delimited by "^^", except for the final initialization string.

```
    Sub ModemList_Click ()
    Dim I As Integer
        ModemConfig.Text = ""
        For I = 1 To Len(Config(ModemList.ListIndex))
            If Mid$(Config(ModemList.ListIndex), I, 2) = _
                    "^^" Then
                ModemConfig.SelText = VbCrLf
                I = I + 1
            Else
                ModemConfig.SelText = _
                    Mid$(Config(ModemList.ListIndex), I, 1)
            End If
        Next I
    End Sub
```

ModemList allows the user to select a modem by manufacturer and model from the associated list boxes. The configuration strings in the associated Config string array are displayed in the ModemConfig textbox. ModemList and the Config array are populated from the MODEM.TXT file in the Form_Load routine.

```
Private Function WaitForResp(SendCommand As String,
Response As String, Timeout As Long) As Integer
Dim RThreshold As Integer
Dim Buffer As String
    If Comm1.PortOpen = False Then Exit Function
    Comm1.Output = SendCommand
    WaitForResp = 2
'assume a timeout error
    Timeout = Timer + 1
'this section of code fails at midnight
    Do Until Timer > Timeout
'so, you may want to enhance it
        DoEvents
        Buffer = Buffer & Comm1.Input
        If InStr(Buffer, Response) Then
            WaitForResp = 0
            'OK!
            Exit Do
        End If
        If InStr(Buffer, "ERROR" & vbCrLf) Then
            WaitForResp = 1
            'Error! This is for a modem, but...
            Exit Do
        End If
    Loop

End Function
```

The WaitForResp function is a variation on the ModemCommand routine that was presented earlier. This routine allows you to specify an expected response and to specify a timeout, perhaps longer than the one second that is built into the ModemCommand function.

This function also is useful for sending strings to non-modem devices because it allows you to specify a response other than "OK". I have included code to detect an "ERROR" response. This extra code might be omitted if you use it with non-modem devices that have no "ERROR" response.

4.5.5   Magstripe Reader

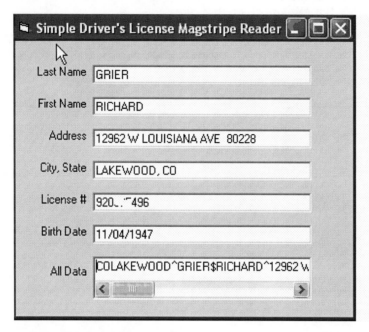

Problem: After I wrote the 3rd Edition of this book, I started an extensive project that relies on serial magstripe readers to decode data from credit and other types of cards that have magstripes. So, how should I present that sort of application here?

Solution: Most US state Driver Licenses have a magstripe where critical license data may be read and decoded. This example illustrates the basic decoding that might be required. It could be extended to other types of cards, though the details required to parse each field will vary from one card vendor to another. There are recording standards, and standard field formats that often (though not universally) are used. See the CD ROM for more details on various formats. Here is a quick description of the format used for most driver's licenses.

The American Association of Motor Vehicles Administrators adopted recommended data formats for the three tracks on Driver Licenses and Identification Cards on 9 September 1992, and most states, except California, follow the AAMVA rules for their DL/ID Cards. AAMVA uses the ISO ALPHA Data Format on Track #1, the ISO BCD Data Format on Track #2, and the ISO ALPHA Data Format on Track #3, with the additional provision that the End Sentinel may be used as a data character on Track #3. This requires the reader to test for the true End Sentinel on Track #3.

Note: I have obscured part of my driver's license number in the image above.

```
Private StartSent As String
Private EndSent As String
Private FieldMarker As String
Private NameMarker As String
```

Open the port and specify the EndSentinal (EndSent), FieldMarker, and NameMarker variables so that they may be used to parse decoded data.

```
Private Sub Form_Load()
    With MSComm1
        .CommPort = 1
        .Settings = "9600, N, 8, 1"
        .PortOpen = True
        .RThreshold = 1
        .DTREnable = True
        .RTSEnable = True
    End With
    EndSent = "?" & vbCr
    FieldMarker = "^"
    NameMarker = "$"
End Sub

Private Sub MSComm1_OnComm()
Static Buffer As String
Dim EndSentPos As Integer
Dim NextField As Integer
Dim NamePos As Integer
```

Append receive data to a string buffer. Scan the buffer for the EndSentinal, which marks the end of the card.

```
    Buffer = Buffer & MSComm1.Input
    EndSentPos = InStr(Buffer, EndSent)
    If EndSentPos > 0 Then
        Buffer = Mid$(Buffer, 2, EndSentPos - 2)
        'now, parse out the data fields
        txtAll.Text = Buffer
        NextField = InStr(Buffer, FieldMarker)
        txtStateCity.Text = Mid$(Buffer, 3, NextField _
- 3) & ", " & Left$(Buffer, 2)
        Buffer = Mid$(Buffer, NextField + 1)
        NamePos = InStr(Buffer, NameMarker)
        txtLastName.Text = Left$(Buffer, NamePos - 1)
        Buffer = Mid$(Buffer, NamePos + 1)
        NextField = InStr(Buffer, FieldMarker)
        txtFirstName.Text = Left$(Buffer, NextField - 1)
        Buffer = Mid$(Buffer, NextField + 1)
        NextField = InStr(Buffer, FieldMarker)
        txtAddress.Text = Left$(Buffer, NextField - 1)
        Buffer = Mid$(Buffer, NextField + 1)
        txtLicense.Text = Mid$(Buffer, 9, 9)
```

```
                   txtBirthdate.Text = Mid$(Buffer, 27, 2) & "/" _
        & Mid$(Buffer, 29, 2) & "/" & Mid$(Buffer, 23, 4)
                   txtAddress.Text = txtAddress.Text & "   " _
        & Mid$(Buffer, 36, 5)
                   Buffer = ""
              End If
         End Sub
```

4.5.6   DMM/LOGGER

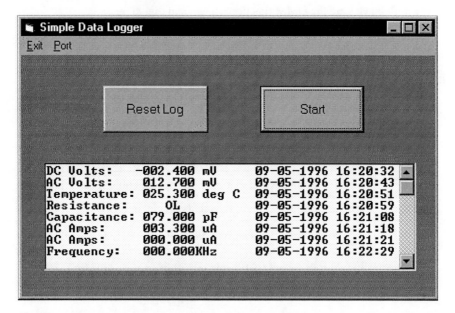

Problem: I have selected a Digital Multimeter (DMM) with an RS-232 interface to instrument a test setup on my client's electronic test bench. I need to log various readings at regular intervals and to display them. Later I will add a way to chart the readings as a function of time and combine that data with information from the function generator that I am using to stimulate the circuit. For now, all I want to do is build the basic interface.

The EXTECH 383273 DMM with PC interface meets my needs. It is inexpensive and flexible. It even comes with a Windows interface program written in Visual Basic. However, the manufacturer did not furnish the source code. The program itself does not work too well and it has several grammatical and spelling errors. I need to code my own.

My client has specified that I provide a VB 6.0 program, so that is what I have shown here. The code could be cut and pasted to VB 3.0; the only change required would be to define a global string variable, vbCrLf = Chr$(13) & Chr$(10).

Note: This example was developed for the 1St Edition of this book. Sufficient time has passed that current DVMs that have a RS-232 interface are less expensive and are simpler than the EXTECJ 38273 DVM. Some cost less than $50. One of these inexpensive DVMs is the MAS-345 from Electronix Express. I provide an example on the CD ROM for that DVM. There is a ReadMe file provided there that has more details.

The manual that is furnished with the EXTECH DMM describes the communications protocol that is used. To request data from the multimeter, send any non-reserved character. Reserved characters are used to configure the DMM. When the DMM receives a non-reserved character, it acquires a new measurement and replies with a five-byte response. The five bytes are the DMM function, range, and digital reading.

To decode the five bytes to a form that can be easily displayed and used, I have to create a parsing tree that extracts data based on the following relationship:

Byte	Description
1	02 (hexadecimal number that marks the start of data)
2	xx (hexadecimal code that indicates function and range)
3	xx (hexadecimal code that indicates reading polarity and data)
4	xx (hexadecimal code of data and decimal point)
5	03 (hexadecimal number that marks the end of data)

The manual provides a table (not repeated here) that shows how to decode the 2nd byte. It can range from &H00 for DC volts in the 200 milivolt range to &HB0 for AC amps in the 200 microamp range.

The 3rd and 4th bytes comprise a bit-mapped word (16-bits) as follows,

Bit	Description
0	Polarity (1=positive, 0=negative), except in frequency measurement (1=MHz, 0=kHz).
1	Most significant digit (0 or 1)
2-5	The second digit (0000-1001)
6-9	The third digit (0000-1001)
10-13	The least significant digit (0000-1001)
14-15	Decimal point position (00=none, 01=between fourth (LSD) and the third digit, 10=between the second and third digit, 11=between the first (MSD) and the second digit).

Solution: Here is the applet.

```
Private Sub Com_Click(Index As Integer)
    Dim I As Integer
    On Error Resume Next
    If Comm1.PortOpen Then Comm1.PortOpen = False
    Comm1.CommPort = Index
    Comm1.PortOpen = True
    If Err Then
        MsgBox "There was an error when this port " & _
            "was selected."
    Else
        For I = 1 To 4
            Com(I).Checked = False
        Next I
        Com(Index).Checked = True
    End If
End Sub
```

This menu-item routine allows the user to select an alternate serial port. If the selected port is not available, a MsgBox is displayed.

```
Private Sub ExitLogger_Click()
    Unload Me
End Sub
Private Sub Form_Load()
    On Error Resume Next
    Comm1.PortOpen = True
    If Err Then
        MsgBox "There was an error opening port " _
            & Format$(Comm1.CommPort) & " ."
    End If
    Com(Comm1.CommPort).Checked = True
End Sub
```

Open the serial port when the program starts. Display a MsgBox if the port is not available.

```
Private Sub Decode(Buffer As String)
Dim RecFunction As String
Dim V As Long
Dim Range As String
Dim VA As String
Dim Digit1 As Integer
Dim State As Integer
Dim DIGIT(14) As String
Dim Digit2 As Integer
Dim Digit3 As Integer
    On Error GoTo ErrorHandle
    DIGIT(0) = "0"
    DIGIT(1) = "8"
    DIGIT(2) = "4"
    DIGIT(4) = "2"
    DIGIT(6) = "6"
    DIGIT(8) = "1"
    DIGIT(9) = "9"
    DIGIT(10) = "5"
    DIGIT(12) = "3"
    DIGIT(14) = "7"

    RecFunction = ""
    Range = ""
    Select Case Asc(Mid$(Buffer, 2, 1))
        Case 0
            RecFunction = "DC Volts:      "
            Range = " mV    "
        Case 1
            RecFunction = "DC Volts:      "
            Range = " V     "
        Case 2
            RecFunction = "DC Volts:      "
            Range = " V     "
        Case 3
            RecFunction = "DC Volts:      "
            Range = " V     "
        Case 4
            RecFunction = "DC Volts:      "
            Range = " V     "
        Case 5
            RecFunction = "Frequency:     "
            Range = " KHz   "
        Case 6
            RecFunction = "DIODE          "
            Range = "       "
        Case 8
            RecFunction = "Resistance:    "
            Range = "Ohms   "
        Case 9
            RecFunction = "Resistance:    "
            Range = "KOhms  "
```

```
Case 10
     RecFunction = "Resistance:   "
     Range = "KOhms "
Case 12
     RecFunction = "Resistance:   "
     Range = "KOhms "
Case 16
     RecFunction = "Resistance:   "
     Range = "MOhms "
Case 17
     RecFunction = "Resistance:   "
     Range = "MOhms "
Case 18
     RecFunction = "Capacitance: "
     Range = " uF    "
Case 20
     RecFunction = "Capacitance: "
     Range = " uF    "
Case 24
     RecFunction = "Capacitance: "
     Range = " nF    "
Case 32
     RecFunction = "Capacitance: "
     Range = " pF    "
Case 33
     RecFunction = "DC Amps:      "
     Range = " A      "
Case 34
     RecFunction = "DC Amps:      "
     Range = " mA     "
Case 36
     RecFunction = "DC Amps:      "
     Range = " mA     "
Case 40
     RecFunction = "DC Amps:      "
     Range = " mA     "
Case 48
     RecFunction = "DC Amps:      "
     Range = " uA     "
Case 64
     RecFunction = "Temperature: "
     Range = " deg F"
Case 65
     RecFunction = "Temperature: "
     Range = " deg F"
Case 66
     RecFunction = "Temperature: "
     Range = " deg C"
Case 68
     RecFunction = "Temperature: "
     Range = " deg C"
Case 128
```

```
                    RecFunction = "AC Volts:        "
                    Range = " mV     "
          Case 129
                    RecFunction = "AC Volts:        "
                    Range = " V      "
          Case 130
                    RecFunction = "AC Volts:        "
                    Range = " V      "
          Case 131
                    RecFunction = "AC Volts:        "
                    Range = " V      "
          Case 132
                    RecFunction = "AC Volts:        "
                    Range = " V      "
          Case 161
                    RecFunction = "AC Amps:        "
                    Range = " A      "
          Case 162
                    RecFunction = "AC Amps:        "
                    Range = " mA     "
          Case 164
                    RecFunction = "AC Amps:        "
                    Range = " mA     "
          Case 168
                    RecFunction = "AC Amps:        "
                    Range = " mA     "
          Case 176
                    RecFunction = "AC Amps:        "
                    Range = " uA     "
          Case 255
                    RecFunction = "WARNING ---HOLD"
    End Select
    Log.SelText = RecFunction
'
'Get the third and fourth bytes only (Multimeter reading)
'
    V = Asc(Mid$(Buffer, 4, 1))
    V = V * 256
    V = V + Asc(Mid$(Buffer, 3, 1))


'
'     Check overload and initial state embedded in byte 3
'     Get the last 6 bits
'
    State = V Mod 64
'     Initial state
    If State = 63 Then Exit Sub
'     Positive overload
    If State = 15 Then
        Log.SelText = "     OL" & "             " & _
                      Date$ & " " & Time$ & vbCrLf
        Exit Sub
```

```
End If
'   Negative overload
If State = 14 Then
    Log.SelText = "    -OL" & "            " & _
                     Date$ & " " & Time$ & vbCrLf
    Exit Sub
End If
If Asc(Left$(Buffer, 1)) <> 2 Or _
      Asc(Right$(Buffer, 1)) <> 3 Then
    Log.SelText = "RECEIVE DATA ERROR!" & _
             "       " & Date$ & "       " & Time$ & vbCrLf
    Exit Sub
End If


'
'    Decode polarity, KHz or MHz
'
If RecFunction = "Frequency" Then
    If V And 1 Then
        Range = "MHz"
        VA = " "
    End If
Else
    If V And 1 Then
        VA = " "
    Else
        VA = "-"
    End If
End If
If V And 2 Then
    VA = VA + "1"
Else
    VA = VA + " "
End If


'
'    4th and 3rd byte assemble a word,
'    bit 2 to 5 : the second digit
'    bit 6 to 9 : the third digit
'    bit 10 to 13 : the least significant digit
'
'    Shift right 4 bits by dividing it by 16 .
'    Get the last 4 bits by moding it by 16 .
'
Digit1 = (V \ 4) Mod 16
VA = VA + DIGIT(Digit1)
Digit2 = (V \ 64) Mod 16
VA = VA + DIGIT(Digit2)
Digit3 = (V \ 1024) Mod 16
VA = VA + DIGIT(Digit3)

Select Case V \ 1024 \ 16
```

```
        Case 2
            VA = Left$(VA, 4) + "." + Right$(VA, 1)
        Case 1
            VA = Left$(VA, 3) + "." + Right$(VA, 2)
        Case 3
            VA = Left$(VA, 2) + "." + Right$(VA, 3)
    End Select

    If Val(VA) < 0 Then Log.SelStart = Log.SelStart - 1
    Log.SelText = Format$(VA, "000.000") & " " & Range _
                & " " & Date$ & " " & Time$ & vbCrLf
    Exit Sub

ErrorHandle:
    Debug.Print "ERROR: " & Err.Description
End Sub
```

The Decode routine implements the serial protocol that was described in the program requirements (the Problem).

A straightforward Select/Case block is used to decode byte-2 of the receive data (function and range).

The only tricky parts of the decode routine are the methods used to extract "packed binary data" (BCD - Binary Coded Decimal) from the receive packet byte-3 and byte-4. Integer divide and modulus operators (\ and mod, respectively) are used to extract four bits at a time from the third and fourth bytes in the receive data packet. These are then reformatted and combined to create a string that can be displayed.

The DMM will send a special bit pattern in byte-3 that is not described in the shortened functional description that I provided in the Problem description. If the DMM is in its Initial State (it has made no measurements yet) or if it is in Positive or Negative Overload, it outputs a unique bit pattern that cannot represent actual data. The 6 LSB are used for this purpose. Initial State is represented by the bit pattern XX111111, Positive Overload is XX001111, and Negative Overload is XX001110. Decoding for this special state is done by comparing the 6-LSB (V& Mod 64) with 65, 15, and 14, respectively.

```
    Private Sub Reset_Click()
        Log.Text = ""
    End Sub
```

Clear the log text box.

```
Private Sub StartStop_Click()
    If StartStop.Caption = "Start" Then
        StartStop.Caption = "Stop"
        Timer1_Timer
        Timer1.Enabled = True
    Else
        StartStop.Caption = "Start"
        Timer1.Enabled = False
    End If
End Sub
```

Enable or disable logging serial data from the DMM.

```
Private Sub Timer1_Timer()
Dim Buffer As String
Dim Timeout As Single
    If Comm1.PortOpen = False Then
        Log.SelText = "Port " & Format$(Comm1.CommPort) & _
            " not available" & "        " & Date$ & "      " _
            & Time$ & vbCrLf
        Beep
        Exit Sub
    End If
    Timeout = Timer + 1
    If Timeout >= 86399 Then Exit Sub    'skip midnight
    Comm1.Output = " "
    StartStop.Enabled = False
    Do Until Comm1.InBufferCount >= 5 Or Timer > Timeout
        DoEvents
    Loop
    StartStop.Enabled = True
    If Timer > Timeout Then
        Log.SelText = "TIMEOUT WAITING FOR DATA" & "      " _
            & Date$ & " " & Time$ & vbCrLf
        Exit Sub
    End If
    Buffer = Comm1.Input
    Decode Buffer
End Sub
```

The Timer control is used to poll the DMM at the interval specified at design-time. A reasonable addition to this project would be to make the polling interval programmable by the user. The maximum rate at which this particular DMM may be polled is roughly 600 mS (that is a higher rate than is specified in the DMM User's Manual).

To poll the DMM a single, non-reserved character is sent from the PC software. I used the space character. When the DMM receives the space character, it performs a conversion and sends a five-byte packet (described elsewhere) in response. The poll routine waits for a response. When five bytes have been received, it parses and displays the data. If data is not received within a specified timeout, a TIMEOUT message is logged.

I simplified the code by skipping any poll that might be sent immediately before midnight. That way I do not have to account for the Timer function midnight rollover from 86399 to 0.

The sort of "packed binary" data that this applet decodes is not uncommon in instrumentation and control systems. So, while your application may differ in detail, the methods for extracting individual bytes and bits from bit-mapped data may be useful anyway.

A simple addition to this applet would be to create a log file that records the data that is displayed. Another addition might be to use DDE or OLE automation to provide this data to other applications.

4.5.7   GPS

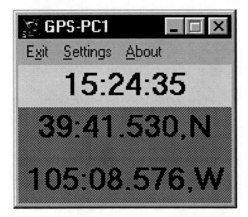

Problem: The Global Positioning System (GPS) constellation of satellites provides an ideal way to locate a receiver's position and to synchronize clocks very accurately. Data acquisition system and network timeservers can benefit from a program that can decode the output of a GPS receiver.

The National Electronic Manufacturers Association (NMEA-0183) protocol provides a standard serial protocol that is supported by quite a few different GPS receiver manufacturers.

Solution: Here is a VB6 applet that has been used with a number of commercial GPS-based products.

A second GPS application is included on the CD ROM that accompanies this book. It decodes additional NMEA-0183 sentences for Altitude, Velocity, Course, and Satellites in View. I will not describe it further here. However, it has some useful features (and a few simplifications) that you may find to be useful.

Note, in 2005 I encapsulated the code that is shown here for both the GPS decode and TimeZone calculations in class modules. These objects provide a substantial improvement in code clarity. I suggest that you use them as the building blocks for future development. See the CD_ROM for the DecodeGPS.CLS and clsTimeZone.cls files, and an example that illustrates their use (GPSClass.vbp).

```
Private WDog As Boolean
Private GPSDate As String
Private GPSStatus As String * 1
Private GPSWidth As Single
Private UTC As String
Private TrueOffset As Integer
Private Sub Comm1_OnComm()
Static Buffer As String
Dim Latitude As String
Dim Longitude As String
Dim commaPosition As Integer
Dim Temp As String
Dim TempString As String
Dim GetNext As Integer
Dim TimeVar As String
Dim Tempx As Integer
Dim TI As String

    On Error Resume Next
    '--- Wait for serial data
    '
    'decode $GPGGA sentence for altitude
    Buffer = Buffer & Comm1.Input
    commaPosition = InStr(Buffer, "GPRMC")
```

```
If commaPosition > 0 And InStr(Buffer, "GPGGA") Then
    Temp = Mid$(Buffer, commaPosition + 6, 2)
    UTC = Mid$(Buffer, commaPosition + 6, 6)
    GPSStatus = Mid$(Buffer, commaPosition + 13, 1)
    If GPSStatus = "V" Then
        lblGPS.BackColor = vbRed
    Else
        lblGPS.BackColor = vbYellow
    End If
    TempString = Mid$(Buffer, commaPosition + 15, _
                    Len(Buffer) - commaPosition - 15)
    For GetNext = 1 To 7
        GPSDate = ExtractString(TempString)
    Next GetNext
    Tempx = Val(Temp) + TrueOffset
    If Tempx > 23 Then Tempx = Tempx - 24
    If Tempx < 0 Then Tempx = Tempx + 24
    If Tempx < 10 Then
        Temp = "0" & Format(Tempx)
    Else
        Temp = Format$(Tempx)
    End If
    Temp = Temp & Mid$(Buffer, commaPosition + 8, 4)
    TimeVar = Left$(Temp, 2) & ":" & _
            Mid$(Temp, 3, 2) & ":" & Right$(Temp, 2)
    If (Len(Temp) = 6 And WindowState = 1 And _
                    Background.Checked = False) Then
        GPS.Caption = TimeVar
    ElseIf Background.Checked = False And _
                                WindowState = 0 Then
        lblGPS.Caption = Trim$(TimeVar)
    End If
    If WindowState = 0 And Background.Checked = _
                                        False Then
        lblGPS.Refresh
    End If
    If GPSStatus = "A" Then
        Buffer = Mid$(Buffer, commaPosition + 15)
        commaPosition = InStr(Buffer, ",")
        Latitude = Left$(Buffer, commaPosition - 7) _
            & ":" & Mid$(Buffer, commaPosition - 6, 8)
        Buffer = Mid$(Buffer, commaPosition + 3)
        If Background.Checked = False And _
                                WindowState = 0 Then
            lblLat.Caption = Trim$(Latitude)
        End If
        commaPosition = InStr(Buffer, ",")
        Longitude = Left$(Buffer, commaPosition - 7) _
            & ":" & Mid$(Buffer, commaPosition - 6, 8)
        If Background.Checked = False And _
                                WindowState = 0 Then
            lblLon.Caption = Trim$(Longitude)
```

```
                          End If
                          'place speed and course decode here ,<7>, <8>
                          TI = Right$(TimeVar, 8)
                    ElseIf Background.Checked = False Then
                          If WindowState = 1 Then GPS.Caption = "? " _
                                                  & GPS.Caption
                          lblLat.Caption = "No Latitude"
                          lblLon.Caption = "No Longitude"
                    End If
                    Buffer = Mid$(Buffer, InStr(Buffer, "GPGGA") + 1)
                    WDog = True
                    If (Comm1.InBufferCount > 1 Or _
                                          Len(Buffer) > 49) Then
                          Comm1_OnComm              'recurse for pending data
                    End If
                    Time$ = TI
              End If
        End Sub
```

The GPS applet decodes NMEA-0183 sentences in the OnComm receive event. This OnComm receive technique is lower overhead. It requires less processor bandwidth than polling MSCOMM in a DoEvents loop. Receive data is buffered in the Static string variable, InString$.

Most of the decoding is done "inline" to make it as responsive as possible. Real-time processing of receive data is required because one of the project requirements is to set PC time to GPS time. This is accurate only if all available receive data is processed every second. This applet might be multitasked with other applications which means that it must be as efficient as possible.

If the applet WindowState = Minimized and Background is checked, there is no need to update the display, only to set time and date. This can further improve speed.

The NMEA-0183 protocol specifies comma delimited, fielded data. The portions of the protocol that are used in this applet are completely described in the appendix. Each "sentence" starts with a $GPRMC or $GPGGA (or other) marker and is terminated by a carriage return and line feed. In between, commas separate data fields. Individual fields may be fixed or variable length as described in the protocol specification.

The ExtractString function is called in the OnComm routine to provide common code for extracting portions of each sentence. It will be described later.

The OnComm code maintains flags that can be polled by an external Timer routine. These flags specify whether or not valid data is being received from the GPS receiver. GPSStatus = "A" indicates valid data while GPSStatus = "V" indicates that the receiver cannot provide a valid navigation solution. If a valid navigation solution is not available, the PC date and time are not set and Daylight Savings time is not checked. The Wdog flag is provided so that the watchdog timer can indicate loss of valid data.

If the code determines that the GPS receiver has a valid navigation fix, the decoded time is certain to be correct. The PC time is set to agree with it.

I should make one more comment about the OnComm code to explain something that is somewhat unusual. After the current data is received and processed, I recursively call OnComm. That is to assure that any pending data that may have been received while the processing the current NMEA-0183 sentence is immediately buffered. This recursive call can improve the real-time performance of the applet. Such a technique is not required for most applications.

```
Function ExtractString (X As String) As String
Dim Position As Integer
    Position = InStr(X, ",")
    ExtractString = Left$(X, Position - 1)
    X = Mid$(X, Position + 1)
End Function
```

This ExtractString function assigns the first field in a comma delimited string to Extract_String and returns the balance of the string in the calling parameter, less the first field and comma. This allows the calling routine to "step-by-step" through a comma-delimited string buffer, one field at a time.

```
Private Sub Form_Load()
Dim Temp1 As String
Dim TZInfo As clsTimeZone
Dim Msg As String
On Error Resume Next
    If App.PrevInstance Then
        Unload Me
    Else
        Set TZInfo = New clsTimeZone
        TZInfo.GetTimezoneInfo
        Select Case TZInfo.TimeZoneInfo
            Case TIME_ZONE_ID_INVALID
                Msg = "The time zone information on " _
                    & "this system is not valid."
            Case TIME_ZONE_ID_UNKNOWN
                Msg = "Time time zone id on this " _
                    & "system is unknown."
            Case TIME_ZONE_ID_STANDARD
                TrueOffset = (TZInfo.TimeZoneBias + _
                            TZInfo.StandardBias) / 60
            Case TIME_ZONE_ID_DAYLIGHT
                TrueOffset = (TZInfo.TimeZoneBias + _
                            TZInfo.DaylightBias) / 60
        End Select
        If Len(Msg) > 0 Then MsgBox Msg, vbExclamation
        Set TZInfo = Nothing
        GPSWidth = GPS.Width
```

```
            lblGPS.Caption = "NONE"
            WDog = False
            Timer1.Interval = 2005
            Timer1.Enabled = True
            Timer2.Interval = 65535
            Timer2.Enabled = True
            ChDrive App.Path
            ChDir App.Path
            Comm1.commport = Val(GetSetting(App.EXEName, _
                              "Port", "Number", "1"))
            Temp1 = GetSetting(App.EXEName, "Run", _
                              "Flag", "False")
            If InStr(UCase$(Temp1), "TRUE") Then
                RunFlag.Checked = True _
            Else
                RunFlag.Checked = False
             End If
            Temp1 = GetSetting(App.EXEName, "Run", _
                              "Minimized", "False")
            If InStr(UCase$(Temp1), "TRUE") Then
                Background.Checked = True
                lblGPS.Caption = "Display Off"
            Else
                Background.Checked = False
            End If
            SetComm
            SOnTop
        End If
    End Sub
```

Code in the Form_Load event does various useful things. The Windows Registry is used to store all applet configuration information. This allows the user to change configuration and have those changes automatically recorded so that they will be used as the startup configuration when the applet is run subsequently.

Thirty-two-bit versions of Windows provide a way to calculate Daylight Savings time based on an API. The version of GPS that is furnished with this edition encapsulates the API in a class module named clsTimeZone.cls. The critical portions of that class follow.

```
Private Declare Function GetTimeZoneInformation _
    Lib "kernel32" (lpTimeZoneInformation As _
    TIME_ZONE_INFORMATION) As Long
Private Type SYSTEMTIME
  wYear As Integer
  wMonth As Integer
  wDayOfWeek As Integer
  wDay As Integer
  wHour As Integer
  wMinute As Integer
  wSecond As Integer
  wMilliseconds As Integer
End Type
Private Type TIME_ZONE_INFORMATION
    Bias As Long
    StandardName(0 To 63) As Byte
    StandardDate As SYSTEMTIME
    StandardBias As Long
    DaylightName(0 To 63) As Byte
    DaylightDate As SYSTEMTIME
    DaylightBias As Long
End Type
Public Enum TimeZoneConsts
    TIME_ZONE_ID_INVALID = &HFFFFFFFF
    TIME_ZONE_ID_UNKNOWN = 0
    TIME_ZONE_ID_STANDARD = 1
    TIME_ZONE_ID_DAYLIGHT = 2
End Enum
Private m_TimeZoneInfo As TimeZoneConsts
Private m_TimeZoneBias As Long
Private m_StandardBias As Long
Private m_DaylightBias As Long
Public Sub GetTimezoneInfo()
Dim TimeZone As TIME_ZONE_INFORMATION
    m_TimeZoneInfo = GetTimeZoneInformation(TimeZone)
    m_TimeZoneBias = TimeZone.Bias
    m_StandardBias = TimeZone.StandardBias
    m_DaylightBias = TimeZone.DaylightBias
End Sub
```

The Public method GetTimeZoneInfo is called in the GPS Form_Load event. This sets
Bias, StandardBias, and DaylightBias properties that will be used in the Comm1_OnComm
event to convert GPS date and time into time and date that accounts for Daylight Savings
time, if enabled by the user.

```
Private Sub Form_Unload(Cancel As Integer)
    SaveSetting App.EXEName, "Port", "Number", _
                        Format$(Comm1.commport)
    If RunFlag.Checked = True Then
        SaveSetting App.EXEName, "Run", "Flag", "True"
```

```
        Else
            SaveSetting App.EXEName, "Run", "Flag", "False"
        End If
        If Background.Checked = True Then
            SaveSetting App.EXEName, "Run", "Minimized", "True"
        Else
            SaveSetting App.EXEName, "Run", "Minimized", "False"
        End If
    End Sub
```

Save all configuration information in the Windows Registry. This saved configuration will be used when the applet is run next.

```
    Private Sub RunFlag_Click()
        RunFlag.Checked = Not (RunFlag.Checked)
        SetComm
    End Sub
```

Open the selected serial port.

```
    Private Sub SetComm()
        '--- 9600 baud, no parity, 8 data and 1 stop bit
        Comm1.Settings = "9600,N,8,1"
        '--- Some GPS receivers default to 4800 bps, so change
    to
        'Comm1.Settings = "4800,N,8,1"
        '---Tell the control to read entire buffer when
        '   the Input property is used.
        Comm1.InputLen = 0
        '--- Open the port
        On Error Resume Next
        If RunFlag.Checked And Comm1.PortOpen = False Then
            GPS.Comm1.PortOpen = True
        End If
        If Err.Number <> 0 Then
            MsgBox "The selected communications port is " & _
                    "not available.  Please select another." _
                    , vbCritical
            RunFlag.Checked = False
        End If
        If RunFlag.Checked = False Then
            If Comm1.PortOpen = True Then Comm1.PortOpen = False
        Else
            Comm1.RThreshold = 1
        End If
    End Sub
```

This code allows the user to select a comm port.

The code is shown to emphasize the point that many off-the-shelf GPS receivers use a serial rate of 4800 bps, not the 9600 bps that is used in this applet. You might customize the speed for your specific application in this routine. You might also specify the Settings that you need in the MSCOMM Properties window at design-time instead of at runtime.

```
Private Sub Timer2_Timer()
Dim Temp As Integer
Dim Yr As String
    On Error Resume Next
    If GPSStatus = "A" Then
        Temp = Val(Left$(UTC, 2)) + TrueOffset
        Yr = Right$(GPSDate, 2)
        If Val(Yr) > 93 Then
            GPSDate = Left$(GPSDate, 4) & "19" & Yr
        Else
            GPSDate = Left$(GPSDate, 4) & "20" & Yr
        End If
        If (Temp > 0 And Temp < 24) Then
            Date$ = Mid$(GPSDate, 3, 2) & "-" & _
                    Left$(GPSDate, 2) & "-" & _
                    Right$(GPSDate, 4)
        End If
    End If
End Sub
```

If valid GPS data is being received, this routine sets the PC date. Timer2.Interval is set to 65535. That means that this routine will run about every 65 seconds. It is not necessary to set the date more frequently than this.

```
'the following code is found in the COMMSEL form.
Private commport As Integer
Private Sub Form_Load()
    ListAvailablePorts
End Sub
Private Sub OKBUTTON_Click()
Dim PortOpen As Boolean
Dim OldPort As Integer
    On Error Resume Next
    With GPS.Comm1
        OldPort = .commport
        PortOpen = .PortOpen
        If PortOpen = True Then .PortOpen = False
        .commport = Val(Trim$(Right$(PortSel.Text, 2)))
        If PortOpen = True Then .PortOpen = True
        If Err.Number > 0 Then
            .commport = OldPort
            .PortOpen = PortOpen
        End If
```

```
            End With
            Unload Me
      End Sub
      Private Sub ListAvailablePorts()
      Dim I As Integer
      Dim PortOpen As Boolean
            On Error Resume Next
            With GPS.Comm1
                  PortOpen = .PortOpen
                  commport = .commport
                  If .PortOpen = True Then .PortOpen = False
                  For I = 1 To 16
                        .commport = I
                        .PortOpen = True
                        If Err.Number = 0 Then
                              PortSel.AddItem "Com " & Format$(I)
                              .PortOpen = False
                        Else
                              Err.Clear
                        End If
                  Next I
                  .commport = commport
                  .PortOpen = PortOpen
            End With
            Me.Caption = "Select From Available Ports (" & _
                        Format$(PortSel.ListCount) & ")"
      End Sub
```

ListAvailable port is called to test each possible serial port and to add any that are found (and are available) to a listbox.

A number of the routines that are actually used in this applet are not reproduced here. They are neither essential to understanding the process of writing an applet that decodes the NMEA-0183 protocol nor to handling time zones and Daylight Savings time. However, you should study the complete code if you intend to use this code in your own application. Refer to the CD ROM.

4.5.8   NIST Automated Computer Time

Problem: I want to reset my computer time and date accurately but I do not want to invest in a GPS receiver. What can I do?

Solution: The National Institute of Standards and Technology provides a dial-up modem timeserver located in Boulder Colorado (there are others in other areas of the country, too). Call the server at 1200 bps. One second after connect, a fixed-length string is returned of this form:

MJD	Y-M-D	H:M:S	DST	LS	DUT1	msADV	OTM
47222	88-03-02	21:39:15	83	0	+.3	045.0	UTC(NIST) *

Figure 4.5.8.1 Data returned from NIST

where MJD = Mean Julian Date, Y = year, M = month, D = day, H = hour, M = minute, S = second, DST = Daylight Savings Time (a flag meaning that a time change is coming). LS = Leap Second (a flag meaning that a leap second is to be added), DUT1 = UT1-UTC (earth rotation time minus coordinated universal time), msADV = milliseconds of advance of time of the marker, OTM = On Time Marker.

To keep the applet simple, the only correction is for local time. Additional changes could be made to correct for Daylight Savings Time. See the GPS applet for possible enhancements.

The only two fields that are extracted from the serial data received from NIST are the date and time fields.

```
Sub Comm1_OnComm ()
    Static Buffer As String
    Dim Position As Integer
    Dim OTM As Integer
    Dim NowDate As String
    Dim NowTime As String
    Dim Hours As String
    Dim Years As String
    Dim Months As String
    Dim Days As String
    Buffer = Buffer & Comm1.Input
    Position = InStr(Buffer, "*")
    If Position > 0 Then
        OTM = InStr(Buffer, "<OTM>")
        NowDate = Mid$(Buffer, OTM + 15, 8)
        Days = Mid$(NowDate, 7, 2)
        NowDate = Months & "-" & Days & "-" & Years
        NowTime = Mid$(Buffer, OTM + 24, 8)
        Hours = Format$(Val(Left$(NowTime, 2)))
        If Val(Hours) <= Offset Then
            'it's yesterday!
            NowDate = Format$(DateValue(NowDate) - 1, _
                "mm-dd-yy")
            Mid$(NowTime, 1, 2) = Format$(Val(Hours) _
                - Offset + 24, "00")
        Else
            Mid$(NowTime, 1, 2) = Format$(Val(Hours) _
                - Offset, "00")
```

```
            End If
            Buffer = ""
            Comm1.RThreshold = 0
            Time$ = NowTime
            Date$ = NowDate
            Label1.Caption = Time$
            Label2.Caption = Date$
        End If
    End Sub
```

This code is another variation on a parser for serial receive data. In this case, data fields are at fixed positions in a fixed-length packet. The start of data is marked by the asterisk of the preceding packet and terminated by the "<OTM>" marker.

Time zone correction is done using inline code. If the local time zone (Mountain Standard time is seven hours earlier than Universal Time) is less than the hours portion of Universal time, the local date is one day earlier than the date received from NIST and an appropriate adjustment is made.

```
    Sub Form_Activate ()
    Dim Ret As Integer
        Offset = 7                         'Mountain Standard Time
        Comm1.PortOpen = True
        Comm1.Output = "ATV1Q0&D2DT1-303-494-4774" & VbCr
        Ret = WaitFor(Comm1, "CONNECT", 30)
        If Ret = 1 Then
            Timer1.Enabled = True
            Comm1.RThreshold = 1
            Wait 15
        ElseIf Ret <> 2 Then
            MsgBox "Did not connect."
            Unload Me
        End If
    End Sub
```

This code dials the NIST number and waits for connect. The routine then waits 15 seconds to allow time and date data to be received and parsed by the OnComm receive routine.

Note that the telephone number specified is long distance. There may be a local telephone number available in certain parts of the country.

This applet uses a hard-coded comm port and the time zone. A more practical program would make these parameters user configurable. I chose to keep the applet as simple as possible. Feel free to enhance it.

The WaitFor routine is similar to the WaitForResp routine that I used in the MODEMCFG applet. The primary difference is that it is somewhat more general purpose and might be used with a modem or to wait for a response from a remote system.

I did not duplicate the WaitFor or Delay routines here. Please refer to the code on CD ROM for a complete listing for this applet

4.5.9 Barcode Reader

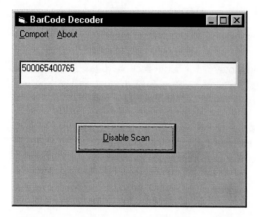

Problem: After the first edition of this book was published, I often received requests from readers for information on barcode readers. My most common response was "This is straightforward. You just connect a serial barcode reader to the PC and use MSComm or similar to receive the data. Then, write a routine to parse the decoded data to access associated information in a database or write a routine to do something else useful with the data." However, simply getting over the initial threshold was a problem in some cases.

Solution: Here is a simple application that receives decoded data from a Worthington Data Solutions serial barcode reader. See Appendix A for additional information on Worthington Data Systems.

Decoded data is written to a standard textbox for display. Obviously, a practical application would use this decoded data in a more useful fashion. However, it suffices to illustrate the issues involved.

```
'Barcode Form code

Private Sub About_Click()
Dim RThreshold As Integer
Dim PortOpen As Boolean
    RThreshold = MSComm1.RThreshold
    PortOpen = MSComm1.PortOpen
    If PortOpen Then MSComm1.PortOpen = False
    MSComm1.RThreshold = 0
    MsgBox "BarCode displays decoded barcodes in a Text
Box." & VbLf _
```

```
                & VbLf & "Copyright (c) 1998 - 2004 by Hard &
    Software."
        MSComm1.RThreshold = RThreshold
        MSComm1.PortOpen = PortOpen
    End Sub
```

Disable MSComm before the MsgBox is displayed and re-enable MSComm after the user dispatches it. This avoids any potential problems with partial scans, as remote as this possibility might be.

```
    Private Sub EnableDisableScan_Click()
        If EnableDisableScan.Caption = "&Enable Scan" Then
            On Error Resume Next
            MSComm1.PortOpen = True
                If MSComm1.PortOpen = False Then
                    MsgBox "Port open failed.  Error: " & Error$
                    Exit Sub
                End If
            EnableDisableScan.Caption = "&Disable Scan"
        Else
            MSComm1.PortOpen = False
            EnableDisableScan.Caption = "&Enable Scan"
            DisplayScan.Text = ""
        End If
    End Sub
```

The Enable/DisableScan command button is used to do just that. It acts as a toggle button. When first clicked, the port is opened and it reports any errors that might be detected. Alternate clicking of the button closes the port to disable scans.

```
    Private Sub MSComm1_OnComm()
    Static Buffer As String
        Buffer = Buffer & MSComm1.Input
        If InStr(Buffer, vbCrLf) Then
            Process Buffer
            Buffer = ""
        End If
    End Sub
```

The MSComm1.RThreshold property was set to 1 at design-time. So, when serial data is received, the OnComm event is called. Barcode readers usually add a CRLF (vbCrLf) at the end of each decoded barcode. This allows easy identification of a complete scan. When complete, the string Buffer contains the barcode and the terminating CRLF. It is then sent to the Process routine.

```
    Private Sub Process(BarCode As String)
        DisplayScan.Text = Left$(BarCode, Len(BarCode) - 2)
```

```
        Beep
End Sub
```

Replace the above code with your own to use the barcode in some useful way!

```
Private Sub SelectComm_Click()
    SelComm.Show vbModal
End Sub
```

Display the SelComm form to select a serial port. This is just one of a number of different ways to select a free port. See the ListAvailablePorts routine on the CD ROM.

```
'SelComm Form code

Private Sub cmdCancel_Click()
    Unload Me
End Sub

Private Sub OKBUTTON_Click()
Dim PortOpen As Boolean
Dim OldPort As Integer
    On Error Resume Next
    With BarCode.MSComm1
        OldPort = .CommPort
        PortOpen = .PortOpen
        If PortOpen = True Then .PortOpen = False
        .CommPort = Val(Trim$(Right$(PortSel.Text, 2)))
        If PortOpen = True Then .PortOpen = True
        If Err.Number > 0 Then
            .CommPort = OldPort
            .PortOpen = PortOpen
        End If
    End With
    Unload Me
End Sub
```

Notes:

The simplicity of this code is evident. However, barcode readers are sometimes used in more complex ways than simply connecting a single reader to a PC. For example, a series of barcode readers may be used in a production line to route components or finished product to destinations in a factory. Another similar use in a production environment would be inventory and warehouse control. In this case, the barcode readers might be all connected to a PC using RS-485 wiring (sometimes called "Daisy Chain" in the manufacturers' documentation). The communications protocol that would be used for this network of readers would be a little more complex than what is shown above. Each reader would be assigned an ID number (a network address), typically between 1 and 255 (1-0FFH). To get barcode data from a reader in the network, a poll command is sent to retrieve data from each reader in sequence. A VB Timer might be used to send these polls. When a barcode reader receives a poll, it responds with a data packet that incorporates its ID number and any pending barcode.

This multipoint protocol often will incorporate error checking to ensure accurate reporting of the decoded data. A checksum is appended to the decoded barcode that can be used to ascertain the accuracy of the received barcode. If the checksum agrees with the barcode, the program would respond by sending an ACK character (Chr$(6)), while an error would cause the program to send a NAK character (Chr$(15)). If the reader receives a NAK, it resends the barcode packet.

The details of the protocol, such as the method for calculating a checksum, will vary between barcode reader manufacturers. However, the actual implementation should be fairly easy. Simply start with a single reader configured with an address and test your code with that reader. After the single reader is working, add more until you have the whole thing working.

4.5.10 XMCommCRC ActiveX Control

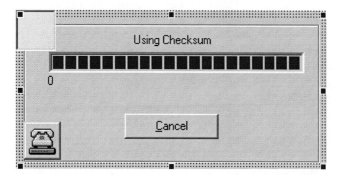

Problem: Quite a few readers of the First Edition of this book asked for 32-bit code that implements XMODEM error-checked file transfers. Although I provided VB3 code (see Appendix B), it used the Windows API and was not as clean as I would have liked. In addition, it would be nice to be able to use MSCOMM features in VBA environments without license issues.

Solution: Here is source code for an ActiveX control. It can be compiled using either VB 5.0 or VB 6.0. XMCommCRC exposes all useful MSCOMM properties and adds properties and methods for XMODEM/checksum file transfers. I have added XMODEM/CRC error-correction in the Third Edition of the book.

XMCommCRC uses three intrinsic VB controls, a Frame, a Label, and a Picture Box, and two controls that are furnished with VB, the Microsoft Communications Control (MSCOMM32.OCX) and the Microsoft Common Controls to provide a ProgressBar. The picture above shows how those controls were placed to create a User Control.

XMCommCRC implementation bends the XMODEM specification that Ward Christianson published. To speed file transfers, it does not process each receive character "on the fly." Rather, it buffers an entire 128-byte receive packet and processes the packet. However, this "improvement" should not affect interoperability with other XMODEM implementations. The transfer throughput, over an error-free, 115.2K bps link, exceeds 50K bps (on a fast computer). This is quite good for a "half-duplex" protocol and far exceeds the maximum expectation for XMODEM, which has often been said to be useful only over fairly slow serial links.

This example also introduces useful VB 5/VB 6 concepts such as the PropertyBag, AmbientProperties, and Extender Objects. Unfortunately, not all ActiveX containers implement the PropertyBag or Extender objects. For that reason, references to the Extender have been removed. Some environments may require that properties be persisted in code – the PropertyBag which provides design-time persistence for VB programmers is not available in all environments.

A Setup program with the XMCOMMCRC.OCX and all other required files can be downloaded from my homepage. See the Introduction for the current URL.

Also included on the CD ROM that accompanies this book are class modules that implement the XMODEM/checksum and CRC protocols. These may be incorporated in a VB program without having to distribute the XMCommCRC OCX.

```
Event OnComm()
Event TransferStatusChange()

Private mCommEvent As Integer
Private m_BeforeTransferMode As Integer
Private m_DisplayStatus As Boolean
Private m_EnableCancel As Boolean
Private m_MaxErrorCount As Integer
Private m_UseCRC As Boolean
Private m_BlockNumber As Byte

Private Const SOH = 1
Private Const ACK = 6
Private Const NAK = &H15
Private Const EOT = 4
```

```
Private Const CAN = &H18
Private Const CTRLZ = &H1A
Private Const CRCSTART = 67
Private Const NAKRECEIVED = -1
Private Const CRECEIVED = -3
Private Const CANRECEIVED = -2
Private Const ACKRECEIVED = 1

Private m_ReceiveBlockNumber As Byte
Private m_PacketTimeout As Single
Private m_ReceiveFilename As String
Private m_SendFileName As String
Private m_PackCharacter As Byte
Private m_Cancel As Boolean
Private m_Filesize As Long
Private m_StartTime As Long
Private m_TransferStatus As Integer
Private m_BytesTransfered As Long
Private m_RTHresholdBeforeTransferMode As Integer
Private m_TransferStarted As Boolean
Private m_Settings As String

Public Enum HandshakeConstants
    comNone = 0
    comXOnXoff = 1
    comRTS = 2
    comXonXoffRTS = 3
End Enum

Public Enum FileTransferStatus
    WAITING = 0  'Transfer not yet started
    TRANSFER_STARTED = 100 'Transfer in process
    PACKET_SENT = 200 'Packet sent
    PACKET_RECEIVED = 300  'Packet received
    NO_START = -100 'File transfer not able to start.
                    'No signal from other computer.
    NO_FILENAME = -200 'No filename selected to receive.
    CANCELLED = -300 'CANcel received
    ERROR_COUNT_EXCEEDED = -400 'Transfer failed.
    FILENAME_NOT_FOUND = -500 'Send file does not exist
    TRANSFER_SUCCEEDED = 1000 'Transfer succeeded!"
End Enum

'--- XMCOMM event constants
Public Enum XMCommConstants
    XMCOMM_EV_SEND = 1
    XMCOMM_EV_RECEIVE = 2
    XMCOMM_EV_CTS = 3
```

```
                XMCOMM_EV_DSR = 4
                XMCOMM_EV_CD = 5
                XMCOMM_EV_RING = 6
                XMCOMM_EV_EOF = 7

        '--- XMCOMM error code constants
                XMCOMM_ER_BREAK = 1001
                XMCOMM_ER_CTSTO = 1002
                XMCOMM_ER_DSRTO = 1003
                XMCOMM_ER_FRAME = 1004
                XMCOMM_ER_OVERRUN = 1006
                XMCOMM_ER_CDTO = 1007
                XMCOMM_ER_RXOVER = 1008
                XMCOMM_ER_RXPARITY = 1009
                XMCOMM_ER_TXFULL = 1010
        End Enum

        Private Declare Function timeGetTime Lib "winmm.dll" () As
        Long
```

First, we need to declare a few constants that will be used often. In addition, all of the
private data that is used in the control is specified along with the timeGetTime API function
declaration.

```
        Private Sub mComm_OnComm()
            If m_TransferStarted = False Or mComm.CommEvent _
                                            = comEvCD Then
                mCommEvent = mComm.CommEvent
                RaiseEvent OnComm
            End If
        End Sub
```

We are using the MSCOMM32.OCX as the primary constituent control. The name we have
selected is mComm. Whenever mComm generates an OnComm event, we echo that event
and the associated CommEvent property to the XMCommCRC OnComm event. An
exception is when a file transfer is in progress. Most normal MSComm OnComm events are
not propagated during a transfer; change of carrier detect state is propagated at all times.

```
        Public Property Get CommId() As Long
            CommId = mComm.CommId
        End Property

        Public Property Get CommEvent() As Integer
            CommEvent = mCommEvent
        End Property

        Public Property Get CommPort() As Integer
            On Error Resume Next
```

```
        CommPort = mComm.CommPort
End Property

Public Property Let CommPort(ByVal NewValue As Integer)
    mComm.CommPort = NewValue
End Property

Public Property Get DTREnable() As Boolean
    DTREnable = mComm.DTREnable
End Property

Public Property Let DTREnable(ByVal NewValue As Boolean)
    mComm.DTREnable = NewValue
End Property

Public Property Get Handshaking() As HandshakeConstants
    Handshaking = mComm.Handshaking
End Property

Public Property Let Handshaking(ByVal NewValue As
HandshakeConstants)
    mComm.Handshaking = NewValue
End Property

Public Property Get InBufferSize() As Integer
    InBufferSize = mComm.InBufferSize
End Property

Public Property Let InBufferSize(ByVal NewValue As Integer)
    mComm.InBufferSize = NewValue
End Property

Public Property Get InBufferCount() As Integer
    InBufferCount = mComm.InBufferCount
End Property

Public Property Let InBufferCount(ByVal NewValue As Integer)
    mComm.InBufferCount = NewValue
End Property

Public Property Get InputLen() As Integer
    InputLen = mComm.InputLen
End Property

Public Property Let InputLen(ByVal NewValue As Integer)
    mComm.InputLen = NewValue
End Property
```

```
Public Property Get OutBufferCount() As Integer
    OutBufferCount = mComm.OutBufferCount
End Property

Public Property Let OutBufferCount(ByVal NewValue As
Integer)
    mComm.OutBufferCount = NewValue
End Property

Public Property Get InputMode() As InputModeConstants
    InputMode = mComm.InputMode
End Property

Public Property Let InputMode(ByVal NewValue As
InputModeConstants)
    mComm.InputMode = NewValue
End Property

Public Property Get NullDiscard() As Boolean
    NullDiscard = mComm.NullDiscard
End Property

Public Property Let NullDiscard(ByVal NewValue As Boolean)
    mComm.NullDiscard = NewValue
End Property

Public Property Get OutBufferSize() As Integer
    OutBufferSize = mComm.OutBufferSize
End Property

Public Property Let OutBufferSize(ByVal NewValue As Integer)
    mComm.OutBufferSize = NewValue
End Property

Public Property Get ParityReplace() As String
    ParityReplace = mComm.ParityReplace
End Property

Public Property Let ParityReplace(NewValue As String)
    mComm.ParityReplace = NewValue
End Property

Public Property Get RThreshold() As Integer
    RThreshold = mComm.RThreshold
End Property

Public Property Let RThreshold(ByVal NewValue As Integer)
    mComm.RThreshold = NewValue
```

```
End Property

Public Property Let RTSEnable(ByVal NewValue As Boolean)
    mComm.RTSEnable = NewValue
End Property

Public Property Get RTSEnable() As Boolean
    RTSEnable = mComm.RTSEnable
End Property

Public Property Let Settings(NewValue As String)
On Error Resume Next
    m_Settings = NewValue
    mComm.Settings = NewValue
    If Err.Number > 0 Then Err.Raise Err.Number
End Property

Public Property Get Settings() As String
    Settings = mComm.Settings
End Property

Public Property Let SThreshold(ByVal NewValue As Integer)
    mComm.SThreshold = NewValue
End Property

Public Property Get SThreshold() As Integer
    SThreshold = mComm.SThreshold
End Property

Public Property Get Break() As Boolean
    Break = mComm.Break
End Property

Public Property Let Break(ByVal NewValue As Boolean)
    mComm.Break = NewValue
End Property

Public Property Get CDHolding() As Boolean
    CDHolding = mComm.CDHolding
End Property

Public Property Get CTSHolding() As Boolean
    CTSHolding = mComm.CTSHolding
End Property

Public Property Get DSRHolding() As Boolean
    DSRHolding = mComm.DSRHolding
End Property
```

```
Public Property Let Output(vNewValue As Variant)
On Error Resume Next
    If mComm.PortOpen And m_TransferStarted = False Then
        mComm.Output = vNewValue
    End If
End Property

Public Property Get PortOpen() As Boolean
    PortOpen = mComm.PortOpen
End Property

Public Property Let PortOpen(ByVal NewValue As Boolean)
    mComm.PortOpen = NewValue
End Property
```

All MSCOMM properties are echoed in the XMCommCRC OCX, with the exception of the Input property. Input is a Visual Basic keyword. Therefore, it cannot be used as a property name. We have decided to use the property name InputData because of its pneumonic value. To avoid errors, data **will not** be returned or transmitted during file transfers.

```
Public Property Get InputData() As Variant

    If m_TransferStarted = False Then
            InputData = mComm.Input
    End If
End Property
```

The additional property Let and Get procedures are used for the XMODEM file transfer features that we have added. The PacketTimeout property specifies how much time the sending and receiving routines will wait for the next acknowledgement or data packet, respectively. NOTE: the PacketTimeout is specified in seconds but is converted to milliseconds by the property procedures.

```
Public Property Get PacketTimeout() As Single
    PacketTimeout = m_PacketTimeout / 1000
End Property

Public Property Let PacketTimeout(ByVal NewValue As Single)
    m_PacketTimeout = NewValue * 1000
    If m_PacketTimeout < 1000 Then
        m_PacketTimeout = 1000
        Beep
    End If
End Property
```

XMODEM data uses fixed-length packets of 128 bytes. This means that if the file length is not a multiple of 128 bytes, the last packet must be padded with data that had no intrinsic meaning. ^Z, Chr$(26)), is often used because it marks the end of text files in DOS (and some other operating systems). However, your specific application may find it worthwhile to use an alternate padding character. Null, Chr$(0), or Chr$(&HFF) might be used.

```
Public Property Get PackCharacter() As Byte
    PackCharacter = m_PackCharacter
End Property

Public Property Let PackCharacter(ByVal NewValue As Byte)
    m_PackCharacter = NewValue
End Property
```

Use the ReceiveFilename property to specify the name, including path if desired, to be used to save the file on the receiving side. This name need not be the same as the source filename on the sending system.

```
Public Property Get ReceiveFilename() As String
    ReceiveFilename = m_ReceiveFilename
End Property

Public Property Let ReceiveFilename(NewValue As String)
    m_ReceiveFilename = NewValue
End Property
```

Use the SendFilename property to specify the name, and path if desired, of the file to be sent.

```
Public Property Get SendFilename() As String
    SendFilename = m_SendFileName
End Property

Public Property Let SendFilename(NewValue As String)
    m_SendFileName = NewValue
End Property
```

Your program can poll the TransferStatus property to obtain the most recent file transfer status. Use of this property is optional, of course. Likewise, the BytesTransfered property will report the current byte transferred total.

```
Public Property Get TransferStatus() As Integer
    TransferStatus = m_TransferStatus
End Property

Public Property Get BytesTransfered() As Long
    BytesTransfered = m_BytesTransfered
End Property
```

The length of the receive file is not included in the XMODEM file transfer protocol. Therefore, the receive status progress bar normally cannot display transferred bytes as a percent of total. Therefore, when receiving a file, the progress bar will display the current receive byte count. However, the program or programmer may have some method to determine the actual size of the file to be received. If this property is set to a value greater than zero, receive file transfer status will be displayed as a percentage.

```
Public Property Let ReceiveFileSize(ByVal NewValue As Long)
    m_Filesize = (NewValue * 128 + 127) \ 128 'adjust for
XMODEM padding
End Property
```

Use the EnableXferCancelButton property to allow or disallow the user to cancel a file transfer after it has started.

```
Public Property Get EnableXferCancelButton() As Boolean
    EnableXferCancelButton = m_EnableCancel
End Property

Public Property Let EnableXferCancelButton(ByVal NewValue As
Boolean)
    m_EnableCancel = NewValue
    If NewValue = False Then
        cmdCancel.Visible = False
    Else
        cmdCancel.Visible = True
    End If
End Property
```

Use the DisplayStatus property to enable or disable the file transfer status window. If disabled, XMCommCRC is not visible at runtime.

```
Public Property Get DisplayStatus() As Boolean
    DisplayStatus = m_DisplayStatus
End Property

Public Property Let DisplayStatus(ByVal NewValue As Boolean)
    m_DisplayStatus = NewValue
End Property
```

The MaxErrorCount property is used to specify the number of errors that will cause a file transfer to be aborted.

```
Public Property Get MaxErrorCount() As Integer
    MaxErrorCount = m_MaxErrorCount
End Property
```

```
Public Property Let MaxErrorCount(ByVal NewValue As Integer)
    If NewValue < 5 Then
        m_MaxErrorCount = 5
    Else
        m_MaxErrorCount = NewValue
    End If
End Property
```

The UseCRC property is used to select CRC error-correction when set to True. When UseCRC is False, checksum error-correction will be used.

```
Public Property Get UseCRC() As Boolean
    UseCRC = m_UseCRC
End Property

Public Property Let UseCRC(ByVal NewValue As Boolean)
    m_UseCRC = NewValue
    If NewValue = True Then
        lblCRC.Caption = "Using CRC"
    Else
        lblCRC.Caption = "Using Checksum"
    End If
End Property
```

Here starts the "nitty-gritty" of the file transfer process. SendPacket accepts an array of type Byte from the SendFile procedure. ByteCount also is specified so that padding characters can be added to bring the packet data size up to 128 bytes for the final packet.

A packet consists of the PacketHeader character, the current BlockNumber, the complement of the current BlockNumber, 128-bytes of data, and the calculated checksum or CRC. The total size of a packet is 132 bytes for checksum error-correction and 133 bytes for CRC error-correction.

If the packet to be sent is a repeat of the last packet, because of either the receipt of a NAK character or because of the failure to receive an ACK, then the BlockNumber is not incremented. If a new packet, then the BlockNumber is incremented. Since the BlockNumber is a single byte, the code forces it to "roll over" to zero after each set of 256 packets has been sent.

```
Private Sub SendPacket(SendBuffer() As Byte, ByVal ByteCount
As Integer, Resend As Boolean)
Dim TempBN As Integer
Dim PacketHeader(2) As Byte
Dim CheckSum As Integer
Dim CS() As Byte
Dim I As Integer
Dim Size As Byte
Dim CRC As Long
```

```
            PacketHeader(0) = SOH
            If Resend = False Then
                m_BlockNumber = (m_BlockNumber + 1) And &HFF
                PacketHeader(1) = m_BlockNumber
                TempBN = &HFF - m_BlockNumber
                PacketHeader(2) = TempBN
            Else
                PacketHeader(1) = m_BlockNumber
                TempBN = &HFF - m_BlockNumber
                PacketHeader(2) = TempBN
            End If
            mComm.Output = PacketHeader
            If UBound(SendBuffer) < 127 Then _
                    ReDim Preserve SendBuffer(127)
            For I = ByteCount + 1 To 127
                    SendBuffer(I) = m_PackCharacter
            Next I
            mComm.Output = SendBuffer
            If m_UseCRC = False Then
                For I = 0 To UBound(SendBuffer)
                    CheckSum = (CheckSum + SendBuffer(I)) And &HFF
                Next I
                ReDim CS(0) As Byte
                CS(0) = CheckSum
                mComm.Output = CS
            Else
                CRC = CRCcalc2(SendBuffer(), UBound(SendBuffer))
                ReDim CS(1) As Byte
                CS(0) = (CRC And &HFFFF) \ &H100
                CS(1) = (CRC And &HFFFF) Mod &H100
                mComm.Output = CS
            End If
            XTransferStatus PACKET_SENT
        End Sub
```

Use the SendFile method to transmit a file. Before this method is called, a valid filename must be assigned to the SendFileName property. Before a file transfer is begun, the SendFile method must save all XMCommCRC properties that may have been set to some value other than those needed for the file transfer. Then, when the transfer is complete or aborted, those properties are restored by calling the ResetMode routine.

The MSCOMM32.OCX binary mode is used for file transfers (comInputModeBinary). This improves performance and allows the control to be used under any version of Windows. Performance is also improved by optimizing data reads from disk. Each time a packet is to be sent, 128 bytes are read. If the code had been written to read a single byte at a time, data throughput would have been less than 50% of the rate that is achieved by reading 128 bytes at a time.

The SendFile method waits for a NAK character from the receiving system. As soon as the initial NAK is received, the first packet is transmitted using the SendPacket routine. After this and each subsequent packet sent, the code must wait for an ACK character from the receiving system. The WaitACK routine is called here. If a NAK is received or if neither an ACK nor NAK is received before the timeout specified by PacketTimeout property, then the last packet is re-transmitted. If the anticipated ACK is received then the next 128 bytes is read from the file and sent. This process continues until all data has been read from the file, sent, and acknowledged. After the final packet has been acknowledged, SendFile sends an EOT character to signal the end of the file transfer.

If the DisplayStatus property has been set to True, the file transfer progress is indicated in a ProgressBar by calling the UpdateStatus routine. Each state that is reached during the file transfer is reported in the TransferStatus property by calling the XtransferStatus routine.

If the user clicks the Cancel button during a file transfer (the DisplayStatus and EnableCancel properties must be True), a CAN character is transmitted and the file transfer is aborted. Likewise, the code may call the CancelTransfer method with the same result.

```
Public Function SendFile() As Boolean
Dim SendBuffer() As Byte
Dim I As Integer
Dim Bytes As Integer
Dim Tries As Integer
Dim EndOfTransfer(0) As Byte
Dim Count As Integer
Dim Filenum As Integer
Dim TotalBytes As Long
Dim Ret As Integer
Dim SendFilename As String
Dim WaitForStart As Integer
    SendFilename = Dir$(m_SendFileName, vbReadOnly _
        Or vbNormal Or vbHidden)
    If SendFilename = "" Then
        XTransferStatus FILENAME_NOT_FOUND
        Exit Function
    End If
    TotalBytes = 0
    m_BeforeTransferMode = mComm.InputMode
    m_RTHresholdBeforeTransferMode = mComm.RThreshold
    mComm.RThreshold = 0
    m_TransferStarted = True
    mComm.InputMode = comInputModeBinary
    m_Cancel = False
    m_StartTime = 0
    If mComm.PortOpen = False Then mComm.PortOpen = True
    mComm.InBufferCount = 0
    If m_DisplayStatus Then
```

```
        m_Filesize = FileLen(m_SendFileName)
        Frame1.Caption = "Waiting for Start"
        Label1.Caption = "0"
        Label2.Caption = Format$(m_Filesize)
    End If
    XTransferStatus WAITING
    Do Until Count >= m_MaxErrorCount Or m_Cancel = True
        Count = Count + 1
        WaitForStart = WaitACK
        If WaitForStart = NAKRECEIVED Or _
        WaitForStart = CRECEIVED Then Exit Do
    Loop
    If m_Cancel = True Then
        XTransferStatus CANCELLED
        ResetMode
        Exit Function
    End If
    If Count >= m_MaxErrorCount Then
        XTransferStatus NO_START
        ResetMode
        Exit Function
    End If
    If m_DisplayStatus Then Frame1.Caption = "Sending..."
    XTransferStatus TRANSFER_STARTED
    m_BlockNumber = 0
    Filenum = FreeFile
    Open m_SendFileName For Binary As Filenum
    Do Until Loc(Filenum) >= LOF(Filenum)
        TotalBytes = TotalBytes + 128
        If Loc(Filenum) <= LOF(Filenum) - 128 Then
            ReDim SendBuffer(127) As Byte
            Bytes = 128
        Else
            ReDim SendBuffer(LOF(Filenum) - Loc(Filenum)) _
                                              As Byte
            Bytes = LOF(Filenum) - Loc(Filenum)
        End If
        Get #Filenum, , SendBuffer
        If Loc(Filenum) >= LOF(Filenum) Then
            If m_DisplayStatus Then Frame1.Caption = _
                      "Send finished with this packet"
        End If
        If m_Cancel = True Then
            Close #Filenum
            EndOfTransfer(0) = CAN
            Delay 2                'wait for receive to clear
            mComm.Output = EndOfTransfer
            ResetMode
```

```
                    Exit Function
            End If
            Do Until Tries > m_MaxErrorCount Or m_Cancel = True
                Tries = Tries + 1
                If Tries = 1 Then
                    SendPacket SendBuffer, Bytes, False
                Else
                    SendPacket SendBuffer, Bytes, True
                End If
                Ret = WaitACK
                If Ret = ACKRECEIVED Then
                    If m_DisplayStatus Then
                        UpdateStatus TotalBytes
                        m_BytesTransfered = TotalBytes
                        Tries = 0
                    End If
                    Tries = 0
                    Exit Do
                ElseIf Ret = CANRECEIVED Then
                    Close #Filenum
                    XTransferStatus CANCELLED
                    ResetMode
                    Exit Function
                End If
            Loop
            If Tries >= m_MaxErrorCount Then
                XTransferStatus ERROR_COUNT_EXCEEDED
                Close #Filenum
                ResetMode
                Exit Function
            End If
            If m_Cancel = True Then
                XTransferStatus CANCELLED
                Close #Filenum
                ResetMode
                Exit Function
            End If
        Loop
    Loop
    Delay 0.5
    EndOfTransfer(0) = EOT
    mComm.Output = EndOfTransfer
    Do Until Count >= 6
        Count = Count + 1
        If WaitACK = 1 Then
            Exit Do
        Else
            mComm.Output = EndOfTransfer
        End If
```

```
        Loop
        If WaitACK <> 1 Then mComm.Output = EndOfTransfer
        XTransferStatus TRANSFER_SUCCEEDED
        If m_DisplayStatus Then Frame1.Caption = _
                                  Format$(Loc(Filenum))
        SendFile = True
        Close #Filenum
        ResetMode
    End Function
```

Use the ReceiveFile method to receive a file. Before this method is called, a valid filename must be assigned to the ReceiveFileName property. Before a file transfer is begun, the ReceiveFile method must save all XMCommCRC properties that may have been set to some value other than those needed for the file transfer. Then, when the transfer is complete or aborted, those properties are restored by calling the ResetMode routine.

ReceiveFile also uses the MSCOMM32.OCX binary mode (comInputModeBinary) for optimum performance.

ReceiveFile initiates a file transfer by transmitting an initial NAK character. It then calls the ReceivePaket routine to wait for data from the sending system. If a valid packet is received, an ACK character is transmitted. Otherwise, a NAK character is transmitted. If an EOT character or CAN character is received then the file transfer process is terminated. If a file transfer is aborted, for any reason, the partial ReceiveFileName file is deleted.

If the DisplayStatus property has been set to True, the file transfer progress is indicated in a ProgressBar by calling the UpdateStatus routine. Each state that is reached during the file transfer is reported in the TransferStatus property by calling the RtransferStatus routine.

At any time during the file transfer, if the user clicks the Cancel button (the DisplayStatus and EnableCancel properties must be True), a CAN character is transmitted and the file transfer is aborted. Likewise, code may call the CancelTransfer method with similar results.

```
    Public Function ReceiveFile() As Boolean
    Dim Ret As Boolean
    Dim I As Integer
    Dim Filenum As Integer
    Dim ACKNAK(0) As Byte
    Dim EOTFlag As Boolean
    Dim RFilenum As Integer
    Dim ErrorCount As Integer
    Dim ReceiveBuffer(127) As Byte
    Dim TotalBytes As Long
        TotalBytes = 0
        m_BeforeTransferMode = mComm.InputMode
        m_RTHresholdBeforeTransferMode = mComm.RThreshold
        mComm.RThreshold = 0
        m_TransferStarted = True
```

```
mComm.InputMode = comInputModeBinary
m_Cancel = False
m_StartTime = 0
'm_Filesize = 0     'file size not known for receive
'however, the calling program may want to set a
'known size
'if m_Filesize > 0, then the ProgressBar will
'indicate actual status
If m_Filesize = 0 Then Label2.Caption = ""
If mComm.PortOpen = False Then mComm.PortOpen = True
mComm.InBufferCount = 0
If m_UseCRC = False Then
    ACKNAK(0) = NAK
Else
    ACKNAK(0) = CRCSTART
End If
mComm.Output = ACKNAK      'start the transfer
RFilenum = FreeFile
If m_ReceiveFilename = "" Then
    XTransferStatus NO_FILENAME
    ResetMode
    Exit Function
End If
XTransferStatus TRANSFER_STARTED
If m_DisplayStatus Then Frame1.Caption = "Receiving..."
Open m_ReceiveFilename For Binary As RFilenum
Do
    If m_Cancel And m_DisplayStatus Then
        ACKNAK(0) = CAN
        mComm.Output = ACKNAK
        Close #RFilenum
        Kill m_ReceiveFilename
        ResetMode
        XTransferStatus CANCELLED
        Exit Function
    End If
    Ret = ReceivePacket(ReceiveBuffer, EOTFlag)
    If Ret Then
        If EOTFlag Then
            ACKNAK(0) = ACK
            mComm.Output = ACKNAK
            Close #RFilenum
            XTransferStatus TRANSFER_SUCCEEDED
            ReceiveFile = True
            Exit Do
        Else
            TotalBytes = TotalBytes + 128
            If m_DisplayStatus Then
```

```
                                UpdateStatus TotalBytes
                                m_BytesTransfered = TotalBytes
                        End If
                        Put #RFilenum, , ReceiveBuffer
                        'Erase ReceiveBuffer
                        ACKNAK(0) = ACK
                        mComm.Output = ACKNAK
                        ErrorCount = 0
                    End If
                Else
                    If EOTFlag = True Then
                        Close #RFilenum
                        Kill m_ReceiveFilename
                        XTransferStatus CANCELLED
                        Exit Do
                    End If
                    ACKNAK(0) = NAK
                    mComm.Output = ACKNAK
                    ErrorCount = ErrorCount + 1
                    If ErrorCount >= m_MaxErrorCount Then
                        On Error Resume Next
                        Close #RFilenum
                        Kill m_ReceiveFilename
                        XTransferStatus ERROR_COUNT_EXCEEDED
                        Exit Do
                    End If
                End If
            End If
        Loop
        ResetMode
    End Function
```

The ReceivePacket function is the complement to the SendPacket function. It is called from ReceiveFile to wait for a packet to be received. Each packet is 132 or 133 bytes, consisting of 128 data bytes and four bytes of header, block number, and checksum or CRC. If 132 bytes are received before the timeout specified by the PacketTimeout property expires, that data is validated. Validation consists of making certain that the block number matches the next number that is expected, and that the calculated checksum matches the checksum byte or the calculated CRC bytes match the CRC that has been received.

If a valid packet has been received, it is returned to the ReceiveFile method. Otherwise, codes are returned to indicate end of file transfer (EOT character received), file transfer aborted by sender (CAN character received), or no valid data received before the specified timeout.

```
    Private Function ReceivePacket(ReceiveBuffer() As Byte,
    EOTFlag As Boolean) As Boolean
    Dim Buffer() As Byte
    Dim CheckSum As Integer
```

```
Dim I As Integer
Dim Timeout As Long
Dim CRC As Long
Dim PacketLength As Integer
    If UseCRC = False Then
        PacketLength = 132
    Else
        PacketLength = 133
    End If
    Timeout = timeGetTime + m_PacketTimeout
    Do Until mComm.InBufferCount >= PacketLength Or _
                            timeGetTime >= Timeout
        DoEvents        'allow messages to be processed
    Loop
    If timeGetTime < Timeout Then
        Buffer = mComm.Input
        If Buffer(0) <> SOH Then
            ReceivePacket = False
            Exit Function
        ElseIf Buffer(1) = m_ReceiveBlockNumber Then
            ReceivePacket = True
            Exit Function
        Else
            For I = 3 To 130
                ReceiveBuffer(I - 3) = Buffer(I)
                If UseCRC = False Then CheckSum = _
                        (CheckSum + Buffer(I)) And &HFF
            Next I
            If UseCRC = False Then
                If CheckSum = Buffer(131) Then
                    m_ReceiveBlockNumber = _
                        (m_ReceiveBlockNumber + 1) And &HFF
                    ReceivePacket = True
                Else
                    ReceivePacket = False
                End If
            Else
                CRC = CRCcalc2(ReceiveBuffer, 127)
                If (CRC And &HFFFF) \ &H100 = _
                        Buffer(131) And (CRC And &HFFFF) _
                        Mod &H100 = Buffer(132) Then
                    m_ReceiveBlockNumber = _
                        (m_ReceiveBlockNumber + 1) And &HFF
                    ReceivePacket = True
                Else
                    ReceivePacket = False
                End If
            End If
        End If
```

```
                End If
        Else
            If mComm.InBufferCount > 0 Then
                Buffer = mComm.Input   'read the receive buffer
                If Buffer(0) = EOT Then
                    ReceivePacket = True
                    EOTFlag = True
                ElseIf Buffer(0) = CAN Then
                    ReceivePacket = False
                    EOTFlag = True
                End If
            End If
        End If
End Function
```

XMODEM/CRC uses a standard 16-bit CRC calculation. That calculation is done here. CRCcalc2 is called from both the ReceivePacket and SendPacket routines.

```
    Private Function CRCcalc2(Buffer() As Byte, Length As
    Integer) As Long
    Dim I As Long
    Dim Temp As Long
    Dim CRC As Long
    Dim J As Integer
        CRC = 0
        For I = 0 To Length
            Temp = Buffer(I) * &H100&
            CRC = CRC Xor Temp
                For J = 0 To 7
                    If (CRC And &H8000&) Then
                        CRC = ((CRC * 2) Xor &H1021&)  _
                                           And &HFFFF&
                    Else
                        CRC = (CRC * 2) And &HFFFF&
                    End If
                Next J
        Next I
        CRCcalc2 = CRC And &HFFFF
    End Function
```

Transfer status changes are indicated by setting the TransferStatus property and by raising the associated TransferStatusChange event.

```
    Private Sub XTransferStatus(ByVal Status As Integer)
        m_TransferStatus = Status
        RaiseEvent TransferStatusChange
    End Sub
```

Call ResetMode to reset all properties and flags when a file transfer is completed or if aborted for any reason.

```
Private Sub ResetMode()
    mComm.InputMode = m_BeforeTransferMode
    mComm.RThreshold = m_RTHresholdBeforeTransferMode
    mComm.InputLen = m_InputLenBeforeTransferMode
    m_TransferStarted = False
    m_Cancel = False
    ProgressBar1.Value = 0
    m_Filesize = 0
End Sub
```

The WaitACK function is called to wait for a response from the receiving system. ACK (acknowledge), NAK (negative acknowledge), and CAN (cancel) characters, and timeout without receiving one of these characters generate return values that are processed by the SendFile method.

```
Private Function WaitACK() As Integer
Dim Timeout As Long
Dim Buffer() As Byte
    Timeout = timeGetTime + m_PacketTimeout
    Do Until Timeout < timeGetTime
        If mComm.InBufferCount > 0 Then
            Buffer = mComm.Input
            If Buffer(0) = ACK Then
                WaitACK = ACKRECEIVED
                Exit Do
            ElseIf Buffer(0) = NAK Then
                WaitACK = NAKRECEIVED
                Exit Do
            ElseIf Buffer(0) = CAN Then
                WaitACK = CANRECEIVED
                Exit Do
            ElseIf Buffer(0) = CRCSTART Then
                WaitACK = CRECEIVED
                Exit Do
            End If
        End If
        DoEvents            'allow messages to be processed
    Loop
    On Error GoTo ExitFunction
    If timeGetTime >= Timeout Then
    'check for cancel following ACK
        If UBound(Buffer) > 0 Then
            If Buffer(1) = CAN Then
                WaitACK = CANRECEIVED
            End If
        End If
    End If
```

```
        End If
ExitFunction:
End Function
```

The UpdateStatus routine is called to display the number of bytes transferred and percentage of total bytes, if appropriate, in a ProgressBar.

```
Private Sub UpdateStatus(ByVal Bytes As Long)
    On Error Resume Next
    If m_StartTime = 0 Then m_StartTime = timeGetTime
    If m_Filesize > 0 Then
        ProgressBar1.Value = (Bytes / m_Filesize) * 100
    Else
        ProgressBar1.Value = ((Bytes / 65536) * 100) Mod 100
        Label1.Caption = "Total bytes: " & Format$(Bytes)
    End If
    Frame1.Caption = Format$(Bytes * 1000 / (timeGetTime -
m_StartTime), "######") & " bytes/s"
    On Error GoTo 0
End Sub
```

The Delay routine can be called by any code that needs to insert a delay to assure synchronization of the sending and receiving systems. Delays are specified in seconds.

```
Private Sub Delay(ByVal Wait As Integer)
Dim TimeDelay As Long
    TimeDelay = timeGetTime + Wait * 1000
    Do Until TimeDelay <= timeGetTime
        DoEvents
    Loop
End Sub
```

When the XMCommCRC control is initialized, both at design-time and at runtime, it is desirable to set certain properties. Do that in the Initialize event.

```
Private Sub UserControl_Initialize()
    m_PacketTimeout = 1000
    m_PackCharacter = CTRLZ
    m_DisplayStatus = True
    m_MaxErrorCount = 10
    m_DisplayStatus = True
    m_EnableCancel = True
End Sub
```

```
Public Sub About()
  MsgBox "The XMComm custom control was written using VB6."
& vbLf _
  & "It encapsulates the functionality of the MSComm custom
control that" & _
```

```
   vbLf & "is furnished with VB6 Professional and Enterprise
Editions." & vbLf _
   & "It also provides XMODEM error-checked file transfers."
& vbLf _
   & vbLf & "            Copyright, Richard Grier, 1998,
1999, 2001"
End Sub
```

XMCommCRC automatically calls the ReadProperties routine when the control is loaded, both at design-time and at runtime. The PropertyBag object is used to store control properties that have been set by the control user at design-time. Data stored in the PropertyBag object is then read to restore these design-time properties.

```
Private Sub UserControl_ReadProperties(PropBag As
PropertyBag)
Me.PacketTimeout = PropBag.ReadProperty( _
                    "PacketTimeout",  1)
   m_PackCharacter = PropBag.ReadProperty( _
                    "PackCharacter", CTRLZ)
   m_MaxErrorCount = PropBag.ReadProperty("MaxErrors", 20)
   Me.DisplayStatus = PropBag.ReadProperty( _
                    "DisplayStatus", True)
   Me.Settings = PropBag.ReadProperty("Settings", _
                        "38400, N, 8, 1")
   Me.CommPort = PropBag.ReadProperty("CommPort", 1)
   Me.Handshaking = PropBag.ReadProperty("Handshaking", 0)
   Me.InBufferSize = PropBag.ReadProperty( _
                    "InputBufferSize", 512)
   Me.InputLen = PropBag.ReadProperty("InputLength", 0)
   Me.InputMode = PropBag.ReadProperty("InputMode", 0)
   Me.NullDiscard = PropBag.ReadProperty( _
                    "NullDiscard", False)
   Me.OutBufferSize = PropBag.ReadProperty(" _
                        OutputBuffer Size", 512)
   Me.ParityReplace = PropBag.ReadProperty( _
                    "ParityReplace", "?")
   Me.RThreshold = PropBag.ReadProperty( _
                    "OnCommReceiveThreshold", 0)
   Me.RTSEnable = PropBag.ReadProperty( _
                    "RTSEnable", False)
   Me.SThreshold = PropBag.ReadProperty( _
                    "OnCommSendThreshold", 0)
   Me.EnableXferCancelButton = PropBag.ReadProperty( _
                    "EnableCancel", True)

   If m_DisplayStatus = True Then
       Picture1.Visible = False
   End If
```

```
End Sub
```

We use the Terminate event to assure that the port is released properly.

```
Private Sub UserControl_Terminate()
    If mComm.PortOpen Then
        mComm.Handshaking = comNone
        Do Until mComm.InBufferCount = 0
            DoEvents
        Loop
        mComm.PortOpen = False
    End If
End Sub
```

XMCommCRC automatically calls the WriteProperties routine when the form that contains the control is saved. The PropertyBag object is used to store control properties that have been set by the control user at design-time. Properties in the PropertyBag object are then read to restore them when the form is loaded subsequently.

```
Private Sub UserControl_WriteProperties(PropBag As
PropertyBag)
    PropBag.WriteProperty "PacketTimeout", Me.PacketTimeout
    PropBag.WriteProperty "PackCharacter", m_PackCharacter
    PropBag.WriteProperty "MaxErrors", m_MaxErrorCount
    PropBag.WriteProperty "DisplayStatus", m_DisplayStatus
    PropBag.WriteProperty "Settings", Me.Settings
    PropBag.WriteProperty "CommPort", Me.CommPort
    PropBag.WriteProperty "Handshaking", Me.Handshaking
    PropBag.WriteProperty "InputBufferSize", Me.InBufferSize
    PropBag.WriteProperty "InputLength", Me.InputLen
    PropBag.WriteProperty "InputMode", Me.InputMode
    PropBag.WriteProperty "NullDiscard", Me.NullDiscard
    PropBag.WriteProperty "OutputBufferSize", _
                                    Me.OutBufferSize
    PropBag.WriteProperty "ParityReplace", Me.ParityReplace
    PropBag.WriteProperty "OnCommReceiveThreshold", _
                                    Me.RThreshold
    PropBag.WriteProperty "RTSEnable", Me.RTSEnable
    PropBag.WriteProperty "OnCommSendThreshold", _
                                    Me.SThreshold
    PropBag.WriteProperty "EnableCancel", m_EnableCancel
    PropBag.WriteProperty "UseCRC", Me.UseCRC
End Sub

Public Sub CancelTransfer()
    m_Cancel = True
End Sub
```

There is one operational item that you should note. Both the SendFile and ReceiveFile methods "block." That is, the routines do not return until a file transfer has been completed or aborted. Both methods use DoEvents inside state loops so that the VB/VBA program that hosts the control can process messages. It might be desirable to alter these methods so that they activate a non-blocking state machine (perhaps using a Timer). This would complicate the code, so I did not attempt it. In addition, since Windows messages may be processed while the file transfer is in progress, a non-blocking routine is not as useful as otherwise it might be.

You can refer to the XMODEM specification that is included in the XMODEM.TXT file on the CD ROM that accompanies this book. Also described in this file are the requirements for the XMODEM 1K and Ymodem protocols.

4.5.10   XMTerm

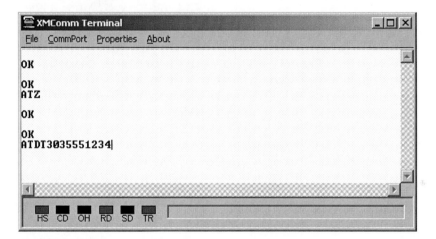

Problem: You have shown me how to write the XMCommCRC ActiveX control. Now, show me how to use it!

Solution: OK. Here is a "repeat" of the VBTerm example code that I presented earlier. This example may be used in the Learning Edition of VB where VBTerm may not be used (MSComm32.ocx is not licensed for the Learning Edition). You may use the VBTerm discussion for most of the code in XMTerm. The exceptions to this are the procedures for sending and receiving files using XMODEM error-correction. I present those routines here.

First let us select a filename that we will receive. The Common Dialog control is used for file selection.

```
Private Sub mnuReceiveFile_Click()
Dim Filename As String
```

```
Dim Ret As Integer
Dim Replace As Integer
On Error Resume Next
    '--- Get File name from the user
    OpenLog.DialogTitle = "Receive a file using XMODEM"
    OpenLog.Filter = "All Files (*.*)|*.*"
    Do
        OpenLog.Filename = ""
        OpenLog.ShowOpen
        If Err.Number = 0 Then
            Filename = OpenLog.Filename
            '--- If file already exists, do we overwrite it?
            Ret = Len(Dir$(Filename))
            If Err Then
               MsgBox Err.Description, vbExclamation
               Exit Sub
            End If
            If Ret > 0 Then
               Replace = MsgBox("Replace existing file - " _
                       & Filename & "?", vbOKCancel)
               If Replace = vbCancel Then Filename = ""
            End If
        Else
            Exit Do            'user pressed Cancel!
        End If
    Loop While Replace <> vbOK And Len(Filename) = 0
    If Len(Filename) > 0 Then
        frmXMComm.Caption = _
                "XMComm Terminal - Receiving: " & Filename
        MOpenLog.Enabled = False
        MSendText.Enabled = False
        MOpen.Enabled = False
```

OK. A filename has been selected and we actually are ready to start the receive process. Set the Filename property and then call the ReceiveFile method. That is about all there is too it.

```
        With XMComm1
            If XMComm1.PortOpen = True Then
                .ReceiveFilename = Filename
                .ReceiveFileSize = 0
                mnuReceiveFile.Enabled = False
                mnuSendFile.Enabled = False
                .Visible = XMComm1.DisplayStatus
                .ReceiveFile
                .Visible = False
                mnuReceiveFile.Enabled = True
                mnuSendFile.Enabled = True
            Else
                MsgBox "Port must be open to receive a
file.", vbExclamation
```

```
                    End If
                End With
                MOpenLog.Enabled = True
                MOpen.Enabled = True
                MSendText.Enabled = True
            End If
    End Sub
```

The other task is to select a file to be sent using XMODEM. This is quite similar to the receive process.

```
    Private Sub mnuSendFile_Click()
    Dim Filename As String
    Dim Ret As Integer
        On Error Resume Next
        MSendText.Enabled = False
        MOpenLog.Enabled = False
        '--- Get file name from the user
        OpenLog.DialogTitle = "Send a file using XMODEM"
        OpenLog.Filter = "All Files (*.*)|*.*"
        Do
            OpenLog.Filename = ""
            OpenLog.ShowOpen
            If Err.Number = 0 Then
                Filename = OpenLog.Filename

                '--- If file doesn't exist, go back
                Ret = Len(Dir$(Filename))
                If Err Then
                    MsgBox Err.Description, vbExclamation
                    MSendText.Enabled = True
                    MOpenLog.Enabled = True
                    Exit Sub
                End If
                If Ret = 0 Then MsgBox Filename & _
                                    " not found!", vbExclamation
            End If
        Loop Until Ret > 0 Or Len(Filename) = 0

        If Ret > 0 Then
            With XMComm1
                If .PortOpen = True Then
                    frmCanSend.Label1.Caption = "Transferring
    file using XMODEM - " & Filename
```

Right. The file to be sent has been selected. Just set the SendFilename property and call the SendFile method. Now, that was not too hard!

```
                    .SendFilename = Filename
                    mnuReceiveFile.Enabled = False
```

```
                              mnuSendFile.Enabled = False
                              .Visible = XMComm1.DisplayStatus
                              .SendFile
                              .Visible = False
                              mnuReceiveFile.Enabled = True
                              mnuSendFile.Enabled = True
                      Else
                              MsgBox "Port must be open to send a
      file.", vbExclamation
                      End If
                 End With
            End If
         MSendText.Enabled = True
         MOpenLog.Enabled = True
      End Sub
```

4.5.11   Host And Remote Using XMCommCRC

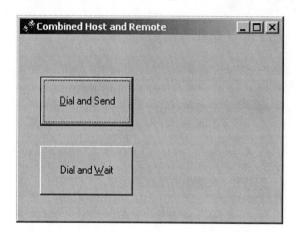

Problem: This picture doesn't look very sexy. You said you are going to offer another example using XMCommCRC.ocx. Is this it?

Solution: Yes! Here is my contribution. Suppose that your design goal is to create a single applet that may be used on either a central computer (Host) or a client computer (Remote) that will automatically send and receive error-checked files. How might this be done?

Actually, this applet is not quite automatic. When it starts up it, assumes Host mode operation. The applet attempts to locate a modem on the computer (see the PortFind example). When a modem is found, it enters "Host" state where it is awaiting a call from another PC. If the user clicks either of the command buttons, "Dial and Send" or "Dial and Wait", the applet suspends "Host" operation and assumes Remote mode operation. After dialing, the applet logs onto a compatible Host and either automatically sends a file or waits to see if a file is available on the host for automatic download.

Let us design a simple table that describes the state machine that the Host and Remote software implement. The actual details will be a little more complex than shown but this will give you an idea of the approach.

Host
WaitForConnect >> Send "LOGIN" >> WaitForUserName >> LookupUserName >> Send "PASSWORD" >> WaitForPassword >> ValidatePassword >> Send "MODE" >> WaitForModeCommand >> ReceiveFile or SendFile (depends on ModeCommand) >> Continue or Disconnect.
Remote
Dial >> WaitForLoginPrompt >> SendUserName >> WaitForPasswordPrompt >> SendPassword >> WaitForModePrompt >> SendModeCommand >> If ModeSending Then SendFile – If ModeWaiting Then WaitForSending >> ReceiveFile. Repeat states as needed.

The frmCommunications.frm code is treated like an object. It remains hidden but provides two facilities. First, it "sites" the XMCommCRC ActiveX control. Next, the code behind this form encapsulates most of the communications functions that we will be using.

Let us begin by looking at the initial operation when the applet starts.

```
With frmCommunications
    .Initialization = "AT&C1&D2"
    .File2Send = "TEST1.TXT"
    .File2Receive = "TEST2.TXT"
    .PhoneNumber = "5551234"
    .UserName = "Dick"
    .Password = "49814981"
    .ParentForm = Me
    Load frmCommunications
    If .ModemFound = True Then
        If .InitModem = False Then
            MsgBox "Modem initialization failed,", _
                                        vbCritical
        Else
            cmdDialSend.Enabled = True
            cmdDialWait.Enabled = True
        End If
    End If
End With
```

As I mentioned, frmCommunications is treated like any other object. It has properties and methods that may be called. The ModemFound function is similar to PortFind from an earlier example (I will not repeat it here; look at the actual code on the CD ROM for details). If a modem is detected, both Dial command buttons are enabled.

```
Private Sub cmdDialSend_Click()
    m_RetryCount = 1
```

```
      With cmdDialSend
          If .Caption = "&Dial and Send" Then
              .Caption = "Cancel &Dial and Send"
              tmrRetry1_Timer
              'this timer retries dialing and file transfer
          Else
              tmrRetry1.Enabled = False
              frmCommunications.HangupReset
              .Caption = "&Dial and Send"
          End If
      End With
  End Sub
```

The DialSend and DialWait command button events are similar. When clicked, they call the RetryTimer (1 or 2, respectively) to actually dial and connect. These routines have two states. If the button is clicked while dialing or communications are in process, they execute code that halts further processing. The applet then returns to Host-mode operation.

```
  Private Sub cmdDialWait_Click()
      m_RetryCount = 1
      With cmdDialWait
          If .Caption = "Dial and &Wait" Then
              .Caption = "Cancel &Dial and Send"
              tmrRetry2_Timer
              'this timer retries dialing and file transfer
          Else
              tmrRetry2.Enabled = False
              frmCommunications.HangupReset
              .Caption = "Dial and &Wait"
          End If
      End With
  End Sub
```

OK. One of the Dial command buttons has been pressed. The tmrRetry_Timer1 code is called to start the dial and send process. If a dialing or connection fails, the Timer is enabled so that it will automatically retry dial, connect and send. The code in Retry_Timer1 implements a state machine that controls the process.

```
  Private Sub tmrRetry1_Timer()
  'This timer activates the send file process.  It also
  'permits the called system to initiate a file transfer
  'after we have sent our file.
  'It will retry up to MAX_RETRIES.
      With frmCommunications
          .EnableAnswer = False
          tmrRetry1.Enabled = False
          If .Dial = True Then
```

The Dial function returns True if dialing results in a "good connection." Look at the Dial code to see what defines a "good connection."

We have designed our applet to send the hard-coded file name "TEST2.TXT". Obviously, a real application design would allow a more flexible file selection method.

```
        If .SendAfterCall("TEST2.TXT") = True Then
            Me.Caption = "File sent."
            If .GetNextFile = True Then
                Me.Caption = "File received."
            End If
            .EnableAnswer = True
            .HangupReset
            cmdDialSend.Caption = "&Dial and Send"
        Else
            If m_RetryCount < MAX_RETRIES Then
                Me.Caption = _
                        "Send failed.  Redialing now."
                tmrRetry1.Enabled = True
                .HangupReset
                m_RetryCount = m_RetryCount + 1
            Else
                Me.Caption = "Max retries exceeded."
                .HangupReset
                cmdDialWait.Caption = "&Dial and Send"
                Me.Caption = _
                            "Dial Failed.  Waiting..."
            End If
        End If
    Else
        If m_RetryCount < MAX_RETRIES Then
            Me.Caption = _
                        "Dial Failed.  Redialing now."
            tmrRetry1.Enabled = True
            m_RetryCount = m_RetryCount + 1
        Else
            cmdDialSend.Caption = "&Dial and Send"
            .HangupReset
            .EnableAnswer = True
            Me.Caption = "Dial Failed.  Waiting..."
        End If
    End If
    End With
End Sub
```

The remaining code in RetryTimer1 is involved with retry states and with notifying the user of the current state.

```
Private Sub tmrRetry2_Timer()
'This timer dials to retrieve any waiting file.  Unlike
' tmrRetry1, it does NOT send a file, only receive one if
' available.  It will retry up to MAX_RETRIES.
Dim Filename As String
```

```
With frmCommunications
    .EnableAnswer = False
    tmrRetry2.Enabled = False
    If .Dial = True Then
        If .WaitAfterCall(Filename) = True Then
```

We have designed our applet to automatically receive a file if the Host program has one waiting. The actual filename to be received is not retrieved from the host, so that our client code needs to know the filename in advance. The state machines that the Host and Remote portions of this applet implement could be expanded to exchange the actual filenames of files to be sent and received.

```
                Me.Caption = "File received: " & Filename
                .EnableAnswer = True
                .HangupReset
                cmdDialWait.Caption = "Dial and &Wait"
            Else
                If m_RetryCount < MAX_RETRIES Then
                    Me.Caption = _
                        "No file received.  Redialing now."
                    tmrRetry2.Enabled = True
                    .HangupReset
                    m_RetryCount = m_RetryCount + 1
                Else
                    Me.Caption = "Max retries exceeded."
                    cmdDialWait.Caption = "Dial and &Wait"
                    .HangupReset
                    Me.Caption = _
                                "Dial Failed.  Waiting..."
                End If
            End If
        Else
            If m_RetryCount < MAX_RETRIES Then
                Me.Caption = _
                            "Dial Failed.  Redialing now."
                tmrRetry2.Enabled = True
                m_RetryCount = m_RetryCount + 1
            Else
                cmdDialWait.Caption = "Dial and &Wait"
                .EnableAnswer = True
                .HangupReset
                Me.Caption = "Dial Failed.  Waiting..."
            End If
        End If
    End With
End Sub
```

The remaining code in RetryTimer2 is involved with retry states and with notifying the user of the current state.

Here are some of the important routines and properties that are in the frmCommunications code module. I have omitted several aspects of the actual code in this module. I do not present the Public Property Len and Get procedures, and have not duplicated the FindModem, SendCommand and other routines that are used in earlier examples. See the actual applet code on the CD ROM for those details.

```
Private m_Password As String
Private m_UserName As String
Private m_PhoneNumber As String
Private m_File2Send As String
Private m_Initialization As String
Private m_AnswerCalls As Boolean
Private m_ModemFound As Boolean
Private m_File2Receive As String
Private Buffer As String
Private m_ParentForm As Object
```

The above are accessed using Public Property procedures.

```
Private Sub Form_Load()
Dim BitRate As Long
Dim PortNum As Integer
    XMCommCRC1.CommPort = "3"
    XMCommCRC1.Settings = "115200, N, 8, 1"
    XMCommCRC1.Handshaking = comRTS
    XMCommCRC1.UseCRC = True
    XMCommCRC1.RThreshold = 0
    XMCommCRC1.InputLen = 0

    PortNum = FindModem(BitRate)
    If PortNum = 0 Then
        MsgBox "No modem detected."
    Else
        m_ModemFound = True
        MsgBox "Modem found.  Port=" & Format$(PortNum) &
".  Supports at least " & Format$(BitRate) & " bps."
        'XMComm has highest supported serial rate and port
set
        EnableAnswer = True
    End If
End Sub
```

This is not much to say about the code in the Form_Load event. It sets some parameters in XMComm that you may want to allow the user to set. I have hard-coded them for convenience. It then calls the FindModem routine. If a modem is found, EnableAnswer is set to True. This places the applet in "Host" mode, ready to receive calls from "Client" or "Remote" versions of the same applet (when the user clicks either Dial command button).

```
Public Function Dial() As Boolean
Dim Rate As Long
```

```
On Error GoTo ErrorHandler
If XMCommCRC1.PortOpen = False Then _
                    XMCommCRC1.PortOpen = True
If ModemCommand("ATDT" & m_PhoneNumber, 55) = _
                    Connect Then Dial = Login
Exit Function
ErrorHandler:
MsgBox "There was an error dialing. Error: " &
Err.Description
End Function
```

If the result of ModemCommand is "Connect" then Dial calls the Login function.

```
Private Function Login() As Boolean
Dim Ret As Integer
Dim FailureString As String
Dim Status As Boolean
Dim RThreshold As String
    FailureString = "Login failed.  "
    Wait 0.5 'wait for things to quiet down
    With XMCommCRC1
        RThreshold = .RThreshold
        .RThreshold = 0
        Ret = WaitForResponse("LOGIN", 5)
```

The state machine implemented by Login assumes that the Host computer sends a text string, "LOGIN". If this string is received then Login next sends the user name stored in the variable m_UserName.

```
If Ret = 0 Then
    Wait 0.5
    .Output = m_UserName & vbCrLf
    Wait 1
    Ret = WaitForResponse("PASSWORD", 5)
```

The next state waits for the text prompt, "PASSWORD" from the Host computer. When the "PASSWORD" prompt is received, we next send the password.

```
If Ret = 0 Then
    Wait 0.5
    .Output = m_Password & vbCrLf
    Wait 3         'adjust this time for the max
req'd to access the pw in the database
    Ret = WaitForResponse("MODE", 5)
```

Here is where we add a little complexity. We need a way to notify the Host computer whether or not a file will be uploaded automatically. When the "MODE" prompt is received from the Host computer, Login is complete.

```
If Ret = 0 Then
        Status = True
    Else
```

```
                        FailureString = FailureString & _
                                    "Mode timeout."
                End If
            Else
                FailureString = FailureString & _
                                    "Password timeout."
            End If
        Else
            FailureString = FailureString & _
                                "Login Timeout."
        End If
        Login = Status
        .RThreshold = RThreshold
    End With
    If Status = False Then
        HangupReset
        m_ParentForm.Caption = FailureString
        Wait 2
        Beep
    End If
End Function
```

After Login has completed satisfactorily, we next call the SendAfterCall or WaitAfterCall routines to implement the next set of states.

```
Public Function SendAfterCall(File2Send As String) As
Boolean
'you can call this function after successful Dial
    Wait 0.5
    XMCommCRC1.Output = "SENDING"
    Wait 1
    XMCommCRC1.SendFilename = App.Path & "\" & File2Send
    XMCommCRC1.SendFile
    If WaitForTransfer = TRANSFER_SUCCEEDED Then
SendAfterCall = True
End Function
```

SendAfterCall is called from the tmrRetry1_Timer routine as a result of the user having pressed the Dial and Send command button.

WaitAfterCall is called from the tmrRetry2_Timer routine as a result of the user having pressed the Dial and Wait command button. WaitAfterCall **also** is called from the tmrRetry1_Timer routine after a file has been sent from the remote to the host. This allows for "bi-directional" file transfers. Thus, if a file is waiting on the host to be sent to the remote, the remote will receive it automatically – even though the original operation was to send a file from remote to host.

```
Public Function WaitAfterCall(File2Receive As String) As
Boolean
'you can call this function after successful Dial/Login
Dim Ret As Integer
Dim RThreshold As Integer
```

```
With XMCommCRC1
    RThreshold = .RThreshold
    .RThreshold = 0
    Wait 0.5
    .Output = "WAITING"
    Wait 1
    Ret = WaitFor2Responses("SENDING", "NONE", 5)
    If Ret = 0 Then
        .ReceiveFilename = File2Receive
        .ReceiveFile
        If WaitForTransfer = TransferStatus_Success
Then
            WaitAfterCall = True
        End If
    End If
    .RThreshold = RThreshold
End With
End Function
```

The NewConnection routine is called when the modem answers in Host-mode operation. It implements the Host logon state machine.

```
Private Sub NewConnection()
'This is generated as soon as the modem connects after
answer
Dim Buffer As String
Dim Timeout As Single
Dim UserNameFound As Boolean
Dim Password As String
Dim Ret As Integer
Dim RThreshold As Integer
    Wait 0.5
    With XMCommCRC1
        RThreshold = .RThreshold
        .RThreshold = 0
        .Output = "LOGIN"
        Wait 1
        Timeout = timeGetTime + 8000
        'Look for the User Name
        Do Until timeGetTime >= Timeout
            Buffer = Buffer & .InputData
            DoEvents
            If InStr(Buffer, vbCr) Then
                UserNameFound = True
                Exit Do
            End If
        Loop
```

We have to make some assumptions about the format of text strings that will be exchanged between Remote and Host. Here we assume that the user name sent from the remote will be terminated by a carriage return character. This allows the user name to be of almost any arbitrary length.

We look up the user name and password. If the user name is found, we next request that the remote send a password string. We look for a match between the password sent from the remote and that in our local database. If it agrees, we proceed with the logon process.

```
'look for the User Name
If UserNameFound = True Then
     Password = LookupName(Buffer)
     If Len(Password) > 0 Then
          Wait 0.5
          .Output = "PASSWORD"
          Wait 1
          'get a password, then validate it
          Ret = WaitForResponse(Password, 5)
```

At this point, the password has been received and is valid.

```
If Ret = 0 Then
     Wait 0.5
     .Output = "MODE"
     Wait 1
     Ret = WaitFor2Responses("SENDING", _
                         "WAITING", 5)
```

We have designed two possible states that may be entered here. First, if the remote is going to send a file, we send a prompt to tell it to start sending. Then we call the ReceiveFile routine to receive that file. After we have received that file from the remote, we synchronize our operation with the remote by waiting for the "WAITING" string. If the variable m_File2Send has a value, we will automatically send that file to the remote computer by first sending the "SENDING" prompt, then calling SendFile.

```
Select Case Ret
     Case 0     'Sending
          ReceiveFile
          Wait 1
          Ret = _
     WaitForResponse("WAITING", 5)
          If Ret = 0 Then
               If m_File2Send <> "" Then
                    Wait 0.5
                    .Output = "SENDING"
                    Wait 1
                    SendFile
                    Wait 0.5
                    HangupReset
               Else
                    Wait 0.5
                    .Output = "NONE"
                    Wait 0.5
                    HangupReset
               End If
          Else
               HangupReset
```

```
                                        End If
```

If the remote user has logged on only to receive any pending file (the user pressed the Dial and Wait command button), we enter this state. If there is a file waiting to be sent to the remote then it is sent.

```
                        Case 1       'Waiting
                            Wait 0.5
                            If m_File2Send <> "" Then
                                .Output = "SENDING"
                                Wait 1
                                SendFile
                                HangupReset
                            Else
                                Wait 0.5
                                .Output = "NONE"
                                Wait 0.5
                                HangupReset
                            End If
                        Case Else 'timeout or other
                            HangupReset
                    End Select
                End If
            End If
        Else
            HangupReset
        End If
    End With
End Sub
```

HangupReset is called to exit from communications by either Remote or Host code. It is the terminal state that is entered if there is either a failure or a satisfactory conclusion. It returns us to the applet startup mode after disconnect.

```
    Public Sub HangupReset()
        On Error Resume Next
        With XMCommCRC1
            .CancelTransfer
            Do Until .TransferStatus <= 0
                DoEvents
            Loop
            .DTREnable = False
            Wait 0.5
            .DTREnable = True
            Wait 0.5
            EnableAnswer = True
            .RThreshold = 1
        End With
    End Sub
```

The LookupName function is trivial for this example. A real-life application almost certainly would look up the user name and password in a database.

```
Private Function LookupName(UserName As String) As String
'This function returns the password associated with a
user.
'Normally, you would use a database to perform this
association.
'Here, we are hard-coding for one user ("Dick").
    If InStr(UserName, "Dick") Then LookupName =
"49814981"
End Function
```

The SendFile and ReceiveFile perform complementary operations. The appropriate
XMCommCRC properties are set to the filename and paths and then the SendFile or
ReceiveFile methods are called. Both of these functions then call WaitForTransfer.

```
Private Function SendFile() As Boolean
    If m_File2Send <> "" Then
        XMCommCRC1.SendFilename = App.Path & "\" &
m_File2Send
        XMCommCRC1.SendFile
        If WaitForTransfer = TransferStatus_Success Then
SendFile = True
    End If
End Function
Private Function ReceiveFile() As Boolean
    XMCommCRC1.ReceiveFilename = App.Path & "\" &
File2Receive
    XMCommCRC1.ReceiveFile
    If WaitForTransfer = TRANSFER_SUCCEEDED Then
ReceiveFile = True
End Function
Private Function WaitForTransfer() As Integer
    With XMCommCRC1
        Do Until .TransferStatus = TRANSFER_SUCCEEDED Or
.TransferStatus < 0
            DoEvents
            If .CDHolding = False Then .CancelTransfer
        Loop
        WaitForTransfer = .TransferStatus
    End With
End Function
```

The GetNextFile function may be called as often as desired to implement multiple file
transfers in one session.

```
Public Function GetNextFile() As Boolean
Dim Ret As Integer
Dim RThreshold As Integer
    With XMCommCRC1
        RThreshold = .RThreshold
        .RThreshold = 0
        Wait 0.5
        .Output = "WAITING"
```

```
            Wait 1
            Ret = WaitFor2Responses("SENDING", "NONE", 5)
            If Ret = 0 Then
                Wait 0.5
                GetNextFile = ReceiveFile
            End If
            .RThreshold = RThreshold
        End With
    End Function
```

Code in the OnComm event is called to parse modem responses. This code is used in Host mode to command the modem to answer and to initiate a host session after connect.

```
    Private Sub XMCommCRC1_OnComm()
    Static Answering As Boolean
        With XMCommCRC1
            If .CommEvent = XMCOMM_EV_RECEIVE Then
                Buffer = Buffer & .InputData
                If InStr(Buffer, "RING") & vbCrLf And _
                                    Answering = False Then
                    Answering = True
                    ModemCommand "ATA", 1
                    Buffer = ""
                ElseIf InStr(Buffer, "CONNECT") And _
                                    Answering = False Then
                    Buffer = ""
                    Answering = False
                ElseIf Answering = True And InStr(Buffer, _
                                    "CONNECT") Then
                    Buffer = ""
                    NewConnection
                ElseIf InStr(Buffer, "NO CARRIER" & _
                                    vbCrLf) Then
                    Buffer = ""
                    Answering = False
                End If
            ElseIf .CommEvent = XMCOMM_EV_CD Then
                If .CDHolding = False Then
                    Answering = False
                    Buffer = ""
                End If
            End If
        End With
    End Sub
```

4.5.12 Flashlite

Problem: Create a Windows-based environment for controlling an embedded PC. There are quite a few embedded PCs available that run DOS. These are easy to program using a DOS-based compiler or built-in BASIC interpreter. Relatively inexpensive controllers for instrumentation, data acquisition and control, security, or other applications may be built using these boards. The features that we want to implement in our applet are:

- error-checked upload and download to and from the board using XMODEM

- a terminal window for interactive control of the embedded PC

- a menu to specify the external DOS compiler that is to be used

- set Date and Time of the embedded PC on startup

- shell the DOS compiler that was previously selected

- print the terminal window.

Solution: JK Microsystems makes a series of small embedded PCs that can be used. These boards are named "Flashlite" because they use flash memory to emulate disk drives. Prior to the 3rd Edition of this book, this applet used PDQComm from Crescent Software. PDQComm is no longer available. It would have been possible to port the code to use Sax Comm Objects, Crystal Comm, or most other commercial communications add-ons. However, XMCommCRC.ocx includes all of the functionality required. So, I have decided to use XMCommCRC. Best of all, it is "free."

```
Private Declare Function timeGetTime Lib "winmm.dll" () As
Long
```

```
Private Enum DataReceived
    NoMatch = 0
    Match = 1
    Error = -100
End Enum

Sub Com_Click (Index As Integer)
Dim I As Integer
Dim SavePort As Integer
    On Error Resume Next
    With XMCommCRC1
        If .PortOpen Then .PortOpen = False
        SavePort = .CommPort
        .CommPort = Index
        .PortOpen = True
        .DisplayStatus = True
        If Err Then
            MsgBox "Comport " & Format$(Index) & _
            " not available.  Select another from Setup."
            .CommPort = SavePort
        Else
            Pace vbCr & "Date " & Date$ & vbCr, 10
            Wait 0.03
            Pace "Time " & Time$ & vbCr, 10
            For I = 1 To 4
                Com(I).Checked = False
            Next I
            Com(Index).Checked = True
            SaveSetting App.EXEName, "Comm", _
                              "Port", .CommPort
        End If
    End With
End Sub
```

Select the comm port to be used to communicate with the Flashlite board. The port number will be saved in the Windows registry to be loaded as the default when the applet is run subsequently.

```
Sub Compile_Click ()
Dim Filename As String
    On Error Resume Next
    Filename = GetSetting(App.EXEName, "DOS", _
                              "Compiler", "")
    If Len(Filename) > 0 Then
        Filename = Dir$(Filename)
        If Err Then
            MsgBox "Cannot access the compiler.  Please
run Setup or make sure that any floppy is accessible."
        Else
            Shell Filename, vbNormalFocus
```

```
            End If
        Else
            MsgBox "No compiler selected.  Please select from
    Setup."
        End If
        TermWindow.SetFocus
    End Sub
```

Run whatever DOS compiler has been previously selected from Compiler under the Setup menu.

```
    Sub Compiler_Click ()
    Dim Compiler As String
        On Error Resume Next
        Compiler = GetSetting(App.EXEName, "DOS", _
                                        "Compiler", "")
        If Dir$(Compiler) <> "" Then
            ChDrive Compiler
            ChDir Compiler
        End If
        dlgFiles.DialogTitle = "Select the DOS compiler you
    want to use"
        dlgFiles.Filter = _
    "Compiler (*.EXE;*.PIF)|*.exe;*.pif|All Files (*.*)|*.*"
        dlgFiles.Filename = Compiler
        dlgFiles.ShowOpen
        If Err.Number = 0 Then
            SaveSetting App.EXEName, "DOS", _
                                        "Compiler", Compiler
            ChDrive App.Path
            ChDir App.Path
        End If
        TermWindow.SetFocus
    End Sub
```

Use the CommonDialog to specify the DOS compiler that will be used to create applications that will be downloaded to the Flashlite board for execution.

```
    Sub Download_Click ()
    Dim Filename As String
    Dim Ret As Integer
    Dim Position As Integer
    Dim I As Integer
    Dim Upfile As String
        On Error Resume Next
        dlgFiles.DialogTitle = "Download File from Flashlite.
    Select the filename and path on your PC."
        dlgFiles.Filter = "All Files (*.*)|*.*"
        dlgFiles.Filename = ""
        dlgFiles.Action = 1
        If Err = 32755 Then
```

```
                  TermWindow.SetFocus
                  Exit Sub
            End If
            Filename = dlgFiles.Filename
            If Len(Filename) <> 0 Then
                  With XMCommCRC1
                        .UseCRC = True
                        For I = 1 To Len(Filename)
                              If Mid$(Filename, I, 1) = "\" Then _
                                                Position = I
                        Next I
                        Upfile = Right$(Filename, Len(Filename) _
                                                - Position)
                        If UCase$(Upfile) = UCase$(Dir$(Filename)) _
                                                Then
                              Ret = MsgBox("This file already exists.
      Are you sure you want to overwrite it?", 36)
                              If Ret = 7 Then Exit Sub
                        End If
                        Upfile = "down" & " " & Upfile & vbCr
                        .RThreshold = 0
                        Pace Upfile, 10
                        Ret = WaitFor("Error opening file", 2)
                        If Ret = Match Then
                              MsgBox "File not found.  You must select
      the SAME name on the PC as the Flashlite."
                              TermWindow.SetFocus
                        Else
                              Wait 0.2
                              .ReceiveFilename = Filename
                              .Visible = .DisplayStatus
                              .ReceiveFile
                              .Visible = False
                        End If
                        .RThreshold = 1
                  End With
            End If
            TermWindow.SetFocus
      End Sub
```

Use the CommonDialog to display files that may be downloaded (received) from the Flashlite board. If that filename already exists on the PC, the user is prompted.

To request a download from the Flashlite board, a serial command of the form "down filename.ext" with a terminating carriage return is sent to the board. If the Flashlite board replies with an "Error opening file" message, the code assumes that the file does not exist.

The Flashlite board has a design limitation that is not uncommon in this type of application. Commands to the Flashlite board must be sent with short pauses between each character. The Flashlite board cannot process commands where the serial data characters are back-to-back. Probably, each character is processed with interrupts disabled. The UART can buffer only a single character. If an additional character comes in before the buffered character is removed, the data will be lost and the command will not be recognized.

The Pace routine sends commands one-character-at-a-time with a short delay between each character. Pace will be described later.

```
Sub Form_Load ()
Dim Bool As Boolean
Dim I As Integer
    With XMCommCRC1
        .RThreshold = 1
        .Settings = "9600, N, 8, 1"
        .DTREnable = True
        .RTSEnable = True
        .DisplayStatus = True
        .Visible = False
        .CommPort = GetSetting(App.EXEName, _
                                "Comm", "Port", 1)
        .PortOpen = True
        If Err Then
            MsgBox "Comport " & CStr(.CommPort) & _
            " not available.  Select another from Setup."
        Else
            For I = 1 To 4
                Com(I).Checked = False
            Next I
            TermWindow.Text = "Flashlite DOS window:"
            Com(XMCommCRC1.CommPort).Checked = True
            Bool = GetSetting(App.EXEName, _
            "Set Date and Time on", "startup", "False")
            If Bool = True Then
                Pace vbCr, 10
                Wait 0.05
                Pace "Date " & Date$ & vbCr, 10
                Wait 0.03
                Pace "Time " & Time$ & vbCr, 10
                Wait 0.03
                Pace "Dir" & vbCr, 10
                SetDateTime.Checked = True
            Else
                .Output = vbCr
            End If
        End If
    End With
    Me.Left = (Screen.Width - Me.Width) \ 2
    Me.Top = (Screen.Height - Me.Height) \ 2
End Sub
```

When the applet first starts, it may have been previously configured to set the Flashlite board's date and time to agree with the PC date and time. This is a very useful option because the Flashlite board does not have a non-volatile RTC (Real Time Clock). This option is loaded from the Windows registry as is the comm port to be used.

```
Private Sub Pace(Buffer As String, Milliseconds As
Integer)
Dim Seconds As Single
Dim I As Integer
    Seconds = Milliseconds / 1000
    For I = 1 To Len(Buffer)
        XMCommCRC1.Output = Mid$(Buffer, I, 1)
        Do Until XMCommCRC1.OutBufferCount = 0
            DoEvents
        Loop
        Wait Seconds
    Next I

End Sub
```

The Pace routine transmits a string with delays between each character in the string for the period of time specified in the Milliseconds parameter. The Wait routine is called to actually implement the specified delay between characters.

If you need to implement a routine like Pace, the actual delay between characters will have to be determined experimentally. You want a sufficiently long delay so that the connected device responds reliably. An unnecessarily long delay will make the program or system seem "unresponsive."

```
Sub PrintScreen_Click ()
    Printer.Print TermWindow.Text
    Printer.NewPage
    Printer.EndDoc
End Sub
```

Here is a feature that Jim Stewart of JK Microsystems requested. He wanted to be able to print whatever is in the terminal window.

```
Private Sub SetDateTime_Click()
Dim Checked As String
    SetDateTime.Checked = Not SetDateTime.Checked
    SaveSetting App.EXEName, "Set Date and Time on" _
                    , "startup", SetDateTime.Checked

End Sub
```

This routine allows the user to specify that the PC date and time will be sent to the Flashlite board each time the applet starts.

```
Private Sub Upload_Click()
Dim Filename As String
```

```
Dim Ret As Integer
Dim Position As Integer
Dim I As Integer
Dim Upfile As String

    On Error Resume Next
    dlgFiles.DialogTitle = "Upload File to Flashlite"
    dlgFiles.Filter = "Executable
(*.EXE;*.COM;*.BAS;*.BAT)|*.exe;*.com;*.bas;*.bat|All
Files (*.*)|*.*"
    dlgFiles.Filename = ""
    dlgFiles.Action = 1
    If Err.Number = 0 Then
        'dialog not canceled
        Filename = dlgFiles.Filename
        With XMCommCRC1
            .UseCRC = False
            .RThreshold = 0
            Upfile = "up" & vbCr
            Pace Upfile, 10
            Ret = WaitFor("Abort..." & vbCr, 10)
            If Ret <> Match Then
                MsgBox "No ACK from the Flashlite"
            Else
                For I = 1 To Len(Filename)
                    If Mid$(Filename, I, 1) = "\" Then _
                                        Position = I
                Next I
                Upfile = Right$(Filename, Len(Filename) _
                                - Position) & vbCr
                Pace Upfile, 10
                Wait 1
                .SendFilename = Filename
                .Visible = .DisplayStatus
                .SendFile
                .Visible = False
            End If
            .Output = vbCr
            Call WaitFor(">", 10)
            .RThreshold = 1
        End With
        Pace "DIR" & vbCr, 10
    End If
    TermWindow.SetFocus
End Sub
```

Use the CommonDialog to display files that may be uploaded (sent) to the Flashlite board.

To request a file upload to the Flashlite board, a serial command of the form "up" with a terminating carriage return is sent to the board. If the Flashlite board replies with an ""Abort..."" message, the code assumes that the Flashlite board did not respond correctly to the upload request.

Like the download process, upload requires that when the "up" command is sent, each character be separated by a short delay (using the Pace routine).

```
Sub Wait (Timeout As Single)
Dim WaitTime As Long
Dim StartTime As Long
    WaitTime = Int(Timeout * 1000) + timeGetTime()
    Do Until timeGetTime() >= WaitTime
        DoEvents
    Loop
End Sub
```

The Wait routine here uses the TimeGetTime API function to provide millisecond-accurate delays.

There are several features in this applet that may be of interest.

First, file upload to the embedded PC uses XMODMEM/Checksum, while download uses XMODEM/CRC. The current versions of the Flashlite boards require this dichotomy. Future versions of Flashlite may use XMODEM/CRC for upload and download.

Second, programmed "keystrokes" that comprise commands to the Flashlite board must have an artificial delay added between characters to simulate keyboard entry. Evidently the serial receive routine in the Flashlite board interprets command characters as they are received. If data arrives too quickly, a UART overrun causes data loss by the Flashlite board. To work around this design limitation, I created the Pace routine. In practice, I found that a 10 millisecond delay between characters is sufficient to avoid data loss by the Flashlite board.

Third, a small delay is required between commands to the Flashlite board in order to allow a previous command time to execute. I found that 30 milliseconds extra delay is sufficient. This keeps the program responsive.

These sorts of compromises and work-arounds are fairly common when dealing with external systems. They usually can only be identified by trial and error, although you can save yourself lots of headaches if you can anticipate some of them.

Most but not all routines in the Flashlite applet are described here. Refer to the code on CD ROM for a complete listing.

Chapter 5   VB.NET

The fall of the year 2000 brought the first beta of the Visual Studio.NET framework and developer suite. Since this chapter was first written, two versions of Visual Studio .NET have been released, Visual Studio 2002 and Visual Studio 2003.

Many of the code examples in this chapter are ports of VB6 code to VB .NET using the Upgrade Wizard. A few changes were made to the code, but for the most part, the Upgrade Wizard works well for serial communications applications.

Why .NET?

Microsoft has started a long journey toward a more distributed, open, and dynamic programming and application environment. This move is away from COM (the Component Object Model) that is the source of much of Windows' power and flexibility.   It has also been the cause of much of Windows' well-recognized reliability problems.

What is .NET?

Ah, the $64,000.09 question. From the Microsoft perspective, .NET is all things to all people. It encompasses operating systems, applications, and development environments. Most of us are interested only in the development environments portion of .NET technologies. Of course, we will restrict ourselves to the even smaller subset, Visual Basic.NET. Here are some features:

- Full object orientation, such as:

    o   Code inheritance

    o   Method overloading

    o   Parameterized constructors

    o   Shared members

- Structured error handling

- New threading model

- Reduced, and in some cases removed, reliance on COM

The .NET framework provides a common forms design environment known as WinForms or Windows Forms. This offers the same Windows GUI capability as older VB forms engines and adds some powerful new capabilities. Fortunately, we do not have to go into detail on most of these new capabilities. Serial communications largely remains the same under VB.NET as it was under earlier versions of VB. There will be extensive coverage of these new design features in other publications. We will limit ourselves to the things that we need to send and receive serial data. Suffice it to say, .NET development is the wave that we will surf over the next few years.

The new .NET design environment has some nice features. We can easily do program development using whatever mix of languages is appropriate. If we need to use VB, C# (pronounced C-sharp), and C++ together, we can. Debugging programs that use multiple languages is straightforward and, by comparison with earlier environments, painless.

.Net is based on a Common Language Runtime (CLR) environment that manages every aspect of code execution. "Common" means that all .NET components, regardless of language used for development, use this runtime. The CLR provides memory management, access to underlying OS services, a secure or safe execution environment, concurrency management, and object-location transparency. Reliance on the CLR creates what is called "Managed Objects or Managed Code." I will leave detailed discussion of the CLR to others.

One key goal of the .NET Framework is to simplify deployment and to eliminate "DLL Hell." There will be some costs and limitations in the attempt to meet this goal. These are topics that others will have to detail for you.

You develop .NET applications under Windows NT 4.0 or Windows 2K (or equivalent later OS). The .NET framework can be installed and executed on Windows 98/Me systems but the development environment cannot. Thus, practical development cannot be done using 98/Me.

All commonly used data types in .NET actually are implemented as objects. Thus, even primitive data types such as Boolean, Byte, Char, DateTime, Decimal, Double, GUID, Int16, Int32, Int64, SByte, and Single, all support the methods: Equals, GetHashCode, GetType, and ToString. These primitive types overlap some of the primitive types from earlier versions of VB. However, **there are differences** that can be important. The most common mistake that might be made (I know from experience) is that the keyword "Integer" refers to an Int32 (32-bit) variable that previously was declared "Long". For clarity in your coding, and to aid debugging, you should adopt the new type naming conventions. Here is an example:

```
Dim iLongVariable As Int32     'good naming convention

Dim iOLongVariable As Integer  'still a 32-bit integer,

                               'but… may be confusing!
```

A vital concept in .NET is that of **namespaces.** Namespaces help organize object libraries and hierarchies, simplify object references, prevent object reference ambiguity, and control the scope of object identifiers.

Class libraries may be referenced before they are used. The reference allows the types to be used in abbreviated form in code instead of requiring a detailed library reference. In VB this is done using the Imports keyword – this is similar to checking a box in the References dialog in earlier versions of VB. For example, a VB.NET form module might have this statement at the beginning:

```
Imports Debug = System.Diagnostics.Debug
```

Later in the code you might use the following statement:

```
Debug.WriteLine(iLongVariable.ToString)
```

These statements in combination first import the System.Diagnostics.Debug object and then assign it to an object variable named Debug. The Debug object may now execute all methods in System.Diagnostics.Debug. The one that we have used here is the WriteLine method. WriteLine requires a String argument. Thus, we have called the ToString method of the iLongVariable object (remember, even primitive data types are objects) to coerce iLongVariable to a string.

Details of .NET program structure that I will not go into here are the assembly (the basic unit of executable code and associated information that represents the application), the manifest, modules, metadata, and the actual compiled byte code (Intermediate Language or IL), or the JIT (Just-In-Time) compiler that converts IL code to native code before execution.

.NET memory management is different than that of earlier versions of VB. The CLR takes care of automatic memory management and garbage collection. The CLR loads components, allocates memory for them, and reclaims memory when those components are no longer referenced, that is, when no longer in scope. Earlier version of VB did this cleanup when the reference count for an object fell to zero. At that time, the Terminate event of the object was called **immediately.** This is known as "deterministic finalization." .NET does not have "deterministic finalization." The CLR does garbage collection only when necessary. Thus, in .NET, it is not possible to know exactly when garbage collection will reclaim unused object resources. The critical aspect of this fact of life is that any code that relied on execution of statements in a Terminate event may need to be manually adjusted to the object lifetime that .NET imposes.

Critical to general acceptance of .NET technologies is the concept of COM (Component Object Mode) Interoperability. COM is the core technology behind ActiveX components. Many millions of lines of code have been written in earlier languages that rely on COM. .NET does not abandon COM (though its use in **new** applications is reduced – ideally eliminated). .NET "Managed Code" may still use the huge set of resources available in "unmanaged code" and COM objects. .NET objects also may be called from unmanaged code by using the TlbExp.exe utility program. Thus, we may continue to rely on COM components in our .NET programs. Several examples provided with the book illustrate ActiveX control use in .NET programs – all made possible through COM Interoperability.

OK. Enough talk. Let us look at some VB.NET code. I will describe one VB.NET example in the text, NETGPS. However, other VB6 examples have been ported and are on the CD ROM that accompanies this book. Notable among these is XMTermNET, a port that employs XMCommCRC.ocx in a .NET terminal program for error-checked file transfers, and NETCommTerm, a terminal program that employs the NETComm.ocx ActiveX control (source code for the control also is included).

The DesktopSerialIO class included in the CD ROM that accompanies this book encapsulates the Windows Serial Communications API to make it accessable to users of Visual Studio 2002 and Visual Studio 2003. I recommend it instead of the other ActiveX solutions that are possible – DesktopSerialIO is simpler to deploy, and provides all of the essential interfaces required for most serial applications. It uses the same basic formula that is shown for CFSerialIO in Chapter 9, so I will not repeat the code here.

A second example on the CD ROM that employs DesktopSerialIO illustrates its use, along with the DecodeGPS class for displaying vehicle location and a moving map. The mapping aspects use Microsoft MapPoint 2004 ™.

Similarly, the XMComNET class provides XMODEM/Checksum and XMODEM/CRC in a native .NET dll that is based on DesktopSerialIO. Find it on the CD ROM, with a simple example that illustrates its possible application.

5.1 NETGPS

Problem: OK. Suppose I want to use MSComm32.ocx in a VB.NET project. Can you show me an example?

Solution: Sure. How about a port of the VB6 GPS program? I can use this to illustrate some of the features of the .NET framework that make things easier for us programmers.

```
Private Sub Comm1_OnComm(ByVal eventSender As _
   System.Object,  ByVal eventArgs As System.EventArgs) _
                                     Handles Comm1.OnComm
     Static Buffer As String
     Dim Latitude As String
     Dim Longitude As String
     Dim commaPosition As Int16
     Dim Temp As String
     Dim TempString As String
     Dim GetNext As Int16
     Dim TimeVar As String
     Dim Tempx As Int16
     Dim TI As String
     Dim Yr As String
     Dim DateStr As String
     Dim UTC As String
     Dim GPSStatus As String
     Dim GPSDate As String
     On Error Resume Next
     '--- Wait for serial data
     '
     'decode $GPGGA sentence for altitude
     Buffer = Buffer & Comm1.Input
     commaPosition = InStr(Buffer, "GPRMC")
     If commaPosition > 0 And InStr(Buffer, "GPGGA") Then
         Temp = Mid(Buffer, commaPosition + 6, 2)
         UTC = Mid(Buffer, commaPosition + 6, 6)
         GPSStatus = Mid(Buffer, commaPosition + 13, 1)
         If GPSStatus = "V" Then lblGPS.BackColor = Red _
                          Else lblGPS.BackColor = Yellow
         TempString = Mid(Buffer, commaPosition + 15, _
                     Len(Buffer) - commaPosition - 15)
         For GetNext = 1 To 7
             GPSDate = ExtractString(TempString)
         Next GetNext
         Temp = Temp & Mid(Buffer, commaPosition + 8, 4)
         TimeVar = VB.Left(Temp, 2) & ":" & Mid(Temp, 3, 2) _
                          & ":" & VB.Right(Temp, 2)
         If WindowState = 0 And Background.Checked = False _
                          Then lblGPS.Refresh()
         If GPSStatus = "A" Then
             WDog = True
             Buffer = Mid(Buffer, commaPosition + 15)
```

```
            commaPosition = InStr(Buffer, ",")
            Latitude = VB.Left(Buffer, commaPosition - 7) _
                & ":" & Mid(Buffer, commaPosition - 6, 8)
            Buffer = Mid(Buffer, commaPosition + 3)
            If Background.Checked = False And WindowState _
                    = 0 Then lblLat.Text = Trim(Latitude)
            commaPosition = InStr(Buffer, ",")
            Longitude = VB.Left(Buffer, commaPosition - 7) _
                & ":" & Mid(Buffer, commaPosition - 6, 8)
            If Background.Checked = False And WindowState _
                = FormWindowState.Normal Then
                            lblLon.Text = Trim(Longitude)
            'place speed and course decode here ,<7>, <8>
        End If
        TI = VB.Right(TimeVar, 8)
        Yr = VB.Right(GPSDate, 2)
        If Val(Yr) > 93 Then GPSDate = VB.Left(GPSDate, 4) _
                    & "19" & Yr Else GPSDate = _
                    VB.Left(GPSDate, 4) & "20" & Yr
        DateStr = Mid(GPSDate, 3, 2) & "-" & _
            VB.Left(GPSDate, 2) & "-" & VB.Right(GPSDate, 4)
        DateString = TimeZone.CurrentTimeZone.ToLocalTime( _
                                    DateStr & " " & TI)
        TimeString = TimeZone.CurrentTimeZone.ToLocalTime( _
                                    DateStr & " " & TI)
        If (Len(Temp) = 6 And WindowState = _
                        FormWindowState.Minimized And _
                        Background.Checked = False) Then
            GPS.DefInstance.Text = _
                    TimeZone.CurrentTimeZone.ToLocalTime( _
                                    DateStr & " " & TI)
        Else
            If Background.Checked = False And WindowState _
                            = FormWindowState.Normal Then
                lblGPS.Text = VB.Right( _
                    TimeZone.CurrentTimeZone.ToLocalTime( _
                            DateStr & " " & TI), 11)
            ElseIf Background.Checked = False Then
                If WindowState = FormWindowState.Minimized _
                        Then GPS.DefInstance.Text = "? " & _
                                GPS.DefInstance.Text
                lblLat.Text = "No Latitude"
                lblLon.Text = "No Longitude"
            End If
            Buffer = Mid(Buffer, InStr(Buffer, "GPGGA") + 1)
            If (Comm1.InBufferCount > 1 Or _
                Len(Buffer) > 49) Then _
                    Comm1_OnComm(Comm1, New System.EventArgs())
                    'recurse to extract any pending
        End If
    End If
End Sub
```

This really does not appear to be very different from the equivalent VB6 code. You can refer to Chapter 4 for comments there.

However, this code does illustrate one of the primary differences between VB.NET and earlier versions of VB. In the VB6 version, we used a class module to wrap the Windows GetTimeZoneInformation API function. This function allowed us to calculate local time from the UTC time decoded from the GPS receiver. The .NET framework provides a powerful set of library functions that expand on this and other API functions and make them more readily available to VB programmers. Here we use the .NET TimeZone.CurrentTimeZone.ToLocalTime method to convert UTC date and time information to local time. This requires that UTC time is converted to a "standard" date and time format that is an acceptable argument for this library function. However, the actual work involved in making this conversion is substantially less than that required for the earlier GPS programs.

Note: In 2005 I wrote a .NET class called DecodeGPS. The source code and dlls for this class are included on the CD ROM that accompanies this book. It provides a superior way to decode and to use the NMEA-0183 data stream from GPS receivers. I suggest that you use it for both desktop/notebook and Compact Framework GPS applications.

5.2 NETSerial

The 3rd Edition included a "native" .NET serial IO class that I named NETSerial. I have modified it slightly for this edition. The text, source code, and embedded inline comments, along with an example that employ it are on the CD ROM. I will not repeat it here, because I want to make space for other information. However, both Visual Studio .NET 2002 and Visual Studio .NET 2003 require such a class (or equivalent functionality via an ActiveX or .NET control). So, it remains of value.

I do want to make some points that I did not cover in the 3rd Edition.

- ActiveX control use requires a form to site the control. When you drop a control on the form from the Toolbox — having added it to the Toolbox using the Components menu/COM tab — .NET Integrated Development Environment (IDE) creates a wrapper DLL (.NET Assembly). The wrapper DLL is the code that .NET actually uses to interface to the ActiveX control. This step is necessarary, but the implication is that the ActiveX control may only be used from a Windows Forms application. Thus, other .NET applications need to use something else, like the NETSerial class. A common example of such an application might be a Windows Service, an application that has no user interface.

- You may either use the NETSerial class by simply incorporating the source code, or you may use reference a .NET DLL compiled from that class. The NETSerialDLL project on the CD ROM does this for you. This DLL may be referenced and used in both VB .NET and C# .NET projects.

- NETSerialDLL does not provide OnComm event notification. To poll you may use a Windows Forms Timer from the toolbox (Windows.Forms.Timer from code) in a Windows Forms project; the Component Timer from the toolbox (Windows.Timers.Timer from code) in a Windows Service or Console application; or you can use a Timer from the Toolbox in an ASP .NET application (server side execution).

- An alternate to the NETSerial class, also included on the CD ROM is the Windows 32-bit version of CFSerialIO. This code is more complete than NETSerial, though it uses a slightly different approach that is optimized for Windows CE applications — it also works well on the desktop. This class provides OnComm and ErrorEvent notification, and has more configuration flexibility.

5.3 DecodeGPS .NET Class

Problem: A preceding example was a simple port of VB6 code to VB .NET. There were a few changes from the equivalent VB6 program, but not very many. Then you explained how to code a serial class in .NET (see section 5.2 and the CD ROM). Can you show me some more VB .NET code?

Solution: Yes. I have some other code to show you. Several sections later in the book deal with other .NET features. Here I would like to illustrate some of the feature that the .NET framework provides that are powerful for parsing and conversion of data. This class takes the NMEA-0183 data stream output by a GPS receiver and decodes it. When the a complete GPS sentence has been decoded, and event is use to notify the calling program that up-to-date GPS data is ready to use.

This class easily could be recompiled for the .NET Compact Framework.

I will not show the code that calls this .NET DLL (Assembly). You may refer to the CD ROM for those details. Add a reference to the DecodeGPS DLL, and then add an Import Statement to use it. Create a New instance of the DecodeGPS class, and pass the receive data stream to the GPSStream method. When a complete NMEA-0183A $GPRMC sentence has been decoded, a GPSDecoded event is generated. See the inline comments for more information.

```
Public Class DecodeGPS
    Private m_Latitude As String
    Private m_Longitude As String
    Private m_UTCTime As String
    Private m_LatitudeHemisphere As String
    Private m_LongitudeHemisphere As String
    Private m_UTCDate As String
    Private m_CourseOverGround As String
    Private m_SpeedOverGround As String
    Private m_MagneticVariation As String
    Private m_MagneticVariationDirection As String
    Private m_LocalTime As String
    Private m_LocalDate As String
```

```
      Private m_Status As String
```

The DecodeGPS object returns data as 13 fields. Two of these fields at calculated from the decoded data, LocalTime and LocalDate. The other 11 fields represent the raw information in each GPRMC field. Thus, you may add additional decoding or processing of this raw data as may be required by your specific application. Refer to Appendix C for a detailed description of the GPS NMEA-0183A serial protocol.

```
   Public Event GPSDecoded(ByVal Status As Boolean)

   Public ReadOnly Property Status() As String
      Get
         Return m_Status
      End Get
   End Property
   Public ReadOnly Property Latitude() As String
      Get
         Return m_Latitude
      End Get
   End Property
   Public ReadOnly Property Longitude() As String
      Get
         Return m_Longitude
      End Get
   End Property
   Public ReadOnly Property UTCTime() As String
      Get
         Return m_UTCTime
      End Get
   End Property
   Public ReadOnly Property LatitudeHemisphere() As
String
      Get
         Return m_LatitudeHemisphere
      End Get
   End Property
   Public ReadOnly Property LongitudeHemisphere() As
String
      Get
         Return m_LongitudeHemisphere
      End Get
   End Property
   Public ReadOnly Property UTCDate() As String
      Get
         Return m_UTCDate
      End Get
   End Property
   Public ReadOnly Property CourseOverGround() As String
      Get
         Return m_CourseOverGround
      End Get
   End Property
```

```
      Public ReadOnly Property SpeedOverGround() As String
         Get
              Return m_SpeedOverGround
         End Get
      End Property
      Public ReadOnly Property MagneticVariation() As String
         Get
              Return m_MagneticVariation
         End Get
      End Property
      Public ReadOnly Property MagneticVariationDirection()
   As String
         Get
              Return m_MagneticVariationDirection
         End Get
      End Property
      Public ReadOnly Property LocalTime() As String
         Get
              Return m_LocalTime
         End Get
      End Property
      Public ReadOnly Property LocalDate() As String
         Get
              Return m_LocalDate
         End Get
      End Property
```

Call GPSStream and pass in receive data. The GPSStream method double-buffers received data. When a complete GPRMC sentence has been received, it will be parsed.

```
      Public Sub GPSStream(ByVal GPSData As String)
         Static Buffer As String

         Dim Delimiter() As Char = ",".ToCharArray
         Dim SplitGPS() As String _
                         Buffer += GPSData
         'Do we have a complete $GPRMC sentence?
         If (Buffer.IndexOf("GPRMC") > 0) And _
            (Buffer.IndexOf(vbCrLf, _
             Buffer.IndexOf("GPRMC")) _
                         > 0) Then
            Dim SubBuffer As String = _
                   Buffer.Substring _
                  (Buffer.IndexOf("GPRMC"))
```

The .NET String.Split method is equivalent to the Visual Basic 6 Split$ function. This gives us an easy way to extract fields from a comma delimited set of data. Each field in the GPRMC sentence will then be found in the SplitGPS string array, and may be copied to private data members that are may be accessed by associated Public Property procedures.

Note; I made the design decision to use separate private data for each property, thus requiring that they be copied from the SplitGPS array. String copy operations in .NET are fairly expensive. Thus, if performance is a factor, it is possible to remove these Private data members, and to change the Property Get procedures to use the SplitGPS array data directly. To do this, on would make SplitGPS() a Private object, instead of local to the GPSStream method.

```
SplitGPS = SubBuffer.Split(Delimiter)
m_UTCTime = SplitGPS(1)
m_Status = SplitGPS(2)
m_Latitude = SplitGPS(3)
m_LatitudeHemisphere = SplitGPS(4)
m_Longitude = SplitGPS(5)
m_LongitudeHemisphere = SplitGPS(6)
m_SpeedOverGround = SplitGPS(7)
m_CourseOverGround = SplitGPS(8)
m_UTCDate = SplitGPS(9)
m_MagneticVariation = SplitGPS(10)
m_MagneticVariationDirection = SplitGPS(11)
```

Time and date from a GPS receiver are Universal Coordinated Time. GPS time and date are not in a format that is useful for display or conversion. So, the first step is to convert these two fields to that used for by a .NET DateTime object.

```
Dim TimeString As String = _
        m_UTCTime.Substring(0, 2)& ":" & _
        m_UTCTime.Substring(2, 2) & ":" _
        & m_UTCTime.Substring(4, 2)
Dim DateString As String = _
        m_UTCDate.Substring(2, 2)& "/" _
        & m_UTCDate.Substring(0, 2) & "/" _
        & "20" & m_UTCDate.Substring(4, 2)
Dim UTCDateTime As DateTime = _
CDate(DateString & " " & TimeString)
```

The .NET framework provides the TimeZone object, which makes conversion of UTC DateTime variables to local DateTime variables easy.

```
m_LocalTime = _
TimeZone.CurrentTimeZone.ToLocalTime _
(UTCDateTime).ToLongTimeString _
m_LocalDate = _
TimeZone.CurrentTimeZone.ToLocalTime _
            (UTCDateTime). _
                ToShortDateString
```

If the GPRMC status fields is "A", a valid navigation solution is present, while a "V" indicates a NAV receiver warning, and all data are suspect. We signal this using the Status variable which is passed as an argument when the GPSDecoded event is raised.

```
If m_Status = "A" Then
```

```
                    RaiseEvent GPSDecoded(True)
            Else
                    RaiseEvent GPSDecoded(False)
            End If
            Buffer = ""
        End If
    End Sub

End Class
```

What is missing from this class? We decode only one sentence, GPRMC, from the NMEA-0183 data stream. There are other sentences that provide information that may be needed. For example, GPGGA sentence contains Antenna Height (altitude above mean sea level), and other potentially inserting fields. The GPGSV sentence has fields that represent the number of satellites in view, and specific elevation and azimuth data for each satellite (useful if you want to map these satellites). Adding decoding for these or other NMEA-0183A sentences would be a straight-forward extension of the code presented here.

The source code included on the CD ROM that accompanies this book for the Compact Framework version of this code uses the `Imports System.Globalization` namespace for localization for different cultures. Refer to it for more information.

5.4 System.IO.Ports

The Visual Studio 2005 release and the .NET Framework improves all part of application development. First, it tailors the software development experience to the needs of the individual developer. This "personalized productivity" will deliver features across the development environment and .NET Framework class libraries to help speed development. Second, Visual Studio 2005 will enable developers to apply existing skills across a broader range of application development scenarios through improved integration with the Microsoft Office System and SQL Server. Finally, it will deliver a new set of tools and functionality that will help satisfy the application development needs of today's enterprises.

This release of Visual Studio 2005 is the first to include serial communications support. The .NET namespace to be used is System.IO.Ports. We will explore it in detail here and will work through an example later.

Let us take a look at the namespace by examining it using the Object Browser, starting with Figure 5.2.1

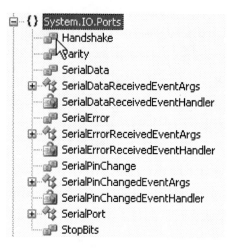

Figure 5.4.1 System.IO.Ports in the Object Browser

The System.IO.Ports class furnishes native .NET access to serial ports. Your code creates an instance of SerialPort, and then it uses that instance to configure the port, to open and close the port, and to use the port to send and receive data, and perhaps to manipulate or read the states of the hardware input and output lines (CTS, CD, DSR, RTS, DTR, and RI).

The SerialPort object naming conventions do a good job of "self-documentation," though the constructor arguments require some elaboration. The New constructors allow use to specify parameters that will be used when the port is opened.   See Figure 5.2.2.1-2.

Figure 5.4.2.1 SerialPort Proprties Methods and Events

Figure 5.4.2.2 SerialPort Properties Methods and Events

New(String, Integer, System.IO.Ports.Parity, Integer, System.IO.Ports.StopBits) is the more complex of the alternative constructors. These arguments in order are:

- PortName

- BaudRate

- Parity

- DataBits

- StopBits

However, each argument also is exposed as a SerialPort property. Use of individual properties often is the best approach to use for port configuration. Thus, the code example in the next section uses this technique.

There are a number of variations for reading and writing data. However, there are two methods for each that I use most frequently, and that I find to be the most useful. The methods that I recommend for writing data are:

- Write(String) — Call write with a String argument. The string contains text to be transmitted.

- Write(Byte(), Offset, Count) — Call write to send binary data. The first argument is an array of type Byte. The second argument is the Offset or starting point in the array, normally 0. The third argument is Count or the number of bytes to be sent, normally the length of the array.

Other methods that might be used to transmit data are Write(Char(), Offset, Count) which is similar to the preceding method except that it sends an array of type Char, and WriteLine(String) which transmits a string buffer (text), while appending a newline terminator (vbCrLf or ControlChars.NewLine – equally ControlChars.CrLf).

The two methods that I use when reading data are:

- ReadExisting(String) — This function returns a string (text) containing all available data.

- Read(Byte(), Offset, Count) — Pass this method an array of type Byte as the first argument, an Offset (normally 0), and Count (normally SerialPort.BytesToRead) to read binary data. For example, a code fragment might look like:

```
Dim Buffer1(SerialPort.BytesToRead - 1) As Byte
Dim NumRead As Integer = SerialPort.Read(Buffer1, _
0, SerialPort.BytesToRead)
```

Read is overloaded, so you also may pass it an array of type Char.

Other methods that might be used for reading data are ReadByte and ReadChar which return a single byte or character, ReadLine which returns a string up to a newline terminator (vbCrLf or ControlChars.NewLine – equally ControlChars.CrLf), and also ReadTo(String) a function that will return a string when the string argument is detected in the receive data stream.

My general approach is **not** to use these alternate methods for receiving data. I prefer to buffer data locally, then to parse it. This allows me to work closely with data as it is received. I find that I seldom understand the actual data patterns (protocol) that I have to implement. Working at a lower level allows me to debug both the code and my understanding of the protocol.

The Open and Close methods are used to open and close the port, respectively.

SerialPort properties closely follow the model provided by MSComm. If you understand MSComm, you will be well on your way to using this class. However, there are a few properties that are new or different. See Figure 5.2.3.

The IsOpen property returns True if the port is open, and False otherwise. This is in line with the Open and Close methods – MSComm used a single property to Open or to Close the port and also to return the port state.

An interesting and very useful method is GetPortNames. The GetPortNames method returns an array of type String that enumerates all installed serial port devices. Here is a code fragment that is used in the Configuration form from the VS2005Term example on the CD ROM. **Note,** The SerialPort control in the Visual Studio Toolbox may be used to create a SerialPort object by drag and drop to the designer surface, or you may instantiate it in code as I do in my example – both approaches provide the same results. I illustrate the pure code technique here.

```
Dim PortNames() As String =
NETSerialTerm.SerialPort.GetPortNames()

    For I = PortNames.Length - 1 To 0 Step -1

        lstCommPort.Items.Add(PortNames(I))

    Next
```

The For/Next loop places each port name into a list box for display and selection. Note, each port name is "friendly." That is, "COMx", where "x" is the port number. Com1 is represented by the string "Com1."

```
    SerialPort.PortName = "Com5"
```

The PortName property is used instead of the MSComm CommPort property. You use the same form described in the preceeding paragraph to set or change the serial port in use. For example,

ReadBufferCount is equivalent to the MSComm InBufferCount property. ReadBufferSize and WriteBufferSize are equivalent to the MSComm InputBufferSize and OutputBufferSize properties, respectively.

A new property is WriteTimeout, which allows us more flexibility than was offered by MSComm. Extend the WriteTimeout value for low data rates, or if you expect to have transmit data flow interrupted by flowcontrol for an extended time period.

MSComm provides a single event, OnComm. The SerialPort event model is more flexible. The ReceivedEvent is generated when the valued of ReceivedBytesThreshold is exceeded. This is equivalent to OnComm when the CommEvent property is comEvReceive. The PinChanged event is equivalent to OnComm when the CommEvent property indicates that one of the hardware input lines changed state (CTS, CD, DSR, or RI). The ErrorEvent is equivalent to OnComm when CommEvent signals a communication error.

5.5   NetSerialTerm

Problem: What is the best way to illustrate the use of the Visual Studio 2005?

Solution: Perhaps another simple terminal emulator applet? This will show how to configure the port, open and close it, and to send and receive data. It also illustrates a new feature of this most recent version of Visual Studio .NET. The feature in question deals with cross-thread data transfer. I will discuss that in more detail in my inline comments in the code.

Concepts like data parsing are covered in the NETGPS example, while logging received data to a file is covered in the examples in the Compact Framework .NET chapter.

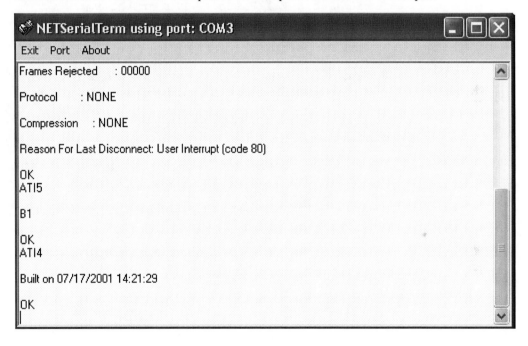

```
Imports System.IO.Ports

Public Class NETSerialTerm
    Inherits System.Windows.Forms.Form

    Public Shared WithEvents SerialPort As SerialPort

    Private Shared m_FormDefInstance As NETSerialTerm
    Private Shared m_InitializingDefInstance As Boolean
```

We need to be able to pass data to and from other forms in our program, specifically so that serial port configuration information may be shared. The DefInstance property is used to create a reference to the terminal form that may be used for this data sharing.

```
Public Shared Property DefInstance() As NETSerialTerm
    Get
        If m_FormDefInstance Is Nothing OrElse _
                m_FormDefInstance.IsDisposed Then
            m_InitializingDefInstance = True
            m_FormDefInstance = New NETSerialTerm
            m_InitializingDefInstance = False
        End If
        DefInstance = m_FormDefInstance
    End Get
    Set(ByVal Value As NETSerialTerm)
        m_FormDefInstance = Value
    End Set
End Property
```

Instantiate a new SerialPort object when the form loads. SerialPort was defined earlier "WithEvents." Thus, the Visual Basic IDE builds the prototypes for all three possible event handlers. We do not have to use the AddHandler method to add these in code; VB takes care of that task for us. This is a feature that C# programmers do not have and one that I miss when I am programming in C#.

```
Private Sub Form1_Load(ByVal sender As _
System.Object, ByVal e As System.EventArgs) _
    Handles MyBase.Load
    SerialPort = New SerialPort
End Sub
```

The menu item Open click event is handled in the next routine. We attempt to open the currently designated serial port here. We display the port status in the NETSerialTerm Title Bar.

```
Private Sub Open_Click(ByVal sender As _
System.Object, ByVal e As System.EventArgs) _
    Handles mnuOpen.Click
    Dim ex As Exception
    With SerialPort
        If .IsOpen = False Then
            Try
                .Open()
            Catch ex
            End Try
        Else
            Try
                .Close()
            Catch ex
            End Try
```

```
                End If
                If .IsOpen = True Then
                    mnuOpen.Checked = True
                    Me.Text = "NETSerialTerm using port: " & _
                                        SerialPort.PortName
                    .RtsEnable = True
                    .DtrEnable = True
                    .ReceivedBytesThreshold = 1
```

.ReceivedBytesThreshold = 1 enables the SerialPort ReceivedEvent.

```
                Else
                    mnuOpen.Checked = False
                    Me.Text = "NETSerialTerm not running"
                End If
            End With
        End Sub
```

Handle the Exit menu click event here.

```
        Private Sub Exit_Click(ByVal sender As _
            System.Object, ByVal e As System.EventArgs) _
                            Handles MenuItem1.Click
    _
            Me.Close()
        End Sub
```

Show the Configuration screen, modally, in the Configure menu click event.

```
        Private Sub Configure_Click(ByVal sender As _
            System.Object, ByVal e As System.EventArgs) _
                            Handles mnuConfigure.Click
            Dim Configuration As New frmConfigScrn
            Configuration.ShowDialog()
        End Sub
```

This is a terminal application. Any keys that are pressed in the terminal window should be transmitted out of the serial port, assuming that the port is open. Since we do not want these keystrokes to be displayed in this window, unless they are echoed by the connected device as receive data, we set the e.Handled = True flag. This notifies the framework that the default textbox display mechanism, which would display the character automatically is to be ignored.

Use the SerialPort.Write method to send the actual data (using the String or text character overload).

```
        Private Sub txtTerm_KeyPress(ByVal sender As _
    Object, ByVal e As _
    System.Windows.Forms.KeyPressEventArgs) _
                        Handles txtTerm.KeyPress
            Dim KeyAscii As Int32 = Asc(e.KeyChar)
            With SerialPort
```

```
          If .IsOpen = True Then
              .Write(Chr(KeyAscii))
          End If
      End With
      e.Handled = True
  End Sub
```

The SerialPort ReceivedEvent handler is called by the framework whenever one or more data bytes are received.

The System.IO.Ports.SerialPort class uses a worker thread to process receive data in the background. Thus, when a ReceivedEvent is generated by SerialPort, the event code executes in that worker thread context, not the STAThread context of the NETSerialTerm form. Therefore, we cannot directly manipulate any UI element, such as the text in the txtTerm textbox. In fact, a new feature of this release of Visual Studio .NET is that any attempt to do such a thing will cause a runtime exception. Earlier versions of Visual Studio handled this problem less gracefully — they failed by erratic operation where the application might simply stop responding to user interaction.

So, we have to marshal data between the SerialPort worker thread and the UI thread. The way to do that is to use Invoke or BeginInvoke on a delegate that executes in the same context as the NETSerialTerm form. DisplayData is that delegate. We pass it the actual data to display, and then allow the .NET framework to use the threading scheduler to dispatch its execution.

```
      Private Shared Sub SerialPort_ReceivedEvent(ByVal _
          sender As Object, ByVal e As _
              System.IO.Ports.SerialReceivedEventArgs) _
              Handles SerialPort.ReceivedEvent
          Dim Buffer As String = SerialPort.ReadExisting()
          DefInstance.txtTerm.BeginInvoke(New _
              DisplayData(AddressOf Display), _
                      New Object() {Buffer})
      End Sub

      Public Delegate Sub DisplayData(ByVal Buffer _
                                      As String)
```

Declare the DisplayData delegate. The delegate signature is the String buffer that contains the data to be displayed.

The .NET Framework calls the actual instance of DisplayData when the UI STAThread is in context. Code in DisplayData does several things.

The first three lines perform some simple filtering that may be needed. Some Unix and mainframe computers use a simple linefeed character (Chr(10)) as a new line terminator. So, when a linefeed is encountered, it is replaced by a carriage return character (Chr(13)). Windows textbox derived controls require a carriage return linefeed character pair (Chr(13) & Chr(10)) to terminate a line. The vbCrLf (or ControlChars.CrLf) constant is terminates a string properly for display in a textbox. Many devices that we use output a CRLF as the newline terminator. The result of the first line will create two carriage return characters in sequence. So, we replace these with the required vbCrLf pair, by first deleting one of the pair, then by replacing all single carriage return characters with vbCrLf.

```
Private Shared Sub Display(ByVal Buffer As String)
    Buffer = Buffer.Replace(vbLf, vbCr)
    Buffer = Buffer.Replace(vbCr & vbCr, vbCr)
    Buffer = Buffer.Replace(vbCr, vbCrLf)
    With DefInstance.txtTerm
```

If a backspace character is received, we want to do just that, erase the character immediately prior to the backspace having been received.

```
If (Buffer.Length = 1) And (Buffer = _
        Chr(8)) Then
    If (.Text.Length > 0) Then .Text = _
        .Text.Remove(.Text.Length - 1, 1)
Else
    .AppendText(Buffer)
End If
```

A standard textbox has a limited number of characters that may be displayed, and too many characters slow update of the display. Thus, the following code limits the maximum length to 8196 characters — when this total is exceeded, only the last 4096 characters are retained. A possible consequence of this operation is that the first character in that is retained is a single linefeed character. This is unusual, but possible. If this happens, then we also delete that character.

```
If .Text.Length > 8196 Then
    .Text = .Text.Remove(0, 4096)
    If Mid(.Text, 1) = vbLf Then _
            .Text = .Text.Remove(0, 1)
    End If
    .SelectionStart = .Text.Length
    End With
End Sub
```

Note, the filtering that we apply here assumes that data may be the result of single events (such as striking the Backspace key on your keyboard, and having that character echoed). If your data come from some non-interactive source (such as a text file dump), and if it contains a sequence of characters that are not normally displayed, the result will be a small rectangular or other characters. For example, ▪•♠♣♦‼☺♀♂♪. Do not worry. However, you may need to examine the actual character value. The text display may not be enough.

It can be important to free up allocated resources. So, when the application ends, we will close the port if necessary.

```
Private Sub NETSerialTerm_Closing(ByVal sender _
    As Object, ByVal e As _
    System.ComponentModel.CancelEventArgs) _
    Handles MyBase.Closing
    If SerialPort.IsOpen Then SerialPort.Close()
End Sub
```

The code in the Configuration form (the frmConfigScrn.vb file). I will not detail that code here. It is pretty straight forward, and you may refer to it when you load the project from the CD ROM.

Chapter 6   Using the Windows API

You can use the Windows API for serial communications in Visual Basic. You do not have to use a DLL or custom control add-on. However, as I have suggested earlier in the book, the API is harder to use than add-ons and, less obviously, the performance using the API is not as good as when an add-on is used.

We can profitably use the API to overcome the maximum speed limitation of the 16-bit versions of the MSCOMM custom control. The 16-bit API can be used to set a non-standard bit rate or non-standard character size, too (see the CD ROM). This second feature can be a lifesaver when communicating with some legacy systems. Do you need to send Baudot data at 50 bps? This is possible using the API.

The 16-bit and the 32-bit communications APIs are substantially different. The 32-bit API also allows us to overcome the speed limitation of the MSCOMM32.OCX that was furnished with VB 4.0. However, the MSCOMM32.OCX that is furnished with VB 5.0 has no such limitation. Just for variety, I have furnished a sample that attacks this problem in a different way.

Here is one caution when using the API for communications. Any 16-bit program that uses OpenComm must use CloseComm to free the port. Windows does this itself if a program ends. However, when you run a program within the Visual Basic design environment (IDE) and you stop its execution, the actual program that owns the port is VB, so Windows does not close the port. You cannot again run the program without first exiting and restarting Visual Basic. The same thing happens with 32-bit programs that use the API. Here the respective functions are CreateFile and CloseFile.

All 16-bit API text and associated code has been moved to the CD ROM

6.1   Non-standard Serial Speed And Character Length

Problem: What techniques can I use so that I can communicate with non-standard communications parameters using MSCOMM (or another communications add-on)?

Solution: Here are code fragments that would be used applying serial communications APIs. I have not included a program on the CD ROM for this because the details are straightforward extensions to the previous applet. This approach is valid for both 16 and 32-bit APIs.

```
Private Declare Function SetCommState Lib "Kernel32" (ByVal
hCommDev As Long, lpDCB As DCB) As Long
Private Declare Function GetCommState Lib "Kernel32"(ByVal
nCid As Long, lpDCB As DCB) As Long


    Private Type DCB  ' Win32API.TXT is incorrect here.
        DCBlength As Long
        BaudRate As Long
        Bits1 As Long
        wReserved As Integer
        XonLim As Integer
        XoffLim As Integer
        ByteSize As Byte
        Parity As Byte
        StopBits As Byte
        XonChar As Byte
        XoffChar As Byte
        ErrorChar As Byte
        EofChar As Byte
        EvtChar As Byte
        wReserved2 As Integer
    End Type


    'This is the code fragment that you would use to set the
    non-standard rate:

    Dim MSCommDCB As DCB
    Dim Port As Integer
    MSComm1.Settings = "9600, N, 8, 1"
    Port = GetCommState(MSComm1.CommID, MSCommDCB)
    MSCommDCB.BaudRate = 7812    'set speed to 7812 bps
    Port = SetCommState(MSComm1.CommID, MSCommDCB)
```

This preceding code fragment allows MSCOMM to be used for unusual speeds. For example, Baudot will often use speeds of 75 or 50 bps.

Some legacy systems, such as some older instruments, require oddball speeds such as 7200 bps. This probably is a result of a design limitation of the legacy system.

```
    Dim MSCommDCB As DCB
    Dim Port As Integer
    MSComm1.Settings = "300, N, 8, 1"
    Port = GetCommState(MSComm1.CommID, MSCommDCB)
    MSCommDCB.BaudRate = 50  'set speed to 50 bps
    MSCommDCB.ByteSize = 5   'five bits per character
    Port = SetCommState(MSComm1.CommID, MSCommDCB)
```

Baudot uses serial data with five bits per character. The preceding fragment sets the character length to five bits. Of course, use of Baudot is quite uncommon. But, if you need it, here it is.

MIDI (Musical Instrument Digital Interface) is a protocol designed for recording and playing back music on digital synthesizers that is supported by many makes of personal computer sound cards. Originally intended to control one keyboard from another, it was quickly adopted for the personal computer. Rather than representing musical sound directly, it transmits information about how music is produced. MIDI uses a serial speed of 31250 bps, so the techniques outlined here should apply to it (untested).

6.2   The 32-bit Windows API

I decided to use the code that Daniel Appleman furnished with his book *Visual Basic Programmer's Guide to the Win32 API*. Dan's kind permission to reprint his program (with some slight modifications, that I have highlighted, and comparable modest UI changes) allowed me to include code that uses a different approach than I might have used.

I do not suggest that VB programmers use the API for any serious communications program. However, you may find a need for it. And this book would not be complete without providing some basic information.

6.2.1   API32Term

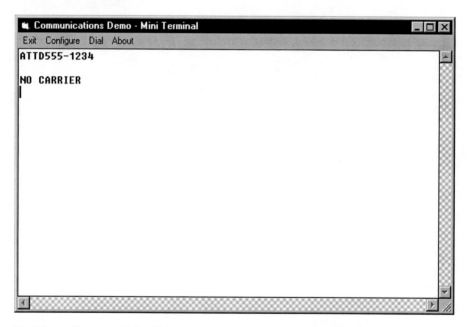

Problem: Get on with it. Show me how to use the 32-bit API.

Solution: Appleman elected to use Visual Basic classes to encapsulate the API functions and a callback to display received data. This approach is beyond reproach. It makes the code more reusable but it does add complexity that is not required for the functions that are provided. Choose your medicine.

```
Private Sub TermText_KeyPress(KeyAscii As Integer)
    If Not (Comm Is Nothing) Then
        Comm.CommOutput (Chr$(KeyAscii))
    End If
    KeyAscii = 0
```

```
End Sub
```

Text that is typed in the TermText textbox is sent using the CommOutput method of the
Comm class module.

```
Public Sub CommInput(thiscomm As dwComm, commdata _
        As String)
    Dim cpos%
    If commdata <> "" Then
        TermText.SelStart = Len(TermText.Text)
        ' Substitute the CR with a CRLF pair, dump the LF
        Do Until Len(commdata$) = 0
            cpos% = InStr(commdata$, VbCr)
            If cpos% > 0 Then
                TermText.SelText = _
                    Left$(commdata$, cpos% - 1) & vbCrLf
                commdata$ = Mid$(commdata$, cpos% + 1)
                cpos% = InStr(commdata$, VbLf)
                If cpos% > 0 Then
                    commdata$ = Mid$(commdata$, cpos% + 1)
                End If
            Else
                cpos% = InStr(commdata$, VbLf)
                If cpos% > 0 Then
                    commdata$ = Mid$(commdata$, cpos% + 1)
                Else
                    TermText.SelText = commdata$
                    commdata$ = ""
                End If
            End If
        Loop
        If Len(TermText.Text) > 4096 Then
            TermText.Text = Right$(TermText.Text, 2048)
        End If
    End If
End Sub
```

This code is similar to the ShowData routine in the VBTerm applet that I discussed earlier.
The fundamental difference is that this routine is called via OLE callback from the
ProcessReadComplete routine in the Comm class module.

I have made modest changes from the code that Appleman furnished with his book which
provide a small performance improvement.

```
Private Sub Timer1_Timer()
    Dim CarrDetect As Boolean
    Static CarrLast As Boolean
    If Not (Comm Is Nothing) Then
        Comm.Poll
```

```
            CarrDetect = Comm.RLSD_ON
            If (CarrDetect = True And CarrLast = False) Then
                Beep
                Me.Caption = "Communications Demo - " & _
                    "Mini Terminal    CD True"
            ElseIf (CarrDetect = False And CarrLast = True) Then
                Beep
                Me.Caption = "Communications Demo - " & _
                    "Mini Terminal"
            End If
            CarrierLast = CarrierDetect
        End If
End Sub
```

Timer1 does two things. First, it calls the Comm.Poll method in the Comm class module. Second, it sets the form caption to provide a "quick and dirty" way to indicate the current status of the modem's carrier detect. Carrier detect status is read from the Comm.RLSD_ON property of the Comm class module.

The following code is found in the COMMCFG form.

```
    Private CurrentBaudSetting&

    ' The port to use and configuration may have changed
    Private Sub CmdOk_Click()
        Dim baudtouse&
        Dim DeviceName$
        ' A more general array approach would also be
        ' easy to implement
        Select Case CurrentBaudSetting
            Case 0
                baudtouse = 1200
            Case 1
                baudtouse = 2400
            Case 2
                baudtouse = 9600
            Case 3
                baudtouse = 19200
            Case 4
                baudtouse = 38400
            Case 5
                baudtouse = 57600
            Case 6
                baudtouse = 115200
        End Select
        ' An approach similar to the baud rate one with indexed
        ' option buttons would be better if you wanted
        ' to support 4 or more comm ports
        If OptionCom1.Value Then
```

```
                DeviceName = "COM1"
        Else
                DeviceName = "COM2"
        End If
        If Not (Comm Is Nothing) Then
            ' Comm is already valid. Have we changed ports?
            If DeviceName <> Comm.DeviceName Then
                ' We're changing device
                ' Note that this also serves to close and
                ' release the previous comm object
                Set Comm = New dwComm
                ' This demo does not use buffer sizes
                Call Comm.OpenComm(DeviceName, CommDemo)
            End If
            ' If device is unchanged, SetCommState is all
            ' that is needed
        Else
            Set Comm = New dwComm
            Call Comm.OpenComm(DeviceName, CommDemo)
        End If
        Comm.DCB.BaudRate = baudtouse
        Comm.DCB.fNull = True
        Comm.DCB.fErrorChar = True
        Comm.DCB.ErrorChar = "~"
        ' Perform any other DCB setting here
        ' Now record the configuration changes
        Call Comm.SetCommState
        Me.Hide
    End Sub
```

I have changed the CmdOk_Click code to allow only modem-supported serial speeds. The original code allowed the user to select 14.4k bps, 28.8k bps and 56k bps. These speeds are not usable with most modems. Actually, the 56k bps speed is converted to 57.6k bps (a legal speed) by the API. However, I changed it for the sake of clarity.

After a serial port and /or speed and other parameters have been selected, the Comm.OpenComm method is called to set the new parameters.

```
    Private Sub Form_Load()
        If Comm Is Nothing Then
            CurrentBaudSetting = 5
        Else
            Select Case (Comm.DCB.BaudRate)
                Case 1200
                        CurrentBaudSetting = 0
                Case 2400
                        CurrentBaudSetting = 1
                Case 9600
                        CurrentBaudSetting = 2
```

```
                Case 19200
                        CurrentBaudSetting = 3
                Case 38400
                        CurrentBaudSetting = 4
                Case 57600
                        CurrentBaudSetting = 5
                Case Else
                        CurrentBaudSetting = 5
        End Select
    End If
    ' Set option button to current device
    OptionBaud(CurrentBaudSetting).Value = True
End Sub
```

I made changes to the Form_Load event code to agree with changes I made to the serial port speeds that can be selected.

```
Private Sub OptionBaud_Click(Index As Integer)
    CurrentBaudSetting = Index
End Sub
'the following code is found in the dwComm.cls module
' dwDCB - Device Communication Block utility class
' Part of the Desaware API Class Library
' Copyright (c) 1996 by Desaware Inc.
' All Rights Reserved
Private Type COMMTIMEOUTS
    ReadIntervalTimeout As Long
    ReadTotalTimeoutMultiplier As Long
    ReadTotalTimeoutConstant As Long
    WriteTotalTimeoutMultiplier As Long
    WriteTotalTimeoutConstant As Long
End Type
Private Type OVERLAPPED
    Internal As Long
    InternalHigh As Long
    offset As Long
    OffsetHigh As Long
    hEvent As Long
End Type
' Private members
Private timeouts As COMMTIMEOUTS
Private handle As Long   ' Comm handle
' Com1, com2 or other compatible comm device
Private devname$
' Public members
Public DCB As dwDCB
' Current state indicators
' Holds output data that arrives while an
' output transfer is in progress
```

```
Private PendingOutput$
' Non zero if events are being watched for
Private CurrentEventMask&
' Buffers for overlapped input and output
' Must take this approach due to VB's ability to
' move strings
Private CurrentInputBuffer&
Private CurrentOutputBuffer&
Private TriggeredEvents&  ' Variable to load with results
' Three overlapped structures,
' 0 = read, 1 = write, 2 = waitevent
Private overlaps(2) As OVERLAPPED
' Indicates background operation is in progress
Private inprogress(2) As Boolean
' Amount of data transferred on write
Private DataWritten&
Private DataRead&
Private CallbackObject As Object
' Declarations
Private Declare Function apiSetCommTimeouts Lib _
    "Kernel32" Alias "SetCommTimeouts" (ByVal hFile As _
    Long, lpCommTimeouts As COMMTIMEOUTS) As Long
Private Declare Function apiGetCommTimeouts Lib _
    "Kernel32" Alias "GetCommTimeouts" (ByVal hFile As _
    Long, lpCommTimeouts As COMMTIMEOUTS) As Long
Private Declare Function CloseHandle Lib "Kernel32" _
    (ByVal hObject As Long) As Long
Private Declare Function CreateFile Lib "Kernel32" _
    Alias "CreateFileA" (ByVal lpFileName As String, _
    ByVal dwDesiredAccess As Long, ByVal dwShareMode _
    As Long, ByVal lpSecurityAttributes As Long, ByVal _
    dwCreationDisposition As Long, ByVal _
    dwFlagsAndAttributes As Long, ByVal hTemplateFile _
    As Long) As Long
Private Declare Function SetupComm Lib "Kernel32" _
    (ByVal hFile As Long, ByVal dwInQueue As Long, _
    ByVal dwOutQueue As Long) As Long
Private Declare Function GetCommModemStatus Lib _
    "Kernel32" (ByVal hFile As Long, lpModemStat As Long) _
    As Long
Private Declare Function GlobalAlloc Lib "Kernel32" _
    (ByVal wFlags As Long, ByVal dwBytes As Long) As _
    Long
Private Declare Function GlobalFree Lib "Kernel32" _
    (ByVal hMem As Long) As Long
Private Declare Function lstrcpyFromBuffer Lib _
    "Kernel32" Alias "lstrcpynA" (ByVal lpString1 As _
    String, ByVal buffer As Long, ByVal iMaxLength As _
```

```
        Long) As Long
Private Declare Function lstrcpyToBuffer Lib _
    "Kernel32" Alias "lstrcpynA" (ByVal buffer As _
    Long, ByVal lpString2 As String, ByVal iMaxLength _
    As Long) As Long
Private Declare Function lstrlen Lib "Kernel32" _
    Alias "lstrlenA" (ByVal lpString As String) As Long
Private Declare Function CreateEvent Lib "Kernel32" _
    Alias "CreateEventA" (ByVal lpEventAttributes As _
    Long, ByVal bManualReset As Long, ByVal bInitialState _
    As Long, ByVal lpName As String) As Long
Private Declare Function WaitForSingleObject Lib _
    "Kernel32" (ByVal hHandle As Long, ByVal _
    dwMilliseconds As Long) As Long
Private Declare Function WriteFile Lib "Kernel32" _
    (ByVal hFile As Long, ByVal lpBuffer As Long, _
    ByVal nNumberOfBytesToWrite As Long, _
    lpNumberOfBytesWritten As Long, lpOverlapped As _
    OVERLAPPED) As Long
Private Declare Function ReadFile Lib "Kernel32" _
    (ByVal hFile As Long, ByVal lpBuffer As Long, _
    ByVal nNumberOfBytesToRead As Long, _
    lpNumberOfBytesRead As Long, lpOverlapped As _
    OVERLAPPED) As Long
Private Const GENERIC_READ = &H80000000
Private Const GENERIC_WRITE = &H40000000
Private Const OPEN_EXISTING = 3
Private Const FILE_FLAG_OVERLAPPED = &H40000000
Private Const INVALID_HANDLE_VALUE = -1
Private Const GMEM_FIXED = &H0
Private Const ClassBufferSizes% = 1024
Private Const ERROR_IO_PENDING = 997 '   dderror
Private Const WAIT_TIMEOUT = &H102&
' GetCommModemStatus flags
Private Const MS_CTS_ON = &H10&
Private Const MS_DSR_ON = &H20&
Private Const MS_RING_ON = &H40&
Private Const MS_RLSD_ON = &H80&
' Error values
Private Const CLASS_NAME$ = "dwComm"
Private Const ERR_NOCOMMACCESS = 31010
Private Const ERR_UNINITIALIZED = 31011
Private Const ERR_MODEMSTATUS = 31012
Private Const ERR_READFAIL = 31013
Private Sub Class_Initialize()
    Dim olnum%
    Set DCB = New dwDCB
    CurrentInputBuffer = _
```

```
        GlobalAlloc(GMEM_FIXED, ClassBufferSizes + 1)
    CurrentOutputBuffer = _
        GlobalAlloc(GMEM_FIXED, ClassBufferSizes + 1)
    ' Create event objects for the background transfer
    For olnum = 0 To 2
        overlaps(olnum).hEvent = CreateEvent(0, True, _
            False, vbNullString)
    Next olnum
End Sub
```

The Class_Initialize routine is called when the statement Set Comm = New dwComm creates a new Comm object.

Vital to use of the Windows 32-bit communications API is maintaining memory buffers that can be used for transmit and receive data. Visual Basic strings should not be used. The WriteFile and ReadFile API functions that are used to actually send and receive serial data use the address of the buffers that are globally allocated in this routine. Windows maintains these buffers at a fixed location in memory while Visual Basic strings are not guaranteed to be in a fixed location in memory.

CreateEvent creates the event objects that are used for the overlapped I/O that is used in the ReadFile and WriteFile API calls.

```
    Private Sub Class_Terminate()
        Dim olnum
        ' Close existing comm device
        Call CloseComm
        ' Dump the event objects
        For olnum = 0 To 2
            Call CloseHandle(overlaps(olnum).hEvent)
        Next olnum
        Set DCB = Nothing    ' Be sure DCB is free
        Call GlobalFree(CurrentInputBuffer)
        Call GlobalFree(CurrentOutputBuffer)
    End Sub
```

The Class_Terminate code is called whenever the Comm object is released. It frees the globally allocated buffers and destroys the event objects. The Class_Terminate routine must be called before the program ends in order to assure an orderly exit and so that memory is not "lost".

```
    Public Function OpenComm(CommDeviceName As String, _
            Notify As Object, Optional cbInQueue, _
            Optional cbOutQueue) As Long
        ' Close an existing port when reopening
        If handle <> 0 Then CloseComm
        devname = "\\.\" & CommDeviceName
        Set CallbackObject = Notify
        handle = CreateFile(devname, _
```

```
            GENERIC_READ Or GENERIC_WRITE, 0, 0, _
            OPEN_EXISTING, FILE_FLAG_OVERLAPPED, 0)
        If handle = INVALID_HANDLE_VALUE Then
            Err.Raise vbObjectError + ERR_NOCOMMACCESS, _
                CLASS_NAME, _
                "Unable to open communications device"
        End If
        ' If the input and output queue size is specified,
        ' set it now
        If Not (IsMissing(cbInQueue) Or _
                IsMissing(cbOutQueue)) Then
            Call SetupComm(handle, cbInQueue, cbOutQueue)
        Else
            Call SetupComm(handle, 4096, 1024)
        End If
        ' Ok, we've got the comm port. Initialize the timeouts
        GetCommTimeouts
        ' Set some default timeouts
        timeouts.ReadIntervalTimeout = 1
        timeouts.ReadTotalTimeoutMultiplier = 1
        timeouts.ReadTotalTimeoutConstant = 1
        timeouts.WriteTotalTimeoutMultiplier = 1
        timeouts.WriteTotalTimeoutConstant = 1
        SetCommTimeouts
        ' Initialize the DCB to the current device parameters
        Call DCB.GetCommState(Me)
        StartInput
    End Function
```

This code calls the CreateFile function to open the comm port. New to the 4th Edition is the setting of the Devname variable with an prefix that permits addressing of all available comports, even those numbered higher than 16. Previous versions were limited to ports between 1 and 16. Next, the input and output buffer sizes that are maintained by the Windows communications driver are set using the SetupComm function. Various timeouts are set in the timeouts structure (all are specified in milliseconds). The Windows communications driver uses the timeouts just set when the SetCommTimeouts function is called. Finally, StartInput is called to begin the receive data process.

```
    Public Function CloseComm() As Long
        ' Already closed, just exit
        If handle = 0 Then Exit Function
        Call CloseHandle(handle)
        handle = 0
    End Function
```

The CloseComm function must be called to release the comm port. Windows releases any open port when the program ends. However, if you run this program in the Visual Basic design environment (IDE, and End a running program from the IDE Run menu or Toolbar, the CloseComm function will not be called. You will not be able to rerun your program unless you exit and then restart Visual Basic. That is because the port is "owned by the VB IDE and not by the program that was running in the IDE.

```
Public Function GetCommState() As Long
    If handle = 0 Then DeviceNotOpenedError
    GetCommState = DCB.GetCommState(Me)
End Function
Public Function SetCommState() As Long
    If handle = 0 Then DeviceNotOpenedError
    SetCommState = DCB.SetCommState(Me)
End Function
Public Property Get CTS_ON()
    Dim modemstatus&
    Dim Res&
    If handle = 0 Then DeviceNotOpenedError
    Res = GetCommModemStatus(handle, modemstatus)
    If Res = 0 Then ModemStatusError
    CTS_ON = (modemstatus And MS_CTS_ON) <> 0
End Property
Public Property Get DSR_ON()
    Dim modemstatus&
    Dim Res&
    If handle = 0 Then DeviceNotOpenedError
    Res = GetCommModemStatus(handle, modemstatus)
    If Res = 0 Then ModemStatusError
    DSR_ON = (modemstatus And MS_DSR_ON) <> 0
End Property
Public Property Get RING_ON()
    Dim modemstatus&
    Dim Res&
    If handle = 0 Then DeviceNotOpenedError
    Res = GetCommModemStatus(handle, modemstatus)
    If Res = 0 Then ModemStatusError
    RING_ON = (modemstatus And MS_RING_ON) <> 0
End Property
Public Property Get RLSD_ON()
    Dim modemstatus&
    Dim Res&
    If handle = 0 Then DeviceNotOpenedError
    Res = GetCommModemStatus(handle, modemstatus)
    If Res = 0 Then ModemStatusError
    RLSD_ON = (modemstatus And MS_RLSD_ON) <> 0
End Property
```

Each of the preceding Public Property Get functions becomes a property of the Comm object that is created when the comm port is opened. I included an example of reading the Comm object properties in the Timer1_Timer event code on the commdemo form. There I use Comm.RLSD_ON to retrieve the current carrier detect state so that it can be displayed in the form caption.

```
Public Function CommOutput(outputdata As String) As Long
    Dim bytestosend&
    Dim Res&
    If handle = 0 Then DeviceNotOpenedError
    PendingOutput = PendingOutput & outputdata
    If inprogress(1) Then ' Write operation is in progress
        CommOutput = True
        Exit Function
    End If
    ' Start a new output operation
    bytestosend = Len(PendingOutput)
    ' No data to send, just exit
    If bytestosend = 0 Then
        CommOutput = True
        Exit Function
    End If
    ' Do not overflow our buffer
    If bytestosend > ClassBufferSizes Then
        bytestosend = ClassBufferSizes
    End If
    If bytestosend > 0 Then
        Call lstrcpyToBuffer(CurrentOutputBuffer, _
            PendingOutput, bytestosend + 1)
    End If
    If bytestosend = Len(PendingOutput) Then
        PendingOutput = ""
    Else
        PendingOutput = Mid(PendingOutput, bytestosend + 1)
    End If
    Res = WriteFile(handle, CurrentOutputBuffer, _
        bytestosend, DataWritten, overlaps(1))
    If Res <> 0 Then
        ProcessWriteComplete
        CommOutput = True
    Else
```

VB5 and VB6 versions of this program require that the Err object be used here. The VB4/32 version of this code called the GetLastError API function, which cannot be used in later versions of VB.

```
        If Err.LastDllError= ERROR_IO_PENDING Then
            inprogress(1) = True
            CommOutput = True
```

```
            End If
         End If
    End Function
```

CommOutput transmits pending data.

Data to be sent is double-buffered because overlapped I/O is used to send data. First, any new data is appended to the string PendingData. If the communications driver is still sending data from a previous call, CommOutput exits. If all previous data has been sent, new data, up to the size of the transmit buffer (4096 characters), is copied to the global memory buffer CurrentOutputBuffer. This copy uses the API function lstrcpyToBuffer. WriteFile is then called to send data from CurrentOutputBuffer.

Overlapped I/O essentially creates a background process that sends data from CurrentOutputBuffer. WriteFile returns immediately so that the remaining code in the function executes immediately. Subsequently, a call to WaitForSingleObject with the parameter overlaps(1) that is returned from WriteFile will determine if all data in CurrentOutputBuffer has been sent.

We might choose not to use Overlapped I/O for WriteFile operations. However, then a call to WriteFile would "block." That is, WriteFile would not return until all data had been sent. This can cause the program to appear unresponsive and it can cause significant performance problems at lower speeds.

The lstrcpyToBuffer function cannot copy strings that contain null characters. Appleman had some extra code to get around this limitation. However, I removed it. There is no way to enter a null character in a textbox. And null characters cannot be received anyway. It seemed superfluous to include the ability to send nulls when they cannot be received.

```
    Public Sub ProcessWriteComplete()
        inprogress(1) = False
        Call CommOutput("")
    End Sub
    ' Called periodically
    Public Sub PollWrite()
        Dim Res&
        If Not inprogress(1) Then Exit Sub
        ' Check the event
        Res = WaitForSingleObject(overlaps(1).hEvent, 0)
        ' If not yet signaled, just exit
        If Res = WAIT_TIMEOUT Then Exit Sub
        ' Data was written - Try writing any pending data
        ProcessWriteComplete
    End Sub
```

The PollWrite routine is called at regular intervals from Timer1_Timer on the commdemo form. It calls WaitForSingleObject to determine if a background WriteFile has completed. If a WriteFile has completed, ProcessWriteComplete is called to complete sending any pending double-buffered data.

```
Private Sub StartInput()
    Dim Res&
    ' Read already in progress
    If inprogress(0) Then Exit Sub
    If handle = 0 Then DeviceNotOpenedError
    Res = ReadFile(handle, CurrentInputBuffer, _
        ClassBufferSizes, DataRead, overlaps(0))
    If Res <> 0 Then
        ProcessReadComplete
    Else
```

VB4/32 version of this code called the GetLastError API function which cannot be used in later versions of VB.

```
        If Err.LastDllError () = ERROR_IO_PENDING Then
            inprogress(0) = True
        Else
            Err.Raise vbObjectError + ERR_READFAIL, _
                CLASS_NAME, _
                "Failure on Comm device read operation"
        End If
    End If
End Sub
```

StartInput reads serial data. ReadFile places any serial data that has been buffered by the communications driver in the global memory CurrentInputBuffer. ReadFile uses overlapped I/O. Unlike the WriteFile operation, there is no real performance advantage to using overlapped I/O for reads. However, the same I/O mode should be used for both reads and writes. So, we will live with the extra code complication here.

```
Public Sub PollRead()
    Dim Res&
    If Not inprogress(0) Then
        StartInput
        Exit Sub
    End If
    ' Check the event
    Res = WaitForSingleObject(overlaps(0).hEvent, 0)
    ' If not yet signaled, just exit
    If Res = WAIT_TIMEOUT Then Exit Sub
    ProcessReadComplete
End Sub
```

The WaitForSingleObject call in PollRead will, almost certainly, return 0. Thus, any pending read will have completed. However, since we are using overlapped I/O for reads, this code is necessary.

```
Public Sub ProcessReadComplete()
    Dim resstring$
    Dim copied&
```

```
        If inprogress(0) Then ' Was overlapped
            DataRead = overlaps(0).InternalHigh
            inprogress(0) = False
        End If
        If DataRead <> 0 Then
            resstring$ = String$(DataRead + 1, 0)
            copied = lstrcpyFromBuffer( resstring, _
                CurrentInputBuffer, DataRead + 1)
            If Not (CallbackObject Is Nothing) Then
                Call CallbackObject.CommInput(Me, resstring)
            End If
        End If
    End Sub
```

ProcessReadComplete finishes the process of reading serial data. Data is copied from the global memory CurrentInputBuffer to the resstring buffer. An OLE callback to CommInput is generated so that the data can be displayed.

The function lstrcpyFromBuffer cannot copy data that contains nulls. Nulls are discarded from receive data anyway because fNull is set to True in the DCB. Binary data cannot be received using this API method.

```
    Public Sub Poll()
        PollWrite
        PollRead
    End Sub
```

The commdemo Timer1_Timer event calls Comm.Poll method at regular intervals.

```
    'the following is found in the dwDCB.cls file
    ' dwDCB - Device Communication Block utility class
    ' Part of the Desaware API Class Library
    ' Copyright (c) 1996 by Desaware Inc.
    ' All Rights Reserved
    Private Type dcbType    ' Win32API.TXT is incorrect here.
        DCBlength As Long
        BaudRate As Long
        Bits1 As Long
        wReserved As Integer
        XonLim As Integer
        XoffLim As Integer
        ByteSize As Byte
        Parity As Byte
        StopBits As Byte
        XonChar As Byte
        XoffChar As Byte
        ErrorChar As Byte
        EofChar As Byte
        EvtChar As Byte
```

```
        wReserved2 As Integer
End Type

Private DCB As dcbType
Private BufferSize As Integer

Private Const ERR_INVALIDPROPERTY = 31000
Private Const CLASS_NAME$ = "dwDCB"

Private Const FLAG_fBinary& = &H1
Private Const FLAG_fParity& = &H2
Private Const FLAG_fOutxCtsFlow = &H4
Private Const FLAG_fOutxDsrFlow = &H8
Private Const FLAG_fDtrControl = &H30
Private Const FLAG_fDsrSensitivity = &H40
Private Const FLAG_fTXContinueOnXoff = &H80
Private Const FLAG_fOutX = &H100
Private Const FLAG_fInX = &H200
Private Const FLAG_fErrorChar = &H400
Private Const FLAG_fNull = &H800
Private Const FLAG_fRtsControl = &H3000
Private Const FLAG_fAbortOnError = &H4000

Private Declare Function apiSetCommState Lib "Kernel32" _
    Alias "SetCommState" (ByVal hCommDev As Long, lpDCB _
    As dcbType) As Long
Private Declare Function apiGetCommState Lib "Kernel32" _
    Alias "GetCommState" (ByVal nCid As Long, lpDCB As _
    dcbType) As Long

Private Sub Class_Initialize()
    ' The structure length must always be set
    DCB.DCBlength = Len(DCB)
    ' Set some default values
    BufferSize = 2048
    fParity = False
    fOutxCtsFlow = True
    fOutxDsrFlow = True
    fDtrControl = 1
    fDsrSensitivity = True
    fTXContinueOnXoff = True
    fOutX = True
    fInX = True
    fErrorChar = True
    fNull = True
    fRtsControl = 1
    fAbortOnError = True
    DCB.XonLim = 100
```

```
            DCB.XoffLim = BufferSize - 100
            DCB.ByteSize = 8
            DCB.Parity = 0
            DCB.StopBits = 0
            DCB.XonChar = 17
            DCB.XoffChar = 19
            DCB.ErrorChar = Asc("~")
            DCB.EofChar = 26 ' ^Z
            DCB.EvtChar = 255
            ' Set some default value
            DCB.BaudRate = 2400
    End Sub
```

The dwDCB class module encapsulates code for manipulating the Device Control Block (DCB). The DCB is used to specify most of the parameters used by the Windows communications driver.

I have included here only those properties and methods that are actually used in this applet. There are many more that might be used in a more sophisticated application. Refer to the code on the CD ROM for the complete listing.

```
    Public Property Let BaudRate(vNewValue As Long)
        Select Case vNewValue
            Case 110, 300, 600, 1200, 2400, 4800, 9600, 14400, _
                  19200, 38400, 56000, 57600, 115200, _
                  128000, 256000
                DCB.BaudRate = vNewValue
            Case Else
                Err.Raise vbObjectError + ERR_INVALIDPROPERTY, _
                    CLASS_NAME, "Invalid baud rate"
        End Select
    End Property
    Public Property Let fErrorChar(vNewValue As Boolean)
        DCB.Bits1 = DCB.Bits1 And (Not FLAG_fErrorChar)
        If vNewValue Then
            DCB.Bits1 = DCB.Bits1 Or FLAG_fErrorChar
        End If
    End Property
    Public Property Let fNull(vNewValue As Boolean)
        DCB.Bits1 = DCB.Bits1 And (Not FLAG_fNull)
        If vNewValue Then DCB.Bits1 = DCB.Bits1 Or FLAG_fNull
    End Property
    Public Property Let ErrorChar(vNewValue As String)
        DCB.ErrorChar = Asc(vNewValue)
    End Property
    Public Function GetCommState(Comm As dwComm) As Boolean
        Dim Res&
        ' Make sure comm device is initialized
        If Comm.hCommDev = 0 Then Exit Function
```

```
        Res = apiGetCommState(Comm.hCommDev, DCB)
        GetCommState = Res <> 0
End Function
' Set the comm device from the current dcb
Public Function SetCommState(Comm As dwComm) As Boolean
    Dim Res&
    ' Make sure comm device is initialized
    If Comm.hCommDev = 0 Then Exit Function
    Res = apiSetCommState(Comm.hCommDev, DCB)
    SetCommState = Res <> 0
End Function
```

The following are in the file COMMDEMO.BAS.

```
' Communication Constants
Global Const NOPARITY = 0
Global Const ODDPARITY = 1
Global Const EVENPARITY = 2
Global Const MARKPARITY = 3
Global Const SPACEPARITY = 4
Global Const ONESTOPBIT = 0
Global Const ONE5STOPBITS = 1
Global Const TWOSTOPBITS = 2
Global Const IGNORE = 0
Global Const INFINITE = &HFFFF
Global Const CE_RXOVER = &H1
Global Const CE_OVERRUN = &H2
Global Const CE_RXPARITY = &H4
Global Const CE_FRAME = &H8
Global Const CE_BREAK = &H10
Global Const CE_CTSTO = &H20
Global Const CE_DSRTO = &H40
Global Const CE_RLSDTO = &H80
Global Const CE_TXFULL = &H100
Global Const CE_PTO = &H200
Global Const CE_IOE = &H400
Global Const CE_DNS = &H800
Global Const CE_OOP = &H1000
Global Const CE_MODE = &H8000
Global Const IE_BADID = (-1)
Global Const IE_OPEN = (-2)
Global Const IE_NOPEN = (-3)
Global Const IE_MEMORY = (-4)
Global Const IE_DEFAULT = (-5)
Global Const IE_HARDWARE = (-10)
Global Const IE_BYTESIZE = (-11)
Global Const IE_BAUDRATE = (-12)
Global Const EV_RXCHAR = &H1
Global Const EV_RXFLAG = &H2
```

```
Global Const EV_TXEMPTY = &H4
Global Const EV_CTS = &H8
Global Const EV_DSR = &H10
Global Const EV_RLSD = &H20
Global Const EV_BREAK = &H40
Global Const EV_ERR = &H80
Global Const EV_RING = &H100
Global Const EV_PERR = &H200
Global Const EV_CTSS = &H400
Global Const EV_DSRS = &H800
Global Const EV_RLSDS = &H1000
Global Const SETXOFF = 1
Global Const SETXON = 2
Global Const SETRTS = 3
Global Const CLRRTS = 4
Global Const SETDTR = 5
Global Const CLRDTR = 6
Global Const RESETDEV = 7
Global Const GETMAXLPT = 8
Global Const GETMAXCOM = 9
Global Const GETBASEIRQ = 10
Global Const CBR_110 = &HFF10
Global Const CBR_300 = &HFF11
Global Const CBR_600 = &HFF12
Global Const CBR_1200 = &HFF13
Global Const CBR_2400 = &HFF14
Global Const CBR_4800 = &HFF15
Global Const CBR_9600 = &HFF16
Global Const CBR_14400 = &HFF17
Global Const CBR_19200 = &HFF18
Global Const CBR_38400 = &HFF1B
Global Const CBR_56000 = &HFF1F
Global Const CBR_57600 = &HFF20
Global Const CBR_128000 = &HFF23
Global Const CBR_256000 = &HFF27
Global Const CN_RECEIVE = &H1
Global Const CN_TRANSMIT = &H2
Global Const CN_EVENT = &H4
Global Const CSTF_CTSHOLD = &H1
Global Const CSTF_DSRHOLD = &H2
Global Const CSTF_RLSDHOLD = &H4
Global Const CSTF_XOFFHOLD = &H8
Global Const CSTF_XOFFSENT = &H10
Global Const CSTF_EOF = &H20
Global Const CSTF_TXIM = &H40
Global Const LPTx = &H80
' Application constants
' The size of the input and output buffers we will use
```

```
Public Const BufferSize% = 2048
Public Comm As dwComm
' The port configuration for the demo
Public Dialing% ' Currently dialing
```

I have made changes to Appleman's code in several areas. I have also made some cosmetic changes. All of these changes agree with my prejudices so do not blame Dan for them.

Figure 6.2.1 is a block diagram that shows the basic steps that might be used (there are other valid variations) to use the Windows 32-bit communications API.

In conclusion, there are some features of this example code that are worth discussing.

- The Windows 32-bit API does not use OpenComm or CloseComm functions. CreateFile and CloseHandle functions replace these.

- WriteFile and ReadFile replace WriteComm and ReadComm functions, respectively.

- The handle returned from the API function CreateFile is not known, a priori. Under the 16-bit API, the equivalent number (the comm port ID) is a function of the port number. Under the 32-bit API, the handle for a serial port is just like any other file handle: it is created from a pool of handles. The actual number returned has no relationship to the port number.

- The ReadFile and WriteFile functions cannot directly read and write VB strings. These functions require a long pointer to a buffer. It is possible to obtain a pointer to a VB string but that pointer is a pointer to a temporary variable that is not valid after the API call. So, the approach taken here is to allocate global memory buffers, copy strings to and from those global buffers, then pass to the global memory buffer a pointer to the ReadFile and WriteFile functions.

- The 32-bit API has a BuildDCB function similar to the 16-bit version. However, the values in the DCB (Device Control Block) can be set manually, to avoid the DOS limitations of the BuildDCB function. Also, the structure of the DCB is different for the 32-bit API than that for the 16-bit API. Make note of the new structure.

- The sample code implements overlapped I/O. This improves performance but at the cost of some extra complexity. Even so, the performance is not very good and I do not know of any good way to improve it substantially. That's why I strongly suggest that any serious application use a communications add-on rather than the API implemented in Visual Basic code.

- An alternative to using the Timer to poll for receive data is to use a Do/Loop. I have tried this and found that it responds to receive data much more quickly because the interval between polls is controlled by the delay in the Sleep API call rather than the 55 (or whatever you select) mS Timer1.Interval. For example,

```
Private Sub Timer1_Timer ()
    Dim CarrDetect As Boolean
```

```
      Static CarrLast As Boolean
      Do
          If Not (Comm Is Nothing) Then
              Comm.Poll
              CarrDetect = Comm.RLSD_ON
              If (CarrDetect = True And CarrLast = False) Then
                  Beep
                  Me.Caption = "Communications Demo - " & _
                      "Mini Terminal CD True"
              ElseIf (CarrDetect = False And _
                      CarrLast = True) Then
                  Beep
                  Me.Caption = "Communications Demo - " & _
                      "Mini Terminal"
              End If
              CarrLast = CarrDetect
              Sleep (2)
              DoEvents
          Else
              Exit Do
          End If
      Loop
    End Sub
```

In this modification, the timer event simply activates the poll loop. If the port is not open (Comm Is Nothing = True), then the loop is not activated. Whether or not this modification is a significant improvement is arguable. I'm offering it here so that you can consider it.

Figure 6.2.2 is a table that describes the Windows 32-bit communications API functions. Those shown with an asterisk (*) are functions that are used for other Windows functions. They are not limited to just serial communications.

I suggest that anyone interested in using the Windows 32-bit API purchase Daniel Appleman's book *Visual Basic Programmer's Guide to the Win32 API*. Appleman's book includes the unmodified version of the code that I have presented in this section. You can compare the differences between his code and mine. More importantly, the book goes into detail on Visual Basic strings and their use with API functions, including the ReadFile and WriteFile functions. This detailed information may be vital, especially if you run into problems. I cannot recommend this book too highly.

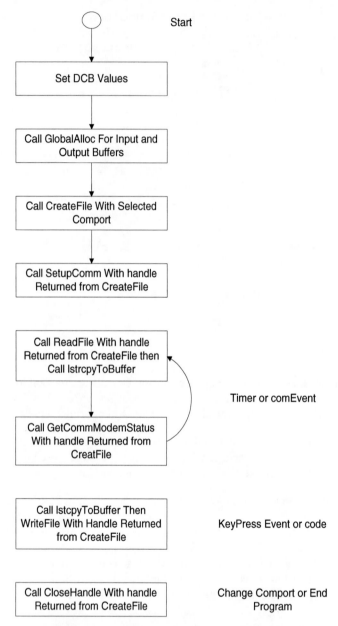

Figure 6.2.1 *Windows 32-bit Communications API*

Function	Description
BuildCommDCB	Modify the DCB (Device Control Block). Specify serial speed, data, start and stop bits, in the DOS MODE command format.
BuildCommDCBAndTimeouts	Set DCB and communications timeouts.
ClearCommBreak	Clear a UART break
* CloseHandle	Close the handle of a resource, including a serial port, that was allocated using CreateFile or CreateFileEX.
CommConfigDialog	Prompt the user for configuration information.
* CreateEvent	Create the event object used by the OVERLAPPED structure.
* CreateFile	Open a resource, such as a serial port.
EscapeCommFunction	Control serial port, such as DTR, RTS, and Break.
GetCommConfig	Returns the current DCB (Device Control Block). Often used prior to making a change in the DCB to avoid having to set unaffected structure elements.
GetCommMask	Returns the current event mask.
GetCommModemStaus	Returns the status of the serial port UART including CTS, RTS, DSR, CD, and RI.
GetCommProperties	Returns various communications properties.
GetCommState	Returns the configuration of a serial port. Like GetCommConfig, this function also returns the current value of the DCB.
GetCommTimeouts	Returns the current COMMTIMEOUTS structure.
* GetLastError	Returns error information, including communications errors.
* GlobalAlloc	Allocate a block of memory and fix its position so that a pointer to it can be used for serial input or output.
* GlobalFree	Remove the block of memory that was allocated using GlobalAlloc.
PurgeComm	Terminate any pending ReadFile or WriteFile operation. This also flushes the input and output buffers.
* ReadFile and ReadFileEX	Read data from a device opened with CreateFile
SetCommBreak	Send a break.
SetCommConfig	Set the configuration of the specified (open) serial port.

SetCommMask	Sets the event mask. Only events that are enabled are signaled.
SetCommState	Sets the configuration of the specified (open) serial port. Changes only values specified in the DCB, unlike SetCommConfig.
SetCommTimeouts	Sets the timeouts for ReadFile and WriteFile functions.
SetupComm	Sets the input and output buffer sizes that are used for ReadFile and WriteFile.
TransmitCommCharacter	Transmit a character ahead of any pending data buffered by WriteFile.
WaitCommEvent	Waits for comm events specified by SetCommMask.
* WaitForSingleObject WaitForSingleObjectEX	Wait for the specified object to be signaled or for a timeout. If the object is signaled, the function returns immediately.
* WriteFile and WriteFileEX	Write data to a device opened with CreateFile

Table 6.2.2 *Windows 32-bit API Communications functions.*

6.3   TAPI

I will present an applet that uses Simple TAPI in this section.

6.3.1   Simple TAPI

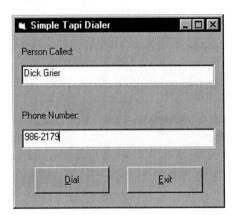

Problem: I would like to use Windows 95/98 and Windows NT 4 Assisted Telephony to create a dialing application.

Solution: Here is a simple application that uses TAPI to dial. It leaves the standard TAPI dialer applet on the toolbar after it completes. The first version uses 16-bit TAPI, so it can be compiled using VB 3.0 or VB 4.0/16. The second uses the 32-bit version, so it can be compiled with VB 4.0/32 or VB 5/6.

```
Sub Dial_Click ()
    DialNum
End Sub
Sub DialNum ()
    Dim AppName As String * 80
    Dim Comment As String * 80
    Dim Ret As Long
    Dim ErrMessage As String
    AppName = App.Title
    Comment = "Test dial"
    Ret = tapiRequestMakeCall(PhoneNumber.Text, _
        App.Title, Party.Text, Comment)
    If Ret <> 0 Then
        Beep
        ErrMessage = "Your request to dial " & Chr$(34) _
```

```
                        & PhoneNumber.Text & Chr$(34)
            ErrMessage = ErrMessage & _
                " could not be fulfilled because "
            Select Case Ret
                Case TAPIERR_NOREQUESTRECIPIENT
                    ErrMessage = ErrMessage & "no Windows " & _
                        "Telephony dialing application is " & _
                        "running and none could be started."
                Case TAPIERR_REQUESTQUEUEFULL
                    ErrMessage = ErrMessage & "the queue of " & _
                        "pending Windows Telephony dialing " & _
                        "requests is full."
                Case TAPIERR_INVALDESTADDRESS
                    ErrMessage = ErrMessage & "the number " & _
                        "you specified is invalid."
                Case Else
                    ErrMessage = ErrMessage & _
                        "of an unspecified error.  Sorry!"
            End Select
            MsgBox ErrMessage
        Else
            Unload Me
        End If
    End Sub
```

The TAPI function tapiRequestMakeCall takes four parameters. First is a string containing the phone number that is to be dialed (called the "destination address"), second is a string containing an application name, third is the name of the party called, and fourth is a comment. The second, third and fourth parameters are optional and might be replaced with an empty string (""). For example, the call might look like this:

```
    Ret = tapiRequestMakeCall(PhoneNumber.Text, "", "", "")

Sub Exit_Click ()
    Unload Me
End Sub
Sub Form_Paint ()
    PhoneNumber.SetFocus
End Sub
Sub PhoneNumber_KeyPress (KeyAscii As Integer)
    If KeyAscii = 13 Then
        DialNum
        KeyAscii = 0
    End If
End Sub
```

If the user presses the enter key, start dialing.

```
Declare Function tapiRequestMakeCall Lib "TAPI.DLL" _
    (ByVal lpszDestAddress As String, ByVal lpszAppName _
    As String, ByVal lpszCalledParty As String, ByVal _
    lpszComment As String) As Long
Global Const TAPIERR_NOREQUESTRECIPIENT = -2&
Global Const TAPIERR_REQUESTQUEUEFULL = -3&
Global Const TAPIERR_INVALDESTADDRESS = -4&

'the following are the 32-bit declarations for that version
'the TAPI dialer
Private Declare Function tapiRequestMakeCall Lib _
    "TAPI32.DLL" (ByVal lpszDestAddress As String, ByVal _
    lpszAppName As String, ByVal lpszCalledParty As _
    String, ByVal lpszComment As String) As Long
```

The 16 and 32-bit versions of SimpleTAPI use only a small portion of TAPI functionality.

VB 5.0 presents the opportunity to expand the use of TAPI without having to use a custom control. That is a subject for the future. The preceding sentence appeared in the first edition of this book. I have taken the opportunity to expand on it by including code on the CD ROM accompanying this book which implements a TAPI subset for data communications in a set of class modules.

Unfortunately, TAPI still is not quite ready for "prime time." TAPI 3.0, already available or soon to be available in Windows 98 and Windows NT 4/2K/XP, will add flexibility and capability. However, TAPI is a work in progress — and it still is not easy to implement, understand, or debug TAPI applications. Visual Studio .NET, including Visual Studio 2005 does not have built-in support for TAPI.

Chapter 7

Using Commercial Communications Add-ons

There is a number of reasons to prefer using commercial communications add-ons instead of MSCOMM, XMCommCRC, or the Windows API.

You may need one or more of the features that these controls provide — features that would be difficult to write in purely Visual Basic code.

You undoubtedly want the performance that commercial add-ons provide — better than you could achieve with VB code alone.

Error-corrected file transfers can be as simple as setting a custom control property or a function call to a DLL. Terminal emulation often needs no extra code beyond a control property or call.

Your design and coding task can be simplified substantially by using a commercial product's example code that duplicates some or all of the functionality that you need.

What if the examples that are furnished do not do what you want? Well, perhaps you are not completely on your own. In this chapter I have included an example that is a practical application using a commercial communications add-on. It does some things that are not really well illustrated in the example code furnished by the vendor.

The example that I have included is a modification of the Host and Remote application (see Chapter 4) that I developed using XMCommCRC. In this example, I use Sax Comm Objects. Sax Comm Objects has a number of features that are not part of XMCommCRC, including TAPI support and the ability to use a number of different error-corrected file transfer protocols. This example does use TAPI and it employs Zmodem file transfers.

Each commercial product has a different set of features that may be useful for your specific project. I'll attempt to outline those features in each of the following sections.

Small modifications would be needed to implement these applets if other vendors' custom controls were used. If you decide to use a DLL-based commercial product, more changes would be required.

The 1st and 2nd Editions of this book included information and examples for communications add-ons that no longer are available. That information has been retained in a .DOC file on the CD ROM.

7.1   Sax Comm Objects

At the writing of the 4th Edition of this book, Sax Comm version 8 was the current ActiveX. Sax Comm Studio also is available. Sax Comm Studio includes native .NET and will be described at the end of this section. Trial version of Sax Comm Objects and Comm Studio are included on the CD ROM that accompanies this book. Thus, you can try the example program here and try the example code that is furnished with Sax Comm without having to download it from the Internet.

Here is a list of Sax Comm Objects features.

- Terminal emulation includes TTY, ANSI, VT-100, and VT-52.

- Two versions of Sax Comm Objects are available, Standard and Enterprise Editions.

- The Standard Edition includes ZMODEM error-checked file transfers. The Enterprise Edition adds these other error-checked file transfer protocols: XMODEM (CRC and Checksum), XMODEM-1K, Ymodem-G, Ymodem-Batch, and Kermit.

- File transfer dialog boxes indicate transfer status. Transfers execute in the background

- The Enterprise Edition includes Sax Zip Objects. Sax Zip Objects add data compression and an extra layer of data protection to your file transfers.

- TAPI support is an important part of the Enterprise Edition. TAPI modem support means that your application can use the modems that are configured in the Windows control panel. There is no need to endure the unpleasant process of configuring modems.

- Another Enterprise Edition feature is Look-up-strings. These provide a very fast way to look for specific strings of text. This can be handy when writing code that automatically logs into a host or looks for patterns from a hardware device.

- Also in Enterprise are many common user interface elements such as toolbars, status bars, and configuration dialogs to make your application look professional with minimal code.

- Sax Comm Objects provide a superset of MSCOMM functions. Any program that can use MSCOMM can also use Sax. However, one place that Sax Comm Objects can be used and where MSComm cannot (easily) is in ASP (Active Server Page).

Extensive debugging features are included. You can Watch incoming and outgoing serial data, you can Watch events and properties, and you can easily add logging and debugging code.

Sax Comm Objects also has dozens of properties and methods that can simplify some programming tasks. Here are just a couple of examples.

OutputLine — send a line, including terminating carriage return and linefeed characters.

LookUp Strings — allow you to trigger events based on receive strings or prompts that you specify. Here is an example,

```
SaxComm.LookUpSeparator = "|"
SaxComm.LookupTimeOut = 30000        'timeout in 30 sec
SaxComm.LookUpInputLen = 0
SaxComm.LookUpText = "Password | Name"
Select Case SaxCommLookUp
    Case 0  'Password
        SaxComm.OutputLine = Password
    Case 1  'Name
        SaxComm.OutputLine = Name
    Case -1 'timeout
        SaxComm.OutputLine = "QUIT"
End Select
```

I need to include one important precaution when you use Sax Comm Objects. Do not call any property or method that attempts to receive data unless there actually is data present. For example, do not use the Input property if InBufferCount = 0. Likewise, do not call Lookup or LookupInput unless InBufferCount > 0.

7.1.1   Host And Remote using Sax Comm Objects

Problem: You have shown me how to use XMCommCRC to send and receive files automatically. However, I need to use TAPI dialing. Also, I want to use Zmodem to improve the performance. Lastly, I don't really want a visible user interface. Can you show me how to use Sax Comm Objects and to NOT use a form to site it?

Solution: Glad to. The first thing that you may notice is that the overall design is very similar to the one that I did for XMCommCRC (why waste good work?). The second thing that you may notice is that there is a **lot** less code required when using Sax Comm. TAPI and other built-in features.

I created a class module to host all of the critical communications functions, expanding on the Sax Comm Objects native methods.   The name of this module is clsSaxComm.cls. Please refer to the equivalent XMCommCRC discussion in Chapter 4 to review the design goals and the control functions (the actual User Interface that is employed in this example applet).

See the actual source code on the CD ROM that accompanies this book for code for the Public property procedures that are used in this class. They are not included in the code that follows.

```
Private m_Password As String
Private m_UserName As String
```

```
Private m_PhoneNumber As String
Private m_File2Send As String
Private WithEvents SaxComm1 As SaxComm
Private Declare Function timeGetTime Lib "winmm.dll" () _
                                              As Long
Public Event CommError(ByVal ErrorString As String)
```

The Dial function is simplified through the use of TAPI enabled dialing. When a TAPI-compliant modem has been selected using the CommPort property, dialing occurs as soon as the PortOpen property is set to True (if the PhoneNumber property has a valid phone number).

There is one addition of note. If an error is encountered, we should not display a MsgBox. Rather, a better design is to raise an event that allows the calling code to respond to the error without interfering with other execution. That is the reason that the CommError event has been included in this class module. I also have omitted a few routines that are duplicated elsewhere in this book.

```
Public Function Dial() As Boolean
On Error GoTo ErrorHandler
    If m_PhoneNumber <> "" Then
        SaxComm1.PhoneNumber = m_PhoneNumber
        SaxComm1.PortOpen = True 'dial now
        If SaxComm1.PortOpen = True Then Dial = Login
    End If
    Exit Function
ErrorHandler:
    RaiseEvent CommError("Dialing error: " & _
                                  Err.Description)
End Function
```

The Login function here is quite similar to that used in the XMCommCRC example. See the discussion there for more details.

```
Private Function Login() As Boolean
Dim Ret As Integer
Dim FailureString As String
Dim Status As Boolean
    FailureString = "Login failed.   "
    Wait 0.5 'wait for things to quiet down
    Ret = WaitForResponse("LOGIN", 5)
    If Ret = 0 Then
        Wait 0.5
        SaxComm1.OutputLine = m_UserName
        Wait 1
        Ret = WaitForResponse("PASSWORD", 5)
        If Ret = 0 Then
            Wait 0.5
            SaxComm1.OutputLine = m_Password
            Wait 3
            'adjust this time for the max req'd to access
            'the pw in the database
```

```
                Ret = WaitForResponse("MODE", 5)
                If Ret = 0 Then
                    Status = True
                Else
                    FailureString = FailureString & _
                                      "Mode timeout."
                End If
            Else
                FailureString = FailureString & _
                                  "Password timeout."
            End If
        Else
            FailureString = FailureString & "Login Timeout."
        End If
        Login = Status
        'If Status = False Then MsgBox FailureString,_
                                      vbCritical

        If Status = False Then
            HangupReset
            RaiseEvent CommError(FailureString)
            Wait 2
            Beep
        End If
    End Function
```

Look at the WaitForResponse and WaitFor2Responses routines. These include conventional VB code that parses responses **plus** commented code that illustrates the same process using built-in Sax Comm Objects methods.

```
    Private Function WaitForResponse(String2Receive As _
                    String, Timeout As Single) As Integer
'       SaxComm1.AutoReceive = False
'       SaxComm1.LookUpText = String2Receive
'       SaxComm1.LookUpInputLen = 0
'       SaxComm1.LookUpTimeOut = Timeout * 1000
'       WaitForResponse = SaxComm1.LookUp
'The above code is equivalent to this more conventional
'routine (Sax routines are easier):
Dim Buffer As String
Dim StopTime As Single
    WaitForResponse = -1
    StopTime = timeGetTime + Timeout * 1000
    Do Until timeGetTime >= StopTime
        Buffer = SaxComm1.Input
        If InStr(Buffer, String2Receive) Then
            WaitForResponse = 0
            Exit Do
        End If
        DoEvents
    Loop
End Function
Private Function WaitFor2Responses(String1 As String, _
```

```
                           String2 As String, ByVal Timeout As Single) As Integer
'        SaxComm1.LookUpSeparator = "|"
'        SaxComm1.LookUpText = String1 & "|" & String2
'        SaxComm1.LookUpInputLen = 0
'        SaxComm1.LookUpTimeOut = Timeout * 1000
'
'        WaitFor2Responses = SaxComm1.LookUp
'The above code is equivalent to this more
'conventional routine:
Dim Buffer As String
Dim TimeoutTime As Long
    TimeoutTime = timeGetTime + Timeout * 1000
    WaitFor2Responses = -1
    Do Until timeGetTime >= TimeoutTime
        Buffer = Buffer & SaxComm1.Input
        If InStr(Buffer, String1) Then
            WaitFor2Responses = 0
            Exit Do
        ElseIf InStr(Buffer, String2) Then
            WaitFor2Responses = 1
            Exit Do
        Else
            DoEvents
        End If
    Loop
End Function
```

The Initialize event executes when the SaxComm class is instantiated as an object by the calling VB code. It sets important properties in Sax Comm Objects. NOTE: the CommPort and Settings properties are "empty". This instructs Sax Comm to use the TAPI settings the user has set in Control Panel.

```
Private Sub Class_Initialize()
'Initialize SaxComm for TAPI, Zmodem, etc.
    Set SaxComm1 = New SaxComm
    With SaxComm1
        .SerialNumber = ""
        'place your serial number between the quotes
        .CommPort = ""
        'Use the first modem available
        .ShutDown = False
        .Settings = ""
        'Use serial speed set by modem settings
        'in the Windows control panel
        .Handshaking = Handshaking_Hardware
        .XferProtocol = Protocol_ZModem
        .RThreshold = 0
        .InputLen = 0
        .AutoReceive = False
    End With
    EnableAnswer = True
End Sub
```

The Terminate event is called when the object is released. We should make sure that Sax Comm Objects shuts down in an orderly fashion before we allow our object to exit.

```
Private Sub Class_Terminate()
    With SaxComm1
        .AbortTransfer
        Do While .XferStatus = 0
            DoEvents
        Loop
        .ShutDown = True
    End With
End Sub
```

The NewConnection routine implements the same state machine that is used in the equivalent XMCommCRC code example. See Chapter 4 for that discussion.

```
Private Sub SaxComm1_NewConnection()
'This event is generated as soon as the modem connects
after answer
Dim Buffer As String
Dim Timeout As Single
Dim UserNameFound As Boolean
Dim Password As String
Dim Ret As Integer
    Wait 0.5
    SaxComm1.Output = "LOGIN"
    Wait 1
    Timeout = timeGetTime + 8000
    'an alternative to this loop would be to use the
    'Sax Comm Objects InputLine property
    'Look for the User Name vvvv
    Do Until timeGetTime >= Timeout
        Buffer = Buffer & SaxComm1.Input
        DoEvents
        If InStr(Buffer, vbCr) Then
            UserNameFound = True
            Exit Do
        End If
    Loop
    'look for the User Name ^^^^
    If UserNameFound = True Then
        Password = LookupName(Buffer)
        If Len(Password) > 0 Then
            Wait 0.5
            SaxComm1.Output = "PASSWORD"
            Wait 1
            'get a password, then validate it
            Ret = WaitForResponse(Password, 5)
            If Ret = 0 Then
                Wait 0.5
                SaxComm1.Output = "MODE"
                Wait 1
```

```
                              Ret = WaitFor2Responses("SENDING", _
                                              "WAITING", 5)
                          Select Case Ret
                              Case 0    'Sending
                                  ReceiveFile
                                  Wait 1
                                  Ret = WaitForResponse( _
                                              "WAITING", 5)
                                  If Ret = 0 Then
                                      If m_File2Send <> "" Then
                                          Wait 0.5
                                          SaxComm1.Output = _
                                                  "SENDING"
                                          Wait 1
                                          SendFile
                                          Wait 0.5
                                          HangupReset
                                      Else
                                          Wait 0.5
                                          SaxComm1.Output = "NONE"
                                          Wait 0.5
                                          HangupReset
                                      End If
                                  Else
                                      HangupReset
                                  End If
                              Case 1    'Waiting
                                  Wait 0.5
                                  SaxComm1.Output = "SENDING"
                                  Wait 1
                                  SendFile
                              Case Else 'timeout or other
                                  HangupReset
                          End Select
                      End If
                  End If
              Else
                  HangupReset
              End If
      End Sub
```

This class module includes Public functions, SendAfterCall, WaitAfterCall, HangupReset, LookupName, SendFile, ReceiveFile, GetNextFile and WaitForTransfer that are equivalent to the similarly named routines in the XMCommCRC example in Chapter 4. I will not repeat that code here. Refer to the discussion there and the actual code on the CD ROM.

7.2   Sax Comm Studio

Sax Comm Studio makes it easy to add scalable serial communications to all your .NET applications.

Sax has combined their knowledge and experience to design a powerful, easy to use object model that is fully integrated with the Microsoft .NET Framework. The result is a class library that combines developer productivity with a high degree of scalability and robustness.

Sax.net Communications can connect to serial devices, modems, as well as legacy computer systems. Industry-standard file transfer protocols including XModem and YModem help you transfer files reliably. The class library supports both RS232C and RS485 serial ports.

- All serial communications components were written using 100% managed C# code. Because serial communications is not a standard part of the Microsoft .NET framework, the code calls into the Windows API. These API calls are limited to a minimum to maximize performance.

- Standard built-in serial ports

- Virtual serial ports on other systems via network software

- Bluetooth and infrafed serial ports

- High performance serial ports using specialized hardware such as Digi boards

- XMODEM, YModem and ZModem file transfer support

- Powerful text scanning an triggering

- Robust terminal emulations

- Built-in diagnostics

- Includes ActiveX components for use in VB6 or other platforms

Go to the Sax Software site, www.sax.net, for up-to-date information on features as they are added to the Sax Comm Studio component.

7.3   Greenleaf CommX

Greenleaf has been a leading vendor of serial communications libraries for C and C++ programmers. In early 1998 they entered the Visual Basic marketplace with CommX. These OCXs have a number of interesting features.

The Port Control OCX, called PortCtl, provides the interface to the serial port hardware. You use this OCX to control port settings (bit rate, port number, parity, handshaking, and other settings). You use it to send and receive raw data. It adds the capability to detect mouse ports, a feature unique to this product.

The File Transfer OCX, called FileXferCtl, provides Zmodem error-corrected file transfers. Unlike most other commercial products, Zmodem is the only supported error-correcting protocol. This OCX also can be used to send and receive ASCII files without error-correction.

The Terminal Control, called TermCtl, provides PC-ANSI and TTY emulation. Unlike other commercial products, it does not directly support other popular terminal emulations. However, this limitation is mitigated by the fact that it provides the building blocks for you to create any other terminal emulation protocol with only a little extra effort. Thus, it is more flexible than most other offerings but does require that the programmer do some extra work to implement a terminal emulation.

7.3   MagnaCarta CommTools

CommTools 3.0 is a mix of DLLs and custom controls. Both 16-bit and 32-bit DLLs and custom controls are included. There are some specialized features and functions that can make this approach to a communications add-on quite attractive.

- Multi-port access allows you to use more than one serial port at a time. The communications driver limits the total number of ports.

- Terminal emulation includes TTY, ANSI and ANSI X3.64, VT-102, VT-52, VT-220, TV-910, TV-912, and TV-925.

- Error-checked file transfer protocols include XMODEM (CRC and Checksum), XMODEM-1K, Ymodem-G, Zmodem, CompuServe B/ B+/QuickB, and Kermit. File transfer dialog boxes indicate transfer status.

- CommTools provides a superset of MSCOMM function in the 16-bit VBX. Any program that can use MSCOMM can also use the VBX. The 16-bit and 32-bit DLLs are powerful but MSCOMM code would have to be ported to account for syntactical and logical differences.

- The CommTools package includes basic example programs.

- CommTools claims CAS fax support. However, I have not used this feature.

- CommTools has built-in support for various modems such as Racal-Vadic and Hayes compatible modems using the ATI6 command. Many modems use equivalent initialization sequences so this can, in some cases, be used in lieu of a modem database.

- Network modem support via NASI/NCSI, INT14 (BIOS), and INT14(EBIOS) is provided.

- Multiport boards that are supported are the DigiBoard PC/Xe and PC/Xi and the Star Gate ACL series.

CommTools has a feature similar to Sax Comm Objects, in the VBX version only. This feature triggers a CommEvent when a string that matches one of a specified set is received. This can simplify writing event driven communications code.

CommTools includes a variety of functions such as timers that make it a flexible way to handle communications problems.

7.4   Crystal CrystalCOMM

CrystalCOMM comes in two versions. One is available for 16-bit applications such as VB 3.0. Another is a 32-bit version for 32-bit Visual Basic under Windows 95/98 or Windows NT. Actually, like other DLL-based communications add-ons, CrystalCOMM will work with any development environment that can call DLLs. At this writing CrystalCOMM for NT/Win95 consists of a number of 32-bit DLLs that provide "thunking layers" for underlying 16-bit code. This means that distribution must include a number of files, so read the documentation carefully before creating a setup disk for a program that uses it.

I will not go into great detail on these communications DLLs. This library has been tested (by Crystal) to operate under C/C++, Delphi, SmallTalk, ACCESS, Paradox, FoxBase, and Gupta SQLBase. Here are a few features that make it attractive.

- CrystalCOMM for NT/WIN95 is thread-safe. It can be called concurrently from multiple threads. CrystalCOMM creates a thread to handle a low-level serial port interface.

- In addition to the conventional error-checked file transfer protocols such as XMODEM checksum and CRC, Ymodem, Zmodem, Kermit (including sliding window), and CompuServe B+, CrystalComm also provides a fast proprietary protocol. CrystalCOMM includes the TAP protocol for alphanumeric paging and the VISA credit card clearing protocol. These two protocols are not included in most communications add-ons.

- Included is a modem database and functions for reading it.

- CrystalCOMM also supports functions for communications security. You can encrypt bytes on the sending system and decode them on the receiving side.

- CrystalCOMM has been tested with a variety of multiport boards including Digiboard and COMMTROL.

- Source code in C is available.

7.7 LUCA

This product has been added to the 3rd and later Editions of the book. It has a feature set that is hard to rival; more than 70 communications protocols and drivers make up LUCA. Here are some of the features it provides. NOTE: a number of the supported protocols are unusual (many are unique to this product) while others are Internet oriented.

- Asynchronous communications
  - TAPI
  - XMODEM
  - Zmodem
  - GPS/NMEA
- Voice support via TAPI
  - DTMF
  - WAV
- Paging/SMS
  - UTC
  - TAP
  - GSM
- PLC
  - 3964 (R)
  - RK512
  - RK512
  - AS511
  - Simatic S7/TCP
  - Allen-Bradley
- FieldBus
  - PROFIBUS DP / FMS
- Synchronous Communications
  - HDLC
  - SDLC
  - BSC
  - Frame Relay
- ISDN
  - CAPI
  - X.75
  - V.110
  - V.120
  - X.31
  - EFT
- TCP/IP
  - TCP
  - UDP
  - FTP

- TFTP
- DNS
- Telnet
- RAS
- Email
  - SMTP
  - POP3
  - UUENCODE
  - Base64
  - MIME

7.8 Hotwind (formerly Allen-Martin)

Hotwind Software (Allen-Martin) tools are TAPI-based ActiveX controls. They consist of four different controls. Here is a description:

Use **TAPIEx (amTapi)** in your application to make and answer telephone calls, detect Caller ID on incoming calls, detect remote party DTMF key press tones, generate DTMF tones, enumerate the installed telephony devices, and translate the telephone number according to the user settings in the Windows dialing rules dialog.

You can use **TAPIEx (amTapi)** to build computer telephony integration systems (CTI), interactive voice response systems (IVR), voice mail, audio text, automated attendants, unified messaging, digital recording, outgoing telemarketing calls, school closing systems, predictive dialers, and call centers and other computer telephony applications. It can be used in conjunction with **amWave** to play and record voice streams or with **amComm** to send and receive data streams.

The **amComm** serial data control provides asynchronous serial data communications for your application either directly using the PC serial Comm ports or in combination with **TAPIEx (amTapi)** to provide serial data communication over a modem. It is compatible with code written for MSComm.

The **WaveEx (amWave)** control is designed principally for use with the **TAPIEx (amTapi)** control. Playback and recording is available for all wave devices installed on the PC including, if supported by the telephony device, playback and recording from the telephone line, the telephone handset, and the telephone headset. The wave ID's are identified by **TAPIEx (amTapi)** when the devices are enabled. For example, when a call is connected, the line wave in and wave out devices become available. The **WaveEx (amWave)** control will playback and record from the sound card as well, allowing the pre-recording of announcement messages and the playback of any recordings made directly from the telephone line.

The Hotwind **ToneEncoder** control allows you to encode variant of tone signals such as DTMF, Caller ID, Users define tones, TTY baudot and so on.

The **amSMS** control allows you to add 'Short Message Service' messaging to your application with very little code. SMS messages are alphanumeric text messages of a few hundred characters or so that can be sent for immediate display on the screens of pagers and cellular telephones. It sends the text messages directly to the cellular phone and pager company service centers by serial data communication over a modem using the TAP protocol. **Note**: Hotwind Software does not provide an equivalent SMS control.

7.9   SuperComm

SuperComm is a DLL-based serial communications toolkit for all versions of Windows. It consists of several modules that can be used according to your needs.

The **ComInt** module forms the core of SuperCom and contains all low level routines. It is used with all other modules

DataLink is a part of the high level interface. It provides routines to transmit and receive data.

The **Transport** module with its packet routines forms the transport part of the high level interface. Transport offers packet routines that transmit and receive data packets.

The **Modem** module is the part of the high level interface that is responsible for the communication through the modem.

The **Terminal** module provides terminal emulations for **ANSI, TTY** and **VT52**.

The **Protocol** adds **ZMODEM, YMODEM, YMODEM/BATCH, XMODEM, XMODEM/CRC,** and **ASCII** file transfer protocols.

Chapter 8   Paging

There are several types of paging.

Numeric paging uses DTMF (Dual Tone Multi-Frequency – normal touch-tone) signaling to send the paging message. You dial the paging service, wait for an answer, and then dial the phone number that you want to have displayed on the pager. Alternately, some paging services may require that you dial the service, wait for the answer, dial an access number, and then dial the number to be displayed on the pager. Usually you will use your modem to do this dialing.

Alphanumeric paging uses modems that connect in data mode to send messages. You dial the paging service and wait for the modem to connect. After connect, a data handshake is exchanged to start the protocol. You then send an access number and wait for a response. If the service acknowledges the access number that you have sent, it will then accept an alphanumeric message. The message length is limited. Some services restrict message length to as few as 80 characters while others accept messages of 400 or more characters. In my experience, 80 character messages are sufficient for most situations. Some paging services allow you to send multiple messages for each connection, one message per access number. I have not found this multiple message feature to be useful in my applications but you may find that you need it.

Alphanumeric paging services send messages to pagers as "send and forget." That is, the message is transmitted. If the pager is out of range or is otherwise out of service (i.e., broken), the message is lost.

Wireless paging is, essentially, two-way messaging. The page works similarly to the alpha page described above but it goes a couple of steps further. The pager must acknowledge the radio message sent to it. The service will resend the message until it is acknowledged. Also, most wireless paging services support email and other two-way communications that go beyond simple paging.

I will not discuss wireless paging here. I have never done anything with it and its requirements go beyond the scope of the book.

8.1 Numeric Paging

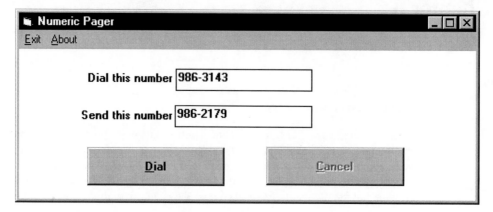

Problem: Create a simple numeric pager.

Solution: Here is one that uses MSCOMM. This could have used the Windows API or other comm add-ons equally well. In fact, the PowerPage DLLs that are used in the next section include numeric paging as a feature.

```
Sub Dial_Click ()
    Dim Buffer As String
    If Len(Num2Call.Text) = 0 Or Len(Num2Send.Text) = 0 Then
        MsgBox "Both numbers must be entered"
        Exit Sub
    End If
    Dial.Enabled = False
    Cancel.Enabled = True
    Comm1.Output = "ATX4V1Q0DT" & Num2Call.Text & "@;" &
vbCr
    'use this alternative, if your modem says "BUSY",
    'when it is not:
    'Comm1.Output = "ATX1V1Q0DT" & Num2Call.Text &
    ' ",,,,,,,,;" & VbCr
    'the number of comma's to be determined by experiment!
    Do Until Dial.Enabled = True
        DoEvents
        Buffer = Buffer & Comm1.Input
        If InStr(Buffer, "BUSY" & vbCrLf) Then
            MsgBox "Telephone Busy.  Try Again later"
            Exit Do
        ElseIf InStr(Buffer, "ERROR" & vbCrLf) Then
            MsgBox "There was a modem error." & _
                " Check the number to dial."
            Exit Do
        ElseIf InStr(Buffer, "OK" & vbCrLf) Then
```

```
                Comm1.Output = "ATDT" & Num2Send & ",;" & vbCr
                Buffer = ""
                Do Until InStr(Buffer, "OK" & vbCrLf
                    DoEvents
                    Buffer = Buffer & Comm1.Input
                    If InStr(Buffer, "ERROR") Then
                        MsgBox "There was a modem error.  " _
                            & "Check the send this number."
                        Exit Do
                    End If
                Loop
                Exit Do
            End If
        Loop
        DialerHangup
        Dial.Enabled = True
        Cancel.Enabled = False
    End Sub
```

Dial the telephone number entered by the user in the Num2Call textbox. There are two methods suggested to assemble the dial string.

The first method terminates the telephone number with an ampersand ("@"). Many modems use an apostrophe in a dial command to "wait for quiet answer." If the modem responds "OK" to a dial string then the assumption is that the number to be paged can be sent. If "BUSY" is detected then a MsgBox is displayed.

The second method inserts after the last number in the dial string a sufficient delay (you have to experiment to get this right) for the paging service to answer and to get past any voice prompts. Each comma in the dial string will cause the modem to delay 2 seconds before issuing the "OK" response.

```
    Sub DialerHangup ()
        Dim Buffer As String
        Comm1.Output = "ATH0" & VbCr
        Do Until InStr(Buffer, "OK" & VbCrLf)
            DoEvents
            Buffer = Buffer & Comm1.Input
        Loop
    End Sub
    Sub Form_Load ()
        Comm1.PortOpen = True
    End Sub
```

This looks simple enough. What's the problem?

Many modems do a mediocre to poor job differentiating voice from busy. So, an answering system with voice prompts that direct entry of a phone number to be paged may confuse your modem. It may output the "BUSY" response when, in fact, a voice is present. Your option, if this is a problem, is to add a fixed delay using commas in the dial string to bypass any voice prompt before the modem sends the number to be paged. For this reason, a computer program for numeric paging may not work very reliably.

This applet uses hard-coded serial port number. A production application would make this user-configurable.

8.2   Alphanumeric Paging

You could use any communications add-on or the API and Visual Basic code to implement the TAP protocol. See Appendix G if you are interested in a detailed explanation of the protocol. Do not confuse TAP (Telocator Alphanumeric Protocol) with TAPI.

The Simple Network Paging Protocol (SNPP) provides features similar to TAP but it uses an Internet connection instead of a modem connection directly to the paging service. I will not discuss SNPP here. X/Pager (mentioned later) provides an implementation of this protocol. See Appendix A for contact information.

There are several Visual Basic add-ons that implement the TAP protocol. CrystalCOMM, Logisoft X/Pager TAPI OCX, LUCA, and Allen-Martin amSMS are all possibilities. See Chapter 7 for more information on CrystalCOMM, LUCA, and amSMS. The PowerPage DLLs that were included with earlier versions of this book no longer are available. The text associated with those DLLs has been moved to the CD ROM.

8.2.2 Logisoft X/Pager TAPI OCX

There is a new alphanumeric pager OCX on the block. It is X/Pager from Logisoft. See Appendix A for contact information. This OCX has a couple of features that make it interesting and valuable for developers.

First, the OCX uses TAPI to handle modem initialization and dialing. This makes it easy to use, eliminating the need to have a modem initialization database. Second, it can send pages using the Internet for any paging service that supports Internet paging, such as Skytel.

Here is how you might use the OCX. For a normal modem connection, set the Pin property with, say, 8697541(the access number for the pager you are calling). Set the ServiceNumber property of the paging phone number, e.g. 18007596366 for Skytel. Set the TAPI DeviceID, perhaps 0 for the first TAPI device, or set the DeviceName, such as Standard Modem 28800

For Internet connections, set the EmailAddress property with, for example, 8697541@skytel.com. Set the RemoteHost property of the SMTP Server, say www.yourserver.com.

Then set the UseInternet property to True or False. Set the Message property to include any text up to MessageLen. Set the MessageLen property using, perhaps, 240. Set the Subject property (optional). Set the UserName and UserCompany properties (optional). Set the Retries property, try 2.

Now, set the Page Methods to start the page. Use the PageQueue method to queue multiple pages at once. Use the PageStart to start the paging process.

You can monitor the progress of a page by using the PageStatus event. There are several methods to display and use standard TAPI dialogs and for error and exception handling.

Chapter 9   .NET Compact Framework for Windows CE

Microsoft Windows Mobile 5, the successor to Windows CE 3.0 and Windows CE 4.x, combines an advanced real-time embedded operating system with powerful tools for rapidly creating the next generation of smart, connected, and small-footprint devices. With a complete operating system feature set and comprehensive development tools, Windows CE .NET contains the features developers need to build, debug, and deploy customized Windows CE .NET-based devices. Windows CE .NET supports Microsoft eMbedded Visual C++ and Microsoft Visual Studio .NET, providing a complete development environment for building applications for the Microsoft .NET Compact Framework, a subset of the Microsoft .NET Framework on the desktop. With these tools, developers can rapidly build designs for applications that use the latest hardware.

At this writing the latest version is Windows Mobile 5, which expands upon the foundation developed in previous Windows CE versions by providing:

- Support for secure and scalable networking.

- Enhanced real-time processing.

- Faster performance.

- Richer multimedia and Web browsing capabilities.

- Greater interoperability with personal computers, servers, Web services, and devices.

Microsoft continues to support Windows CE 3.0 and derivative versions, such as Microsoft Pocket PC 2000, 2002, 2003 and the newly released Windows Mobile 5. Development for these platforms may be done using Microsoft eMbedded Tools 3.0 (Visual C++ or Visual Basic), which is freely available from Microsoft. Pocket PC 2000/2002/2003 software may also be developed using Microsoft Visual Studio .NET and the Compact Framework, though this later developer environment **cannot** be used for Windows CE 3.0 or earlier.

This chapter is devoted to using Visual Studio .NET and the Compact Framework to implement serial communications applications. The Compact Framework 1.0 was released as part of Visual Studio .NET 2003 (Professional and higher Editions). The .NET Compact Framework 1.0, like Visual Studio .NET 2003, has no built-in serial support. Thus, we must write our own serial code that relies on Platform Invoke (P/Invoke) to call the underlying Windows CE serial communications APIs.

At this writing, the Compact Framework 2 has been released with Visual Studio 2005. Visual Studio 2005, Standard Editions and higher provide serial support through the System.IO.Ports namespace, and its usage is the same as that for the desktop .NET Framework that is discussed in Chapter 5.

The Compact Framework 1.x using Visual Studio 2003 as a development platform requires the use of the Windows serial communications APIs, called through Platform Invoke (P/Invoke).

The first step is to create a new Smart Device Application in Visual Studio .NET 2003. This invokes the Smart Device Wizard to create a Compact Framework application or library. In our case we will select a Class Library so that we can create a .NET Assembly (DLL) that may be referenced by other Compact Framework/Smart Device applications. Select a target device, either Pocket PC or Windows CE (though, in general, such a DLL will be usable with both Pocket PC and Windows CE applications – as long as they employ the same processor instruction set). I have selected the name CFSerialIO for this class.

You may be interested in an alternative serial IO implementation. One was developed by the OpenNETCF.org group. It is written in C#, and I have included it on the CD ROM. Please read the accompanying license information for its use. I contributed to this project, and it works well. Both C# and VB examples are provided.

9.1   Windows CE Device Control Block

I will borrow and adapt a quote from Hamlet, "To DCB or not to DCB?" What do I mean? The following image reflects the Windows CE documentation for the Device Control Block (DCB) structure that is used in several critical Windows CE API function calls.

Note the graphic nature of the circle-slash. This DCB documentation is not correct. The Windows CE DCB actually is **exactly the same** as the one used in 32-bit versions of Windows. You may refer to Chapter 6, or examine the source code on the CD ROM for the correct structure and field definitions.

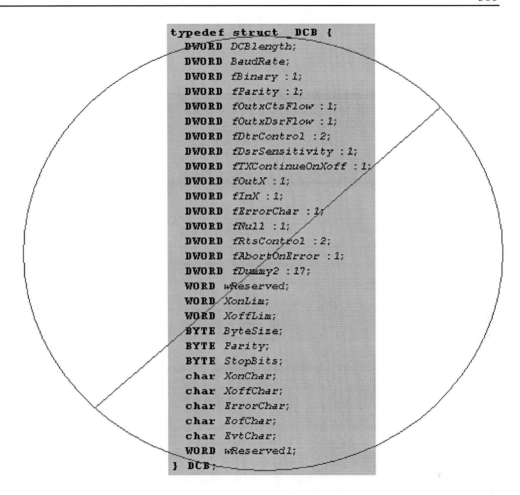

```
typedef struct _DCB {
    DWORD DCBlength;
    DWORD BaudRate;
    DWORD fBinary : 1;
    DWORD fParity : 1;
    DWORD fOutxCtsFlow : 1;
    DWORD fOutxDsrFlow : 1;
    DWORD fDtrControl : 2;
    DWORD fDsrSensitivity : 1;
    DWORD fTXContinueOnXoff : 1;
    DWORD fOutX : 1;
    DWORD fInX : 1;
    DWORD fErrorChar : 1;
    DWORD fNull : 1;
    DWORD fRtsControl : 2;
    DWORD fAbortOnError : 1;
    DWORD fDummy2 : 17;
    WORD wReserved;
    WORD XonLim;
    WORD XoffLim;
    BYTE ByteSize;
    BYTE Parity;
    BYTE StopBits;
    char XonChar;
    char XoffChar;
    char ErrorChar;
    char EofChar;
    char EvtChar;
    WORD wReserved1;
} DCB;
```

Figure 9.1 – Do not use the DCB shown in the Microsoft Windows CE documentation!

9.2 CFSerialIO Class

```
Imports System.Text
Imports System.Runtime.InteropServices
Imports System.ComponentModel
Imports System.Threading
```

The CFSerialIO class uses threading to optimize the processing of receive data.

```
Namespace SerialIO

    Public Class SerialPort
        Implements IDisposable
        Private m_Handle As Integer = INVALID_HANDLE_VALUE
        Private m_Port As Integer = 1
        Private m_Parity As Byte = NONE
        Private m_DataBits As Byte = 8
        Private m_StopBits As Byte = _
                        CByte(StopbitValue.STOPBITS_1)
        Private m_BaudRate As Integer = RATE_9600
        Private m_TXBufferSize As Integer = 512
        Private m_RXBufferSize As Integer = 512
        Private m_Timeout As Integer = 100
        Private m_RxBuffer As Byte()
        Private m_EnableOnComm As Boolean
        Private m_RTSEnable As Boolean
        Private m_DTREnable As Boolean
        Private m_EnableHardwareFlowcontrol As Boolean
        Private m_EnableSoftwareFlowcontrol As Boolean
```

These private variables are used for parameters and data associated with the serial port.

Next we define a method that will execute as a free thread. ReadThread reads receive data and handles communication errors.

```
        Private ReadThread As Thread
```

Two events are defined. The OnComm event may be generated when one or more bytes of data are received. The CommError event will be generated if any one of several communications errors is detected.

```
        Event OnComm()
        Event CommError(ByVal ErrorFlag As CommErrorFlags)
```

There are four Regions that I will not expand here. You may refer to the actual source code on the CD ROM. The Constants Region contains constants used, the Structures Region contains various data structures required by the serial communications APIs, the Enums Region contains Public enumerated constants that make the interface provided by CFSerialIO more "programmer friendly," and the API functions Region contains all of the Windows CE API function declarations — the functions that are the heart of our work.

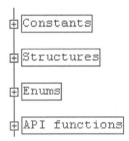

The CTSHandshaking property is use to enable CTS/RTS flow control.

```
Public Property CTSHandshaking() As Boolean
    Get
        Return m_EnableHardwareFlowcontrol
    End Get
    Set(ByVal Value As Boolean)
        m_EnableHardwareFlowcontrol = Value
        If Me.PortOpen = True Then
            UpdateCommState()
        End If
    End Set
End Property
```

The EnableOnComm property enables OnComm receive event generation. This is equivalent to setting the MSComm Rthreshold property to a value of 1 (best practice, in my opinion).

```
Public Property EnableOnComm() As Boolean
    Get
        Return m_EnableOnComm
    End Get
    Set(ByVal Value As Boolean)
        m_EnableOnComm = Value
    End Set
End Property
```

Use the BitRate property to set the serial port speed.

```
Public Property BitRate() As Integer
    Get
        Return m_BaudRate
    End Get
    Set(ByVal Value As Integer)
```

```
                        If Value = RATE_110 Or Value = RATE_300 _
                        Or Value = RATE_600 Or Value = RATE_1200 _
                        Or Value = RATE_2400 Or Value = RATE_4800 _
                        Or Value = RATE_9600 Or Value = _
                        RATE_14400 Or Value = RATE_19200 Or Value _
                        = RATE_38400 Or Value = RATE_56000 Or _
                        Value = RATE_57600 Or Value = RATE_115200 _
                        Or Value = RATE_128000 Or Value = _
                                                RATE_256000 Then
                            m_BaudRate = Value
                            If m_Handle <> INVALID_HANDLE_VALUE _
                                            Then UpdateCommState()
                        Else
                            Throw New Exception _
                                    ("Invalid serial port speed.")
                        End If
                    End Set
                End Property
```

The CommID property returns an open serial port handle. This might be important to allow the user of the class to call some Windows CE API that needs the serial port handle. This would be unusual, but I do not want to limit this sort of capability.

```
                Public ReadOnly Property CommID() As Integer
                    Get
                        Return m_Handle
                    End Get
                End Property
```

The CommPort range for current Windows CE devices is between 1 and 8. This may change in the future.

```
                Public Property CommPort() As Integer
                    Get
                        Return m_Port
                    End Get
                    Set(ByVal Value As Integer)
                        Dim ReOpen As Boolean
                        If m_Handle <> INVALID_HANDLE_VALUE _
                                        Then ReOpen = True
                        m_Port = Value
                        If m_Port < 1 Then
                            Throw New Exception _
                                    ("Invalid port number")
                            m_Port = 0
                        Else
                            If ReOpen = True Then Close()
                            m_Handle = CreateFile("COM" & _
                            m_Port.ToString & ":", GENERIC_READ _
                            Or GENERIC_WRITE, 0, 0, OPEN_EXISTING, _
                                                        0, 0)
                            If m_Handle = INVALID_HANDLE_VALUE Then
```

```
                            Throw New Exception _
                 ("Port already is in use or is not
available.")
                    ElseIf ReOpen = False Then
                        m_Handle = Close()
                    End If
                End If
            End Set
        End Property
```

The DataBits property has a range of 5-8.

```
        Public Property DataBits() As Integer
            Get
                Return CInt(m_DataBits)
            End Get
            Set(ByVal Value As Integer)
                Select Case Value
                    Case 5
                        m_DataBits = CByte(Value)
                    Case 6
                        m_DataBits = CByte(Value)
                    Case 7
                        m_DataBits = CByte(Value)
                    Case 8
                        m_DataBits = CByte(Value)
                    Case Else
                        Throw New Exception _
                        ("Invalid number of data bits.")
                End Select
                If m_Handle <> INVALID_HANDLE_VALUE _
                                Then UpdateCommState()
            End Set
        End Property
```

The MaxInputLen property defines the maximum number of bytes that may be buffered by the ReceiveThread process. This property is equivalent to the MSComm InBufferSize property.

```
        Public Property MaxInputLen() As Integer
            Get
                Return m_RXBufferSize
            End Get
            Set(ByVal Value As Integer)
                m_RXBufferSize = Value
            End Set
        End Property
```

The MaxOutputLen property defines the maximum number of bytes that may be transmitted in any single call to the Output method. This is equivalent to the MSComm OutBufferSize property.

```
            Public Property MaxOutputLen() As Integer
                Get
                    Return m_TXBufferSize
                End Get
                Set(ByVal Value As Integer)
                    m_TXBufferSize = Value
                End Set
            End Property
```

Use the Parity property to define the parity to be used when sending and receiving data.

```
            Public Property Parity() As String
                Get
                    Select Case m_Parity
                        Case EVEN
                            Return "E"
                        Case MARK
                            Return "M"
                        Case NONE
                            Return "N"
                        Case ODD
                            Return "O"
                        Case SPACE
                            Return "S"
                    End Select
                End Get
                Set(ByVal Value As String)
                    If Len(Value) > 1 _
                        Then Value = Value.Substring(0, 1)
                    Value = UCase(Value)
                    Select Case Value
                        Case "E"
                            m_Parity = EVEN
                        Case "M"
                            m_Parity = MARK
                        Case "N"
                            m_Parity = NONE
                        Case "O"
                            m_Parity = ODD
                        Case "S"
                            m_Parity = SPACE
                        Case Else
                            Throw New Exception _
                                ("Invalid parity.")
                    End Select
                    If m_Handle <> INVALID_HANDLE_VALUE _
                        Then UpdateCommState()
                End Set
            End Property
```

Use the PortOpen property either to open a serial port or to return the state of the port.

```
              Public Property PortOpen() As Boolean
                  Get
                      If m_Handle = INVALID_HANDLE_VALUE Then
                          Return False
                      Else
                          Return True
                      End If
                  End Get
                  Set(ByVal Value As Boolean)
                      If Value = True Then
                          Open()
                      Else
                          m_Handle = Close()
                          If m_Handle = INVALID_HANDLE_VALUE Then
                              Thread.Sleep(10)
                              ReadThread = Nothing
                          End If
                      End If
                  End Set
              End Property
```

Under normal circumstances StopBits should be set to 1. Some Windows CE serial ports support **only** 1 stop bit, and using either 1.5 or 2 stop bits will fail.

```
              Public Property StopBits() As Single
                  Get
                      Select Case m_StopBits
                          Case CByte(StopbitValue.STOPBITS_1)
                              Return 1
                          Case CByte(StopbitValue.STOPBITS_15)
                              Return 2
                          Case CByte(StopbitValue.STOPBITS_20)
                              Return 3
                      End Select
                  End Get
                  Set(ByVal Value As Single)
                      Select Case Value
                          Case 1
                              m_StopBits = _
                                  CByte(StopbitValue.STOPBITS_1)
                          Case 1.5
                              m_StopBits = _
                                  CByte(StopbitValue.STOPBITS_15)
                          Case 2
                              m_StopBits = _
                                  CByte(StopbitValue.STOPBITS_20)
                          Case Else
                              Throw New Exception _
                              ("Invalid number of stop bits.")
                      End Select
                      If m_Handle <> INVALID_HANDLE_VALUE _
                                          Then UpdateCommState()
```

```
        End Set
    End Property
```

The Timeout property defines the time that the underlying Windows CE API will buffer transmit data. If data cannot be sent because the flow control state, either hardware CTS, or software Xoff, is signaled false – and if the value of this timeout is exceeded, then that data will be discarded. Our code is designed to use a free thread (ReadThread) to continuously buffer receive data. Thus, this property will not affect receive data.

```
    Public Property Timeout() As Integer
        Get
            Return m_Timeout
        End Get
        Set(ByVal Value As Integer)
            m_Timeout = Value
            If m_Handle <> INVALID_HANDLE_VALUE _
                               Then UpdateCommState()
        End Set
    End Property
```

Two data types may be used to send and receive data. The InputArray function returns receive data as an array of type Byte.

It is important to note that calling either InputArray or InputString returns **all** available receive data. This is equivalent to setting the MSComm InputLen property to 0 (this is the MSComm default). This optimizes performance and simplifies the code in this class.

```
    Public Function InputArray() As Byte()
        If Not (m_RxBuffer Is Nothing) Then
            Dim Buffer(m_RxBuffer.Length - 1) As Byte
            Monitor.Enter(ReadThread)
                m_RxBuffer.CopyTo(Buffer, 0)
                Erase m_RxBuffer
            Monitor.Exit(ReadThread)
            Return Buffer
        End If
    End Function
```

InputString returns receive data as a String. Also see the comment above.

```
    Public Function InputString() As String
        Dim AsciiEncoding As New ASCIIEncoding
        Dim Buffer() As Byte
        If Not (m_RxBuffer Is Nothing) _
                    AndAlso m_RxBuffer.Length > 0 Then
            Monitor.Enter(ReadThread)
            ReDim Buffer(m_RxBuffer.Length - 1)
            m_RxBuffer.CopyTo(Buffer, 0)
            Erase m_RxBuffer
            Monitor.Exit(ReadThread)
            Return AsciiEncoding.GetString _
                        (Buffer, 0, Buffer.Length)
```

```
            End If
        End Function
```

The following New method is the default constructor for the CFSerialIO object. Unlike
MSComm, I believe that it is worthwhile to set the default for the state of RTS to True,
when the port is open. There may be some situations where this is not desired — in that
case, simply change the two NEW constructors to set m_RTSEnable = False. An example
of such an application might be one that communicates using RS-485, though simply setting
the RTSEnable property to False in code will usually be sufficient in this case.

```
        Public Sub New()
            MyBase.new()
            m_RTSEnable = True
            m_DTREnable = True
        End Sub
```

New is an overloaded constructor that may be called when the CFSerialIO object is
instantiated. The following method allows you to specify other than default parameters
when it is created.

```
        Public Sub New(ByVal port As Integer, ByVal BitRate _
            As Integer, ByVal parity As String, ByVal _
            databits As Integer, ByVal stopbits As Single)
            MyBase.new()
            m_Port = port
            m_BaudRate = BitRate
            Select Case parity
                Case "E"
                    m_Parity = EVEN
                Case "M"
                    m_Parity = MARK
                Case "O"
                    m_Parity = ODD
                Case "S"
                    m_Parity = SPACE
                Case Else
                    m_Parity = NONE
            End Select
            Select Case databits
                Case 5
                    m_DataBits = CByte(databits)
                Case 6
                    m_DataBits = CByte(databits)
                Case 7
                    m_DataBits = CByte(databits)
                Case Else
                    m_DataBits = CByte(databits)
            End Select
            Select Case stopbits
                Case 1.5
                    m_StopBits = _
                        CByte(StopbitValue.STOPBITS_15)
```

```
            Case 2
                m_StopBits = _
                    CByte(StopbitValue.STOPBITS_20)
            Case Else
                m_StopBits = _
                    CByte(StopbitValue.STOPBITS_1)
        End Select
        m_RTSEnable = True
        m_DTREnable = True
    End Sub
```

The Output method is overloaded to allow data of type String or an array of type Byte to be transmitted. The signature of the parameter used to call Output determines which of these two overloaded methods are selected at runtime.

```
    Public Overloads Sub Output(ByVal Value As String)
        Dim TxBuffer() As Byte
        Dim Encoder As New ASCIIEncoding
        If m_Handle <> INVALID_HANDLE_VALUE Then
            ReDim TxBuffer(Value.Length - 1)
            TxBuffer = Encoder.GetBytes(Value)
            Me.Output(TxBuffer)
        End If
    End Sub

    Public Overloads Sub Output(ByVal Value() As Byte)
        Dim intResult As Integer
        Dim NumberOfBytesWritten As Integer = 0
        If m_Handle <> INVALID_HANDLE_VALUE Then
            intResult = WriteFile(m_Handle, Value, _
                Value.Length, NumberOfBytesWritten, _
                                        Nothing)
        End If
    End Sub
```

The InBufferCount property returns the number of bytes that have been buffered by the ReadThread process.

```
    Public ReadOnly Property InBufferCount() As Integer
        Get
            If (Me.PortOpen = True) And _
                    (Not IsNothing(m_RxBuffer)) Then
                Monitor.Enter(ReadThread)
                InBufferCount = m_RxBuffer.Length
                Monitor.Exit(ReadThread)
            Else
                InBufferCount = 0
            End If
        End Get
    End Property
```

Open is an internal (Private) method that is called by code in the PortOpen property to open a serial port. When a port is successfully opened, a new ReadThread is created to process receive data and communications errors.

```
Private Function Open() As Boolean
    If m_Port > 0 Then
        If m_Handle > 0 Then Close()
        m_Handle = CreateFile("COM" & _
            m_Port.ToString & ":", _
            GENERIC_READ Or GENERIC_WRITE, _
                0, 0, OPEN_EXISTING, 0, 0)
        If m_Handle <> INVALID_HANDLE_VALUE Then
            ReadThread = New Thread(AddressOf _
                                    ReadThreadProc)
            ReadThread.Start()
            UpdateCommState()
            If m_DTREnable = True Then _
                                    DTREnable = True
            If m_RTSEnable = True Then _
                                    RTSEnable = True
            Dim errCode As Integer
            Dim StatStruct As New COMSTAT
            ClearCommError(m_Handle, errCode, _
                                    StatStruct)
            Return True
        Else
            Return False
        End If
    End If
End Function
```

The ReadThreadProc is the ReadThread object that is created when a port has been opened. ReadThread is a polling routine. As long as the port is open (a closed port is signaled by the port handle being set to -1 (INVALID_HANDLE_VALUE) as a result of Close having been called), the thread calls the ReadFile API function. If data is returned, it is buffered. After data has been buffered, an OnComm event may be generated. Near the end of the loop, the ClearCommError API routine is called. Any communication errors that are were detected by Windows CE result in generation of the CommError event.

An alternative to using a loop in the ReadThreadProc would have been to setup a state machine that employs the SetCommMask , WaitCommEvent and WaitForSingleObject APIs. However, the code that I have selected seems to have two advantages. First, and primary, is that it actually executes more quickly than the alternative. Second, this code is simpler (perhaps this is the reason that the execution efficiency is better?).

```
Private Sub ReadThreadProc()
    Dim Result As Integer = -1
    Dim BytesActuallyRead As Integer = 0
    Dim Buffer(m_RXBufferSize - 1) As Byte
    Dim SentXOFF As Boolean
    Dim LoweredRTS As Boolean
```

```
         Do Until m_Handle = INVALID_HANDLE_VALUE
           If Buffer.Length < m_RXBufferSize _
               Then ReDim Buffer(m_RXBufferSize - 1)
           Result = ReadFile(m_Handle, Buffer, _
               m_RXBufferSize, BytesActuallyRead, _
                                           Nothing)
```

OK. We have received data. Now, what do we do with it? The goal here is to create a circular buffer to retain the data. Such a buffer acts like a FIFO (First In First Out), with a strictly limited capacity. The capacity is the value set using the MaxInputLen property.

We also have to handle flow control. Why? The Windows CE API implements flow control. Why do we have to deal with it here? The answer is simple. We have created a free thread for reading and buffering data. Thus, the receive buffers that Windows maintains, and monitors for flow control, will never cause receive flow control to be signaled. Thus, we have to take over that part of the job.

```
     If BytesActuallyRead > 0 Then
        Monitor.Enter(ReadThread)
        Dim OldLength As Integer = 0
        If Not (IsNothing(m_RxBuffer)) _
            Then OldLength = m_RxBuffer.Length
        Dim OverLimit As Integer = _
                OldLength + BytesActuallyRead
        If (m_EnableHardwareFlowcontrol = _
            True) Then
            If (OverLimit > 0.75 * _
                m_RXBufferSize) Then
                EscapeCommFunction(m_Handle, _
                    ControlLines.ClearRTS)
                LoweredRTS = True
            ElseIf (LoweredRTS = True) _
                AndAlso (m_RTSEnable = True) _
                AndAlso (OverLimit < 0.5 _
                * m_RXBufferSize) Then
                EscapeCommFunction(m_Handle, _
                    ControlLines.SetRTS)
                LoweredRTS = False
            End If
        End If
        If (m_EnableSoftwareFlowcontrol = True) _
            Then
            If (OverLimit > 0.75 * _
                m_RXBufferSize) AndAlso _
                    (SentXOFF = False) Then
                SentXOFF = True
                Output(XOFF)
            ElseIf (SentXOFF = True) AndAlso _
                (OverLimit < 0.5 * _
                        m_RXBufferSize) Then
                SentXOFF = False
                Output(XON)
```

```
                    End If
            End If
            If OverLimit > m_RXBufferSize Then
                    RaiseEvent CommError _
                            (CommErrorFlags.RXOVER)
                    Dim BytesToCopy As Integer = _
                        m_RXBufferSize _
                                - BytesActuallyRead
                    Array.Copy(m_RxBuffer, OldLength - _
                        BytesToCopy, m_RxBuffer, 0, _
                                BytesToCopy)
                    ReDim Preserve m_RxBuffer _
                                (m_RXBufferSize - 1)
                    Array.Copy(Buffer, 0, m_RxBuffer, _
                    BytesToCopy - 1, BytesActuallyRead)
            Else
                    ReDim Preserve m_RxBuffer(OldLength _
                            + BytesActuallyRead - 1)
                    Array.Copy(Buffer, 0, m_RxBuffer, _
                            OldLength, BytesActuallyRead)
            End If
            Monitor.Exit(ReadThread)
            If m_EnableOnComm = True Then _
                                RaiseEvent OnComm()
        Else
            If (m_EnableHardwareFlowcontrol = True) _
                Then
                If (m_RTSEnable = True) AndAlso _
                            (LoweredRTS = True) Then
                    EscapeCommFunction(m_Handle, _
                            ControlLines.SetRTS)
                End If
            End If
            If (m_EnableSoftwareFlowcontrol = True) _
                    AndAlso (SentXOFF = True) Then
                SentXOFF = False
                Output(XON)
            End If
        End If
        Dim errCode As Integer
        Dim StatStruct As COMSTAT
```

This is the place to handle communication errors. ClearCommError does two things. First, it clears all pending errors, so that new errors can be detected. Second, it returns a value (errCode) that may be parsed to determine which errors might have been detected by Windows CE since ClearCommError last was called.

```
        ClearCommError(m_Handle, errCode, StatStruct)
        If errCode > 0 Then
            If (errCode And CommErrorFlags.BREAK) = _
                        CommErrorFlags.BREAK Then
                RaiseEvent CommError _
```

```
                              (CommErrorFlags.BREAK)
                    End If
                    If (errCode And CommErrorFlags.FRAMING) _
                            = CommErrorFlags.FRAMING Then
                        RaiseEvent CommError _
                                (CommErrorFlags.FRAMING)
                    End If
                    If (errCode And CommErrorFlags.OVERRUN) _
                            = CommErrorFlags.OVERRUN Then
                        RaiseEvent CommError _
                                (CommErrorFlags.OVERRUN)
                    End If
                    If (errCode And CommErrorFlags.RXOVER) _
                            = CommErrorFlags.RXOVER Then
                        RaiseEvent CommError _
                                (CommErrorFlags.RXOVER)
                    End If
                    If (errCode And _
                        CommErrorFlags.RXPARITY) _
                            = CommErrorFlags.RXPARITY Then
                        RaiseEvent CommError _
                                (CommErrorFlags.RXPARITY)
                    End If
                    If (errCode And CommErrorFlags.TXFULL) _
                            = CommErrorFlags.TXFULL Then
                        RaiseEvent CommError _
                                (CommErrorFlags.TXFULL)
                    End If
                End If
```

There is one thing more to be done in the loop. Call Threed.Sleep to release time so that any other pending threads, such as the UI thread, may execute. While not mandatory, this statement is a good design element. It helps to ensure optimal performance.

```
                Thread.Sleep(0)
            Loop
        End Sub
```

The Private UpdateCommState method is called from any of the Public Property methods that can cause any port parameters to change. It also is called from Open to make sure that the current serial port parameters are used.

```
        Private Sub UpdateCommState()
            Dim intResult As Integer
            Dim Result As Integer
            Dim DCB As DCB
            Dim CommTimeouts As COMMTIMEOUTS
            Result = PurgeComm(m_Handle, PURGE_RXCLEAR Or _
                                            PURGE_TXCLEAR)
            intResult = GetCommState(m_Handle, DCB)
            SetBits(DCB)
            DCB.StopBits = m_StopBits
```

```
                    DCB.ByteSize = m_DataBits
                    DCB.Parity = m_Parity
                    DCB.BitRate = BitRate
                    DCB.XonChar = Chr(17)
                    DCB.XoffChar = Chr(19)
                    DCB.XonLim = 100
                    If m_RXBufferSize > 20000 Then
                        DCB.XoffLim = 19900
                    Else
                        DCB.XoffLim = CShort(m_RXBufferSize * 0.75)
                    End If
                    intResult = SetCommState(m_Handle, DCB)
                    intResult = SetupComm(m_Handle, m_RXBufferSize, _
                                                   m_TXBufferSize)
                    CommTimeouts.ReadIntervalTimeout = 0
                    CommTimeouts.ReadTotalTimeoutMultiplier = 0
                    CommTimeouts.ReadTotalTimeoutConstant = Timeout
                    CommTimeouts.WriteTotalTimeoutMultiplier = 10
                    CommTimeouts.WriteTotalTimeoutConstant = 100
                    intResult = SetCommTimeouts(m_Handle, _
                                                   CommTimeouts)
            End Sub
```

There are several bit-mapped fields in the DCB (Device Control Block structure) that cannot be set directly. The Private SetBits method is use to mask off the individual bits that are affected, and then to manipulate them.

```
        Private Sub SetBits(ByRef DCB As DCB)
            With DCB
                If m_Parity = 0 Then
                    .Bits = .Bits And (FPARITY Xor &HFFFF)
                Else
                    .Bits = .Bits Or FPARITY
                End If
                If m_EnableHardwareFlowcontrol = False Then
                    .Bits = .Bits And (FOUTXCTSFLOW Xor _
                                                    &HFFFF)
                    .Bits = .Bits And (FRTSCONTROL Xor _
                                                    &HFFFF)
                Else
                    .Bits = .Bits Or FOUTXCTSFLOW
                    .Bits = .Bits Or FRTSCONTROL
                End If
                If m_EnableSoftwareFlowcontrol = False Then
                    .Bits = .Bits And (FOUTX Xor &HFFFF)
                    .Bits = .Bits And (FINX Xor &HFFFF)
                Else
                    .Bits = .Bits Or FOUTX
                    .Bits = .Bits Or FINX
                End If
            End With
        End Sub
```

Software flowcontrol is enabled using the XonXoffHandshaking property.

```
Public Property XonXoffHandshaking() As Boolean
    Get
        Return m_EnableSoftwareFlowcontrol
    End Get
    Set(ByVal Value As Boolean)
        m_EnableSoftwareFlowcontrol = Value
        If Me.PortOpen = True Then
            UpdateCommState()
        End If
    End Set
End Property
```

Request To Send (RTS) is set or reset using the RTSEnable property.

```
Public Property RTSEnable() As Boolean
    Get
        Return m_RTSEnable
    End Get
    Set(ByVal Value As Boolean)
        If m_Handle <> INVALID_HANDLE_VALUE Then
            If Value = True Then
                EscapeCommFunction(m_Handle, _
                        ControlLines.SetRTS)
            Else
                EscapeCommFunction(m_Handle, _
                        ControlLines.ClearRTS)
            End If
        End If
    End Set
End Property
```

Data Terminal Ready (DTR) is set or reset using the RTSEnable property.

```
Public Property DTREnable() As Boolean
    Get
        DTREnable = m_DTREnable
    End Get
    Set(ByVal Value As Boolean)
        If m_Handle <> INVALID_HANDLE_VALUE Then
            If Value = True Then
                EscapeCommFunction(m_Handle, _
                        ControlLines.SetDTR)
            Else
                EscapeCommFunction(m_Handle, _
                        ControlLines.ClearDTR)
            End If
        End If
    End Set
End Property
```

The state of Clear To Send (CTS) is read using the CTS property.

```
Public ReadOnly Property CTS() As Boolean
    Get
        If m_Handle <> INVALID_HANDLE_VALUE Then
            Dim ModemStatus As Integer
            GetCommModemStatus(m_Handle, ModemStatus)
            CTS = CBool(ModemStatus And CTS_ON)
        End If
    End Get
End Property
```

The state of Data Set Ready (DSR) is read using the DSR property.

```
Public ReadOnly Property DSR() As Boolean
    Get
        If m_Handle <> INVALID_HANDLE_VALUE Then
            Dim ModemStatus As Integer
            GetCommModemStatus(m_Handle, ModemStatus)
            DSR = CBool(ModemStatus And DSR_ON)
        End If
    End Get
End Property
```

The state of Carrier Detect (CD or CxR) is read using the CD property.

```
Public ReadOnly Property CD() As Boolean
    Get
        If m_Handle <> INVALID_HANDLE_VALUE Then
            Dim ModemStatus As Integer
            GetCommModemStatus(m_Handle, ModemStatus)
            CD = CBool(ModemStatus And RLSD_ON)
        End If
    End Get
End Property
```

The state of Ring Indicate (RI) is read using the RI property.

```
Public ReadOnly Property RI() As Boolean
    Get
        If m_Handle <> INVALID_HANDLE_VALUE Then
            Dim ModemStatus As Integer
            GetCommModemStatus(m_Handle, ModemStatus)
            RI = CBool(ModemStatus And RING_ON)
        End If
    End Get
End Property
```

One implication of the use of free threading is that we have to terminate threads that we have created properly. The failure to do so will cause the Windows CE/Pocket PC hardware to act erratically (read: lock up, requiring a soft reset). Close closes an open port by calling the CloseHandle API **and** it signals this to the ReadThread, so that ReadThread may exit gracefully.

```
Private Function Close() As Integer
    Dim intResult As Integer
    If m_Handle <> INVALID_HANDLE_VALUE Then
        Dim errCode As Integer
        Dim StatStruct As COMSTAT
        ClearCommError(m_Handle, errCode, StatStruct)
        intResult = CloseHandle(m_Handle)
        If intResult < 0 Then
            Throw New Exception _
                ("Unable to close serial port.")
        End If
    End If
    Return INVALID_HANDLE_VALUE
End Function

⊞ IDisposable implementation

⊞      Protected Overrides Sub Finalize() ...
    End Class
End Namespace
```

There is an issue that I did not cover in the inline comments in the preceding code. The ReceiveThread is a "free thread." It buffers receive data to an array of type Byte. This data, and the associated InBufferCount property, may be accessed by a separate thread, with no inherent synchronization between the threads. Thus there is the potential for a "race condition." That is, the ReadThread process may attempt to change buffered data at the same time that an outside thread attempts to consume that data. Something needs to be done to eliminate this potential contention for the shared data. The .NET threading classes provide a couple of mechanisms for this, and the one that I have used is the System.Threading.Monitor class. Monitor.Enter acquires and exclusive lock for a region of code, and Monitor.Exit releases the lock. While a lock is in place, all variables that are within the scope of the lock can be used without fear that some other contending thread can cause an error. This or other equivalent synchronization mechanism is vital to the safe and effective use of threading.

Synchronization mechanisms are not free. That is, there is an impact on performance. When a block of code has been locked and another other thread that attempts to access the shared resource, that code simply must wait until the lock is released. This "wait" imposes a processing overhead that can easily be measured. For this, and other good reasons, threading should be used to solve real problems (as in this class), but should not be used in every instance. Use it with care, and recognize the consequences of its use. See the Resources Appendix for my recommendation on a book that covers .NET threading in depth.

Some developers want to use a serial class for desktop applications that has the same API as the one used for Pocket PC/Windows CE applications. The CFSerialIO class has been ported to the desktop (see the DesktopSerial folder).

Now we have a serial class. What shall we do with it? There are several programs that illustrate the use of CFSerialIO on the CD ROM. These are CFNETTerm, ImageTransfer, TestPulse, and DataLogger.

9.3   CFNETTerm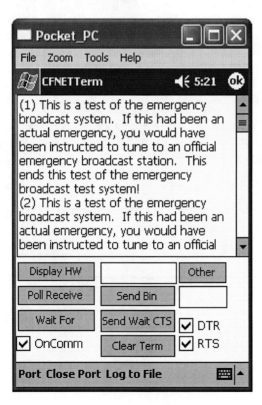

Problem: There are a number of features of the CFSerialIO class that want illustration. How do you select and configure ports, display data, poll and wait for data, display the status of the hardware input lines, handle binary data, and write receive data to a file on the device?

Solution: This example illustrates these functions (and a few others). I will let you refer to the source code on the CD ROM for many of the details. However, there are a few things that require some discussion, so I will do that next.

The ReceiveEvent subroutine is the event handler for the OnComm event. The code in this routine does two things. First, if logging is enabled (see the LogFile code later for more on this), then it writes receive data to a file. Second, if logging is not enabled, receive data is prepared for display — however, receive data is not displayed immediately. Instead, a separate Handler is called using the Invoke method. The ReceiveEvent handler executes in the context of the ReceiveThread in the SerialClass. To interact with the UI, which executes in the STAThread context of the form, data must be marshaled. This marshaling and transfer of thread context is done by using Invoke.

Another thing to note is that receive data is buffered in by using the Append method of the StringBuilder named ScreenBuff. Appending data to a StringBuilder can offer better performance than appending data to a String. In many cases, using StringBuilder is to be preferred. However… If receive data must be parsed (which often requires a String), the advantage of StringBuilder is lost. Thus, the recommendation that often is given in .NET books to use StringBuilder when appending data, instead of String may or may not be valid. If receive data must be parsed, a string and the overhead for appending data to the string, may provide the best performance.

Note: see the CD ROM in the VisualStudio2005 folder for a Compact Framework 2.0 example that uses the built-in System.IO.Ports namespace. I suggest using the built-in serial class for Windows Mobile 5 applications.

```
Private Sub ReceiveEvent() Handles SerialPort.OnComm
        If LogToFile = True Then
            Dim Buffer As String = SerialPort.InputString
            myWriter.Write(Buffer)
        Else
            ScreenBuff.Append(SerialPort.InputString)
            'This is FAR from the best way to display (or
manage) this data.  It is "quick and dirty."
            Me.Invoke(New EventHandler(AddressOf
DisplayData))
        End If
    End Sub
```

The DisplayData event hander executes on the STAThread context.

```
    Private Sub DisplayData(ByVal sender As Object, _
ByVal e As EventArgs)
```

```
                        'This marshals receive data from the receive
            thread context (OnComm or CommError) to the Windows Form
            STAThread context
                    With txtTerm
                        .Text = ScreenBuff.ToString
                        .SelectionStart = .Text.Length
                        If (ErrorMessage.Length > 0) Then
                            .SelectedText = ErrorMessage
                            ErrorMessage = ""
                        End If
                    End With
                    With ScreenBuff
                        If .Length > 0.8 * .Capacity Then
                            ScreenBuff.Remove(0, CInt(0.6 _
            * .Capacity))
                        End If
                    End With
                End Sub
```

The LogFile click event starts and stops writing receive data to a file. The ImageTransfer example on the CD ROM provides and example that writes binary data to a file.

```
                Private Sub LogFile_Click(ByVal sender As
            System.Object, ByVal e As System.EventArgs) _
            Handles LogFile.Click
                    If SerialPort.PortOpen = True Then
                        If LogFile.Text = "Log to File" Then
                            LogFile.Text = "Stop Logging"
                            With SaveFileDialog1
                                .FileName = LogfileName
                                .ShowDialog()
                                If .FileName <> "" Then
                                    LogfileName = .FileName
                                    MsgBox("Logging will start when
            this dialog is closed. Receive data will NOT be displayed.
            Click again to stop logging.", MsgBoxStyle.Information)
                                    fs = New _
                            System.IO.FileStream(LogfileName, _
                            System.IO.FileMode.CreateNew) _
                                    myWriter = New _
                                        System.IO.StreamWriter(fs)
                _                   SerialPort.EnableOnComm = True
                                    chkEnableOnComm.Checked = True
                                    LogToFile = True
                                Else
                                    With LogFile
                                        .Enabled = True
                                        .Text = "Log to File"
                                    End With
                                End If
                            End With
                        Else
```

```
                    With LogFile
                        .Enabled = True
                        .Text = "Log to File"
                    End With
                    Try
                        myWriter.Flush()
                        fs.Flush()
                        fs.Close()
                    Catch ex As Exception
                    End Try
                End If
            Else
                MsgBox("Port must be open before logging.",
MsgBoxStyle.Exclamation)
            End If
        End Sub
```

Chapter 10 — Direct Port I/O

There will be times when Windows gets in your way. You may want to do something that is not supported by the Windows API. You certainly should not use direct I/O to read and write serial data. Although this is technically possible, it would bypass the interrupt driven routines that Windows provides and would be inefficient in the extreme.

Direct access to computer I/O ports is not supported by Visual Basic. However, you can use add-on DLLs or kernel-mode drivers and OCXs to read and write to I/O ports. This can allow you to do some useful things. I will show examples of some of the possibilities in the following sections. 16-bit text and code examples have been moved to the CD ROM.

10.1   WIN95IO and DriverLINX PortIO

Thirty-two-bit programs that access I/O ports require a 32-bit DLL or a 32-bit OCX. Included on the CD ROM that accompanies this book is WIN95IO.DLL. It is a simple, freeware DLL that contains two functions: ReadPort and WritePort. It is limited to use under Windows 9x/Me.

Also included on the CD ROM is free software from Scientific Software Tools. DriverLINX PortIO is similar to WIN95IO.DLL. It is a simple user-mode DLL and NT kernel driver to allow direct hardware I/O access. It can be used on any 32-bit Windows platform. Read the license information that is provided with it on the CD ROM. This device driver and associate files is located in the 3rdEdition/Chapter10 folder.

10.1.1   CDMonitor (32-bit)

Problem: Show me a 32-bit applet that will monitor modem carrier detect.

Solution: OK, here it is. The applet provided with earlier editions of this book used WIN95IO.DLL. That version is on the CD ROM. This version uses the Scientific Software Tools DriverLINX PortIO add-on. It will work with any standard internal or external modem. However, it **will not** work with a WinModem or other modem that does not use a standard UART.

```
Option Explicit
Private Declare Function DlPortReadPortUchar Lib _
        "dlportio.dll" (ByVal Port As Long) As Byte
Dim Addr As Integer
Private Sub Com_Click(Index As Integer)
```

```
        Dim I As Integer
        If Index < 0 Then Exit Sub
        For I = 0 To 3
            Com(I).Checked = False
        Next I
        Com(Index).Checked = True
        SaveSetting App.ProductName, "Settings", "Comm port", _
            Format$(Index + 1)
        Select Case Index
            Case 0
                Addr = &H3FE
            Case 1
                Addr = &H2FE
            Case 2
                Addr = &H3EE
            Case 3
                Addr = &H2EE
        End Select
    End Sub
```

The I/O address of the UART modem status register is the UART I/O base address + 6. The Com routine assigns that address to the Addr variable and saves the port number in the registry for use when the program next runs.

```
    Private Sub Form_Load()
    Dim ComPort As Integer
        ComPort = GetSetting(App.ProductName, "Settings",
    "ComPort", "1")
        Com_Click (Val(ComPort) - 1)
        Me.Show
        Timer1_Timer            'execute carrier detect code
    now
    End Sub
```

If a comm port had previously been selected then it is used for the current run. Timer1 is enabled to sample the state of modem carrier detect once per second.

```
    Private Sub Form_Unload(Cancel As Integer)
        'Place any code that you need to record the total
        'connect time in this routine.  This normally would be
        ' saved in a disk or database, with a date and time
        'stamp.
    End Sub
    Private Sub Timer1_Timer()
        Static Connected As Double
        Dim CD As Byte
        CD = DlPortReadPortUchar(Addr) And &H80
        If CD <> 0 Then
            Me.Icon = Image1.Picture
```

```
                Connected = Connected + 1
                Me.Caption = "Connected for " & _
                    Format$(Connected / 100000, "hh:mm:ss")
            Else
                Me.Icon = Image2.Picture
            End If
    End Sub
```

Mask the UART modem status register bit 7. If the result is a 1, carrier detect is true. Total connect time is incremented when CD is true.

Both this and the preceding applet will not work if the UART registers are virtualized. This happens under (at least) two circumstances. First, Windows virtualizes the serial port hardware for DOS Virtual Machines (VMs). So, you cannot use a DLL to access the UART I/O ports if a DOS application is using the serial port. This method can only be used to monitor Windows applications. Second, some replacement drivers for Windows 95/98, such as TurboCom 95, virtualizes the UART, even for Windows applications. So, these apps will not work with TurboComm95 or equivalent replacement drivers.

10.1.2   HangUp

Problem: I need to create an application that forces the modem to disconnect. Unfortunately, the modem in question is not in use by my application. Another application is using the serial port/modem. An additional complication is that this must run under both Windows 95/98 and Windows NT/2K/XP.

Solution: We could use a DLL like WIN95IO.DLL if this did not need to run under Windows NT/2K/XP. So, this applet uses the Scientific Software Tools DriverLINX PortIO add-on.

```
Option Explicit
Private Declare Sub DlPortWritePortUchar Lib "dlportio.dll" _
                    (ByVal Port As Long, ByVal Value As Byte)
Public Sub Main()
Dim PortNum As String
Dim Addr As Long
    PortNum = Command$
    Select Case PortNum
        Case "1", ""
            Addr = &H3FC
        Case "2"
            Addr = &H2FC
        Case "3"
            Addr = &H3EC
        Case "4"
            Addr = &H2EC
    End Select
    DlPortWritePortUchar Addr, 0
```

```
End Sub
```

You can force DTR false by setting the UART modem control register Data Terminal Ready control bit to zero. This EXE can be run with a command-line argument that specifies the port to force disconnect (lower DTR).

See the documentation that is furnished with PortIO for distribution and installation instructions.

I have included an example called RasDisconnect on the CD ROM. The RAS API can be used to force an external RAS dialup application to disconnect.

10.1.3   SendData

Problem: I want to allow another application to open a serial port. Normally, this means that I cannot **also** use the same port to send data. How can I overcome this limitation?

Solution: Let us use Scientific Software Tools DriverLINX PortIO add-on to write directly to the UART Tx register.

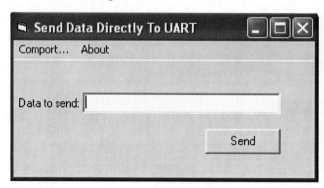

```vb
Option Explicit

Private Declare Function DlPortReadPortUchar Lib _
"dlportio.dll"(ByVal Port As Long) As Byte

Private Declare Sub DlPortWritePortUchar Lib
"dlportio.dll"(ByVal Port As Long, ByVal Value As Byte)

Private Declare Sub Sleep Lib "kernel32" (ByVal _
dwMilliseconds As Long)

Private Addr As Integer

Private Sub SendByte(ByVal Data As Byte)
Dim LineControlRegister As Byte
Dim TEMT As Byte
'     'get the Line Status Register -- if Busy, then wait
    TEMT = DlPortReadPortUchar(Addr + 4)
    Do Until TEMT <> &H40
        DoEvents
        TEMT = DlPortReadPortUchar(Addr + 4) And &H40
    Loop
'The above didn't seem to work, so add a delay after
sending character
    'get the Line Control Register
    LineControlRegister = DlPortReadPortUchar(Addr + 3)
    'reset bit 7
    DlPortWritePortUchar Addr + 3, LineControlRegister And
&H7F
    'place data in the Tx Holding Register
    DlPortWritePortUchar Addr, Data
    'add a delay to allow UART to clear -- the delay may
need to be extended substantially
    Sleep 1          'good for 115200 bps
    'Sleep 30         'good for 300 bps
End Sub

Private Sub cmdSend_Click()
Dim I As Integer
    For I = 1 To Len(txtData2Send.Text)
      SendByte CByte(Asc(Mid$(txtData2Send.Text, I, 1)))
    Next I
End Sub

Private Sub Form_Load()
    Addr = &H3F8
End Sub

Private Sub mnuCom_Click(Index As Integer)
Dim I As Integer
    For I = 1 To 4
```

```
            mnuCom(I).Checked = False
        Next I
        mnuCom(Index).Checked = True
        Select Case Index
            Case 1
                Addr = &H3F8
            Case 2
                Addr = &H2F8
            Case 3
                Addr = &H3E8
            Case 4
                Addr = &H2E8
        End Select
End Sub
```

Chapter 11 Debugging Communications Applications

What do you do when things go wrong?

It is human nature to assume that there is a fault with something other than the code in your program or your design logic — this includes both software and hardware. If you are having trouble with a modem, blame the modem. If you are using the MSCOMM custom control, or some other communications add-on, then blame it. As a last resort, blame Windows.

About one time out of ten, this natural assignment of blame is correct. Unfortunately, nine times out of ten we should look at our design or implementation first, something that is hard to do.

So, the best advice that I can offer is to, "Take three steps backwards, and look around." This means that you should get a sound understanding of whatever communications add-on that you are using, study the manual of each device that is part of the communications link, and review the logic of your design in light of these things. I also hope that you review some of the comments that I have made in earlier chapters about the limitations of various approaches.

What I will discuss here are ways to help you out when you've lost patience with the foregoing and want to get down to the nitty-gritty.

11.1   Hardware Assisted

A mandatory tool for any serious serial communications development that uses RS-232 ports is a breakout box. These are available from a number of manufacturers. A list of breakout boxes and manufacturers is in the Resources appendix.

Breakout boxes have two LEDs to indicate the state of selected serial lines (or two-color LEDs). They provide switches to interrupt a standard path and patch wires to re-route any signal path to any other. This allows you to construct almost any serial connection that you might need and to monitor signals for their status.

Of course, some signals change state too quickly for the breakout box LEDs to give a visible indication. However, these compact testers are almost indispensable.

Alternatives to a breakout box are an RS232 Mini-Tester (must have LEDs) and a separate test card that allows signal patching. These can be purchased for $10 or less. Radio Shack sells a handy Mini-Tester (as do others in the Resources appendix).

RS-232 or other external serial interfaces give you the chance to use the most powerful debugging tools that are available. These are variously called Serial Data Analyzers, Protocol Analyzers, Data Scopes, or RS-232 Data Monitors. These range in complexity and flexibility (and cost). They will help you debug the most intractable problems because they do not interfere with the operation of any other part of the communications system. They are completely passive when used for monitoring serial data, so they do not slow data transfer or induce any latency or delays.

Agilent Technologies (Hewlett Packard), for example, makes powerful units called protocol analyzers that can monitor almost any serial link, either synchronous or asynchronous. The analyzers look somewhat like oscilloscopes with keyboards. They provide onscreen display of real-time data and the state of the hardware lines. These units also allow you to simulate protocols, record and store data to disk, and provide for remote coordination and control of the analyzers.

Other manufactures make simpler but still powerful units. Some are small, hand-held units with LCD for display of data. The one that I use is a model DLM200 from Benedict Computer (see the Resources appendix). It has a 32-charcter X 8-line display and comes with software that can be used to display monitored data on the PC. DLM stands for "Data Line Monitor." It connects to the PC using the PC printer port, so it can be used on the same PC that is executing the serial application that is being debugged (some performance caveats apply). Some others use an external PC to display data (not the PC that you are trying to debug). An example of this is the Comscope Protocol Analyzer by Telebyte, also in the Resources appendix.

Still another alternative in this area is mostly a software solution. Well, not completely software, but software that executes on a standard, separate PC. The software executes on a separate computer so I have placed this in the hardware section. The PC provides two serial ports for monitoring data and it provides the data display. A special cable is furnished that connects to the system being monitored. Examples of this are Comscope II Protocol Analyzer from Telebyte, BreakOut-II by Lifeboat Publishing (the one that I use), Frontline System's STAsync+Spy (described elsewhere in this chapter), and a similarly featured product from sysFire called ViewComm. See the appendix for contract information for ViewComm. I have included a simpler, less powerful Visual Basic code SerialMonitor applet in the next section.

Suppose that you do not have the time or resources to use a protocol analyzer. Is there anything that the technicians in the audience can do? Well, I would not have asked the question if I did not have some sort of answer. If you are not too critical, you can build your own serial monitor in Visual Basic. Remember, Windows is not real-time OS, so any display of the hardware lines (CTS, RTS, DSR, CD, DTR, etc.) status will be comparatively crude.

Figure 9.1 is a simple adapter that you can build that allows you to use a terminal program to monitor both transmit and receive data from another serial port. You could build it in an hour or two on a serial test card such as the Patton Electronics Co. Model 3TC. If you were to combine this with a Model 3 data tap to connect your test circuit to the RS-232 that you are monitoring, total cost would be about $35. I have never seen an adapter like this described anywhere else.

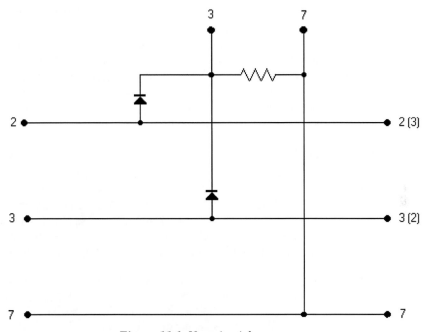

Figure 11.1 Vampire Adapter

The left-most connector connects to the PC serial port that is under test. The right-most connector connects to the other device. If the device you are connecting to is a DCE, such as a modem, there is no crossover. If it is a DTE, such as another computer, then signals are crossed-over. The crossover becomes a null modem connection. Do not crossover if you are using this adapter with a null modem cable or adapter.

The connection shown at the top of the figure connects to the serial port that you are using for your monitor program.

The two diodes isolate transmit and receive data so that both signals can be monitored by a single serial port. The type of diodes is not critical — it might be 1N4148 or 1N4001. This connection is known as a wired-OR. The resistor is optional, although it is a good idea. If used, 20K Ω is a reasonable value.

This simple adapter allows you to monitor, with a single serial port, both transmit and receive data using a simple terminal program like the VBTerm program that was shown in an earlier chapter. If the terminal program runs on the same PC as the program that you are debugging then you must, of course, have two separate RS-232 ports. And, of course, this can only be used to monitor an RS-232 port. It cannot be used with an internal modem.

Of course, there are some problems with this simple adapter. First, if data is transmitted and received simultaneously, the data will collide. This will cause the monitor to display garbage for each transmit character that overlaps a receive character. Often this is not a problem (such as sending commands to modems and viewing responses) but occasionally it can be. Second, transmit and receive data is intermixed with no discrimination. This can be a little (or very) confusing. Third, the terminal program that you use as the monitor will display ASCII text fine but binary data needs a better display method. See the next section on Software Assisted debugging for one way to enhance the display of VBTerm in order to make it more useful for debugging binary serial data.

11.2   Software Assisted

It would be really valuable if you could execute a program that would intercept and display all serial data and also display the status of the serial port handshaking lines. Then you would not need the external hardware devices I discussed in the previous section. Such a tool would be especially valuable when used with an internal modem. Well, that goal is not impossible — it's just hard. And, when it is done, it has some significant negative features.

Under Windows 3.x, it is possible, although not possible from Visual Basic alone, to intercept communications messages. The handle that is used for open serial ports is a constant for each port. Refer to the Windows API chapter for information on this. Under Windows 32, the API assigns the handle to an open serial port from a pool of handles. The 32-bit handle is not normally available outside the application that opens the port.

I know of two Windows 3.x programs that allow you to display, transmit, and receive data.

One is Dean Software's InfoSpy (version 2.61 is available at this writing). InfoSpy is shareware and has some nice features, beyond communications, for spying Windows messages. It runs under Windows 3.x or Windows 95/98. However, the communications spy only works with 16-bit communications applications.

The other is a commercial product from Pacific CommWare called Turbo Commander Pro. This product provides a very useful "serial analyzer" type display. Unfortunately, it works only under Windows 3.1, not Windows 95/98/NT. *Figure 9.2.1* shows TurboCommander Pro in action.

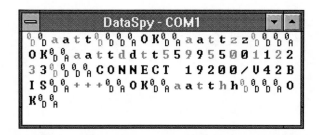

Figure 11.2.1 *Pacific CommWare TurboCommander Pro*

Pacific CommWare sells a new product, TurboCom95, for Windows 95. This provides a monitor that shows the state of hardware handshaking. It allows logging of communications errors. And it displays real-time data throughput. However, as of this writing, it does not have the DataSpy feature of TurboCommander Pro.

There are two products that I have found that incorporate **both** hardware assisted **and** software assisted monitoring. On top of that, the software monitoring features work under Windows 95/98. Also, in this area (see Serial Communications Products/HardAndSoftware on the CD ROM) is my own **VirtalNullModem and DataMonitor**. Check it out.

Frontline Test System's SerialTest ST Async+Spy (SerialTest) product is a well thought out and reasonably priced product.   It has features like those of high priced protocol analyzers and data scopes. Hardware monitoring uses external cables that are furnished with the product. These allow you to use one or two free comports on your PC to monitor serial data.

ST Async+Spy also has a graphical breakout box display that shows the status of all useful hardware handshaking lines (DTR, CD, DSR, CTS, RTS, RI, etc.). The current status is shown as is a "logic analyzer" style display of these lines. This is a powerful tool for analyzing communications problems.

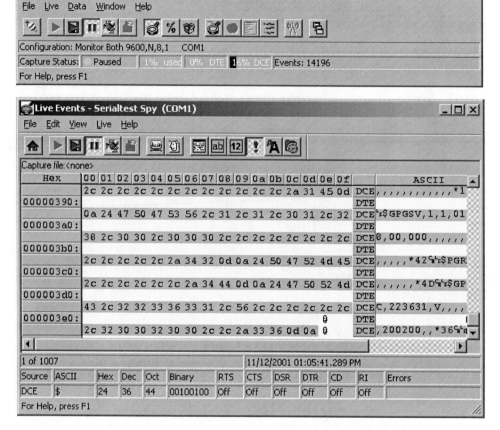

Figure 11.2.2 *Frontline Test System SerialTest ST* Async+Spy

SerialTest also has a Spy Mode with features similar to those that would be available if you used a single external serial port with a cable. However, this mode allows you to "trap" serial traffic internal to the PC so that it is usable with an internal modem. The Com Spy mode replaces the Windows 95/98/NT/2K serial driver with one of its own (similar to Pacific CommWare's TurboCommander for Windows 3.x).

Both hardware and Com Spy modes offer a variety of tools for analyzing and decoding serial data. Data can be logged to a disk file for later analysis (and so that known good data can be compared with new data). SerialTest decodes various protocol stacks such as async PPP, HDLC, ICMP, IP, SLIP, TCP, UDP, VJC, VDU, and X25. It displays decoded data as ASCII, 7-bit ASCII, EBCDIC, and Baudot. Captured data also can be printed.

ST Async+Spy supports serial speeds ranging from 50 bits per second (bps) to 115,200 bps. Parity choices are N, O, E, M and S. Word Length choices are 5, 6, 7, and 8 bits. Stop Bit choices are 1 and 2 bits for word lengths of 6, 7 and 8, and 1 and 1.5 bits for word lengths of 5.

Serial ports with 16650 and 16750 UARTs are not supported. However, 16550 and 8250 UARTS are fully supported. This covers more than 99% of the serial ports that are in use. One caution — ST Async+Spy cannot be used with software drivers that emulate hardware UARTs. Thus, WinModems, USB serial port adapters, etc. are not valid targets for use with the Spy mode of operation. Of course, it will work when connected via cables to external RS-232 devices. So, it can be used with USB serial port adapters when it is used to **physically** monitor them.

Advanced Serial Port Monitor 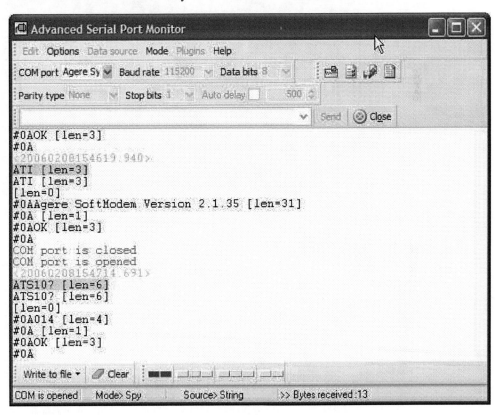 from AGG Software was new when the 4th Edition was first printed. As of 2006, it has over two years of use, and had become invaluable to me. It provides many useful features, and its Spy/Sniffer mode supports USB serial ports. Please study the documentation on the CD ROM that accompanies this book, and give the demo version that is included a try. Here is a screen shot of ASP in action.

Figure 11.2.3 Advanced Serial Port Monitor

All software-based data spies impose some performance penalty. The data sent and received must be intercepted, interpreted, displayed, and then sent on to its destination. All of this extra processing can reduce throughput. So, if you are trying to debug a real-time problem, this extra latency will have to be considered. If you use a spy, make certain that it is removed before you try to measure throughput.

Even the simple monitor applets that follow can impose significant overhead. Serial communications, data manipulation, and display can be slooowwww.

11.2.1   HexTerm

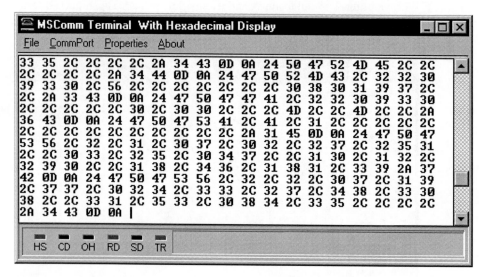

Problem: Binary data is not displayed satisfactorily using a conventional terminal emulation program. What can I do to make debugging easier?

Solution: This modification of the VBTerm applet presented earlier allows you to select a conventional display or to display data in hexadecimal format. Both displays are active, although only one is visible, so the user can toggle back and forth between the two. This facilitates viewing mixed ASCII and binary data. It is not perfect but anything can be helpful.

This applet might be used with the Vampire Adapter in the previous section. If you use the adapter, you can only monitor. Anything that you type will not be displayed. This applet also can be useful as a terminal program for occasional display of binary or ASCII data.

```
Sub HexEnabled_Click ()
    HexEnabled.Checked = Not HexEnabled.Checked
    If HexEnabled.Checked Then
        Term.Text = ""
        Term.Visible = False
        HexDisp.Visible = True
```

```
        Else
            Term.Visible = True
            HexDisp.Text = ""
            HexDisp.Visible = False
        End If
    End Sub
```

The HexEnabled menu selection will enable display of hexadecimal formatted data when checked. The display will be normal text format when HexEnabled is not checked.

```
Sub MSCOMM1_OnComm ()
    Dim EVMsg$
    Dim ERMsg$
    Dim Ret As Integer
    Dim Buffer As String
    Dim Nd As Integer, I As Integer
    Dim Temp As String, Temp1 As String
    If MSCOMM1.CommEvent = MSCOMM_EV_RECEIVE Then
        Buffer = MSCOMM1.Input
        LED2(3) = True              'use Modem Lights
        'If HexEnabled.Checked Then
        If HexDisp.SelStart >= 8192 Then
            HexDisp.Text = Right$(HexDisp.Text, 4095)
        End If
        For I = 1 To Len(Buffer)
            Temp = Hex$(Asc(Mid$(Buffer, I, 1))) & " "
            If Len(Temp) = 2 Then
                Temp = "0" & Temp
            End If
            Temp1 = Temp1 & Temp
        Next I
        HexDisp.SelText = Temp1
        'Else
        Nd = Len(Term.Text)
        If Nd >= 8192 Then
            Term.Text = Right$(Term.Text, 4096)
            Nd = 4096
        End If
        Term.SelStart = Nd
        Do
            I = InStr(Buffer, Chr$(8))
            If I Then
                If I = 1 Then
                    Term.SelStart = Nd - 1
                    Term.SelLength = 1
                    Buffer = Mid$(Buffer, I + 1)
                Else
                    Buffer = Left$(Buffer, I - 2) _
                        + Mid$(Buffer, I + 1)
```

```
                End If
            End If
        Loop While I
        Do
            I = InStr(Buffer, VbLf)
            If I Then
                Buffer = Left$(Buffer, I - 1) _
                    + Mid$(Buffer, I + 1)
            End If
        Loop While I
        I = 1
        Do
            I = InStr(I, Buffer, VbCr)
            If I Then
                Buffer = Left$(Buffer, I) + VbLf _
                    + Mid$(Buffer, I + 1)
                I = I + 1
            End If
        Loop While I
        Term.SelText = Buffer
        'End If
        If hLogFile Then
            I = 2
            Do
                Err = 0
                Put hLogFile, , Buffer
                If Err Then
                    I = MsgBox(Error$, 21)
                    If I = 2 Then
                        MCloseLog_Click
                    End If
                End If
            Loop While I <> 2
        End If
    End If
End Sub
```

The code shown above includes only that which differs from the VBTERM applet discussed earlier. What I have illustrated here is also abbreviated — I have not shown code that is unchanged from the VBTerm OnComm routine. Other forms and modules are the same as in the MSTERM section in the chapter covering MSCOMM. See the code on the CD ROM for a complete listing.

This code is optimized for performance over that in the MSTERM example. Code optimization is discussed later in this chapter. The highlighted MSCOMM1_OnComm subroutine will be discussed there.

11.2.2   SerialMonitor

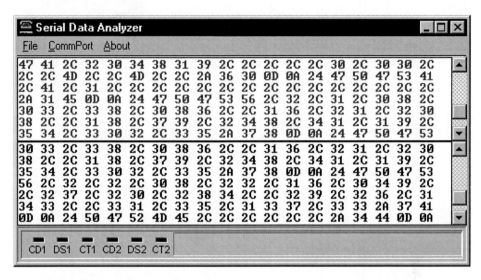

Problem: How about a final applet that displays two channels of serial data and works a little like a serial analyzer.

Solution: Here it is. It is derived from the HexTerm applet. To use it, you will need a serial cable or adapter wired like *Figure 9.2.1*

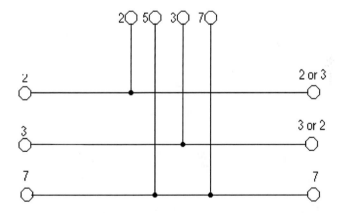

Figure 11.2.1 Passive Monitor Adapter

The left and right connectors connect to the serial devices that are being monitored. The top connectors connect to **two serial ports** on the computer that runs the SerialMonitor applet. Connect pins 2 and 5 to Com1 and pins 3 and 7 to Com2. The pin numbering for Com2 assumes that it uses a 9-pin connector while Com1 uses a 25-pin connector.

You usually will run the SerialMonitor applet on a different computer than the application that you are trying to monitor/debug.

Previous editions of this book provided both VB3 and VB5/6 versions of the SerialAnalyzer program. Those programs still are on the CD ROM. An alternate VB6 version also is provided here. I have added a new feature to SerialMonitor in the 3rd edition. You now can timestamp data at regular intervals.

```
Private ByteCount As Integer
Private StartTime As Long

Private Const TIMESTAMP_COUNT_THRESHOLD = 500

Private Declare Function timeGetTime Lib "winmm.dll" () As
Long
Private Sub MOpen_Click()
    On Error Resume Next
    MOpen.Checked = Not (MOpen.Checked)
    If MSComm1.PortOpen = True Then
        MSComm1.PortOpen = False
        MSComm2.PortOpen = False
    End If
    If MOpen.Checked = True Then
        MSComm1.PortOpen = True
        If Err.Number > 0 Then
            MsgBox Err.Description & " 1st comm port.", _
                                        vbExclamation
            MOpen.Checked = False
            If MSComm2.PortOpen Then MSComm2.PortOpen = _
                                        False
            Exit Sub
        Else
            MSComm2.PortOpen = True
            If Err.Number > 0 Then
                MsgBox Err.Description & _
                    " 2nd comm port.", vbExclamation
                MOpen.Checked = False
                If MSComm1.PortOpen Then _
                                MSComm1.PortOpen = False
            End If
        End If
    End If

End Sub
```

Two serial ports are used in this application. MOpen opens and closes both ports at the same time.

```
Private Sub MSComm1_OnComm()
Dim Nd As Integer
Dim I As Integer
Dim Buffer As String
```

```
Dim BufferArray() As Byte
Dim ASCIIChar As Integer
Dim Temp As String
Dim NowTime As Long
    If MSComm1.CommEvent = comEvReceive Then
        If TimeStampEnabled.Checked Then
            ByteCount = ByteCount + MSComm1.InBufferCount
            If ByteCount >= TIMESTAMP_COUNT_THRESHOLD Then
                NowTime = timeGetTime - StartTime
                Term.SelText = Chr$(169) & _
                            Format$(NowTime / 1000, _
                            "########.000") & Chr$(170)
                Term2.SelText = Chr$(169) & _
                            Format$(NowTime / 1000, _
                            "########.000") & Chr$(170)
                ByteCount = ByteCount - _
                            TIMESTAMP_COUNT_THRESHOLD
            End If
        End If
        If Term.SelStart >= 8192 Then
            Term.Text = Mid$(Term.Text, 4095)
            Nd = Len(Term.Text)
            Term.SelStart = Nd
        End If
        If HexEnabled.Checked Then
            BufferArray = MSComm1.Input
            For I = 0 To UBound(BufferArray)
                Buffer = Buffer & Format$(_
                            BufferArray(I), "00") & " "
            Next I
            Term.SelText = Buffer
        Else
            BufferArray = MSComm1.Input
            For I = 0 To UBound(BufferArray)
                ASCIIChar = BufferArray(I)
                If ASCIIChar < &H20 Or ASCIIChar > _
                                        &H80 Then
                    Temp = Temp & Chr$(174) & _
                            Format$(ASCIIChar) & Chr$(175)
                Else
                    Temp = Temp & Chr$(ASCIIChar)
                End If
            Next I
            Term.SelText = Temp
        End If
    Else
        DisplayCommMessage MSComm1.CommEvent, "1"
    End If
End Sub
```

The subroutine for MSComm2_OnComm replicates that of MSComm1. I will not repeat it here. See the actual source code on the CD ROM.

```
    Private Sub DisplayCommMessage(ByVal CommEvent As _
                                Integer, PortNumber As String)
Dim EVMsg As String
        Select Case CommEvent
            Case comEvCTS
                EVMsg = "Change in CTS Detected Port " _
                                & PortNumber
            Case comEvDSR
                EVMsg = "Change in DSR Detected Port " _
                                & PortNumber
            Case comEvCD
                EVMsg = "Change in CD Detected Port " _
                                & PortNumber
            Case comEvRing
                EVMsg = "Ring Indicate Port " & _
                                PortNumber
            Case comEventBreak
                EVMsg = "Break Received Port " & _
                                PortNumber
            Case comEventFrame
                EVMsg = "Framing Error Port " & _
                                PortNumber
            Case comEventOverrun
                EVMsg = "Overrun Error Port " & _
                                PortNumber
            Case comEventRxOver
                EVMsg = "Receive Buffer Overflow Port " _
                                & PortNumber
            Case comEventRxParity
                EVMsg = "Parity Error Port " & _
                                PortNumber
        End Select
        Status.Caption = EVMsg
    End Sub
```

Error and other status messages are displayed using the DisplayCommMessage routine.

The MSComm32.ocx furnished with VB5/6 allows binary data to be handled differently than previous versions of MSCOMM. An array of Bytes is used to store binary data when the MSCOMM1.InputMode = Binary. Often this will provide faster processing of binary data. More importantly, it eliminates the possibility that a DBCS-enabled (Double Byte Character String) OS will cause incorrect mapping of binary characters to string data.

This code example has one change from that published in the First Edition of this book. I changed the scope of the receive byte array (BufferArray) to local to OnComm instead of using a form-level array. This change removes the requirement to ReDim BufferArray for each OnComm. This should speed processing in the OnComm event.

I have removed the LED status display from the applet . Of course, you can add it back in if you find it to be valuable.

See the code on the CD ROM for a complete listing of this applet.

11.2.3 PortMon

Portmon is a GUI/device driver combination that monitors and displays all serial and parallel port activity on a system. It has advanced filtering and search capabilities that make it a powerful tool for exploring the way NT works, seeing how applications use ports, or tracking down problems in system or application configurations. Portmon works on NT 4.0, Windows 2000 (Win2K), Windows 95 and Windows 98. PortMon was written by Mark Russinovich and is supplied by permission of SysInternals. See Appendix A for contact information.

Remote monitoring: Capture kernel-mode and/or Win32 debug output from any computer accessible via TCP/IP - even across the Internet. You can monitor multiple remote computers simultaneously. Portmon will even install its client software itself if you are running it on a Windows NT/2K system and are capturing from another Windows NT/2K system in the same Network Neighborhood.

Most-recent-filter lists: *Portmon* has been extended with powerful filtering capabilities. It remembers your most recent filter selections and it has an interface that makes it easy to reselect them.

Clipboard copy: Select multiple lines in the output window and copy their contents to the clipboard.

Highlighting: Highlight debug output that matches your highlighting filter. And even customize the highlighting colors.

Log-to-file: Write debug output to a file as its being captured.

Printing: Print all or part of captured debug output to a printer.

#	Time	Process	Request	Port	Result	Other
423	4:44:23 PM	tapisrv.exe	IRP_MJ_WRITE	Serial0	SUCCESS	Length 68: ~...IE..<*... ......P......
424	4:44:23 PM	tapisrv.exe	IOCTL_SERIAL_WAIT_ON_MASK	Serial0	SUCCESS	
425	4:44:23 PM	tapisrv.exe	IRP_MJ_READ	Serial0	SUCCESS	Length 8: ~...IE...
426	4:44:23 PM	tapisrv.exe	IOCTL_SERIAL_WAIT_ON_MASK	Serial0	SUCCESS	
427	4:44:23 PM	tapisrv.exe	IRP_MJ_READ	Serial0	SUCCESS	Length 60: <.o..>..@9..P.......Q\...
428	4:44:23 PM	tapisrv.exe	IRP_MJ_WRITE	Serial0	SUCCESS	Length 72: ~...IE..@+....h.............
429	4:44:23 PM	tapisrv.exe	IOCTL_SERIAL_WAIT_ON_MASK	Serial0	SUCCESS	
430	4:44:23 PM	tapisrv.exe	IRP_MJ_READ	Serial0	SUCCESS	Length 8: ~...IE..
431	4:44:23 PM	tapisrv.exe	IOCTL_SERIAL_WAIT_ON_MASK	Serial0	SUCCESS	
432	4:44:23 PM	tapisrv.exe	IRP_MJ_READ	Serial0	SUCCESS	Length 209: ..p..>..?..P.......5........
433	4:44:23 PM	tapisrv.exe	IRP_MJ_WRITE	Serial0	SUCCESS	Length 68: ~...IE..<...............
434	4:44:24 PM	tapisrv.exe	IRP_MJ_WRITE	Serial0	SUCCESS	Length 68: ~...IE..<..... ....
435	4:44:25 PM	tapisrv.exe	IRP_MJ_WRITE	Serial0	SUCCESS	Length 68: ~...IE..<.... .. ..
436	4:44:26 PM	tapisrv.exe	IRP_MJ_WRITE	Serial0	SUCCESS	Length 71: ~...IE..?/....d........P...
437	4:44:26 PM	tapisrv.exe	IOCTL_SERIAL_WAIT_ON_MASK	Serial0	SUCCESS	
438	4:44:26 PM	tapisrv.exe	IRP_MJ_READ	Serial0	SUCCESS	Length 8: ~...IE..
439	4:44:26 PM	tapisrv.exe	IOCTL_SERIAL_WAIT_ON_MASK	Serial0	SUCCESS	
440	4:44:26 PM	tapisrv.exe	IRP_MJ_READ	Serial0	SUCCESS	Length 469: ..q..>..>..P.......5........
441	4:44:26 PM	tapisrv.exe	IRP_MJ_WRITE	Serial0	SUCCESS	Length 68: ~...IE..<0... Z.............
442	4:44:26 PM	tapisrv.exe	IOCTL_SERIAL_WAIT_ON_MASK	Serial0	SUCCESS	
443	4:44:26 PM	tapisrv.exe	IRP_MJ_READ	Serial0	SUCCESS	Length 8: ~...IE..
444	4:44:26 PM	tapisrv.exe	IOCTL_SERIAL_WAIT_ON_MASK	Serial0	SUCCESS	
445	4:44:26 PM	tapisrv.exe	IRP_MJ_READ	Serial0	SUCCESS	Length 60: <.y...5.......M\...

Use

Portmon understands all serial and parallel port I/O control (IOCTLs) commands and will display them along with interesting information regarding their associated parameters. For read and write requests, Portmon displays the first several dozen bytes of the buffer while using '.' to represent non-printable characters. The Show Hex menu option lets you toggle between ASCII and raw hex output of buffer data.

How it Works: WinNT

The Portmon GUI is responsible for identifying serial and parallel ports. It does so by enumerating the serial ports that are configured under HKEY_LOCAL_MACHINE\Hardware\DeviceMap\SerialComm and the parallel ports defined under HKEY_LOCAL_MACHINE\Hardware\DeviceMap\Parallel Ports. These keys contain the mappings between serial and parallel port device names and the Win32-accessible names.

When you select a port to monitor, Portmon sends a request to its device driver that includes the NT name (e.g. \device\serial0) that you are interested in. The driver uses standard filtering APIs to attach its own filter device object to the target device object. First, it uses **ZwCreateFile** to open the target device. Then it translates the handle it receives back from **ZwCreateFile** to a device object pointer. After creating its own filter device object that matches the characteristics of the target, the driver calls **IoAttachDeviceByPointer** to establish the filter. From that point on, the *Portmon* driver will see all requests aimed at the target device.

Portmon has built-in knowledge of all standard serial and parallel port IOCTLs. They are the primary way that applications and drivers configure and read status information from ports.

How it Works: Win9x

On Windows 95 and 98, the *Portmon* GUI relies on a dynamically loaded VxD to capture serial and parallel activity. The Windows VCOMM (Virtual Communications) device driver serves as the interface to parallel and serial devices, so applications that access ports indirectly use its services. The *Portmon* VxD uses standard VxD service hooking to intercept all accesses to VCOMM's functions.

11.3 Software Breakout Box

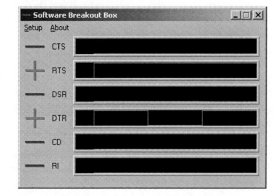

Problem: How can I view the status of the serial control lines for both internal modems and external serial devices?

Solution: Well, here is one way. This application monitors the UART modem status register and modem control register at regular intervals. It displays the results on screen. The current state is shown in the + (True) or – (False) column. A history of changes of state is displayed in "logic analyzer" form. This program, with further documentation, is furnished on the CD ROM.

The version furnished here is a "Demo." That means that it runs for two minutes and then times-out with a message asking that it be upgraded to the "Full" version. The "Full" version has no time limitation. Readers of this book can request the "Full" version from me by email. The source code for this also is available. However, it uses a commercial add-on from National Instruments for the timing-state display (Measurement Studio CWGraph.ocx).

11.4 Code Optimization

This section is a series of "rules-of-thumb," comments, and suggestions to improve communications performance.

Most computer programs can benefit from code optimization. Performance can often be improved by code optimization. However, some optimization involves making code less modular and therefore less maintainable. So, you should always be aware of the tradeoffs involved. I'll try to make comments about when code should be optimized and to point out those tradeoffs.

Serial communication programs are subject to some Visual Basic bottlenecks that can/may affect other type of programs to a lesser degree than they do communications.

For example, string functions (InStr, Left$, Right$, and Mid$) are often needed to extract individual characters from a data stream. These functions are not especially fast. If the program has a serious real-time requirement, these functions should be used with care. The InStr function is **much** faster than using a For/Next loop and the Mid$ function to scan a string for a match. That seems an obvious comment but who knows? It may be worth repeating here.

NOTE: in the preceding paragraph I used the string specific functions for string handling. That brings up variable typing, one of the most important optimization issues. You really must declare the variables that you use. **Do not** rely on the Variant variables that Visual Basic will assign if you do not make an explicit variable declaration.

To achieve optimum performance, use Integer variables for loop counters that match the word size of the target processor. For example, Long (Int32 in VB.NET) variables provide faster access than (short) Integer (Int16) variables. Probably this fact will continue to be true for code executing on future processors that provide 64-bit or wider architectures.

Variant variables require more memory than explicitly declared variables. Using them in functions requires extra time because Visual Basic has to resolve their implicit type at runtime. Visual Basic can also make an error by "deciding" that a Variant variable is a type different than your design might have expected. This can cause obscure, hard to find, errors.

By the same token, you should use explicit string functions (Left$, Right$, and Mid$) and not the Variant versions of the same (Left, Right, and Mid). The reasons are the same as I outlined in the previous paragraph. However, this paragraph does not apply to VB.NET. The $ data-type suffix is no longer used. Variant variables and conversion are no longer an issue.

In a similar vein, you **should not** use control default properties for assignment. For example, the default property of a text box is the Text property. So, these two statements are equivalent:

```
TextBox1 = SomeString$
TextBox1.Text = SomeString$
```

However, the second will execute faster than the first. As a bonus, the second is easier to read, and not much harder to write. What other things might you consider? Well, one thing might be the structure of your code. A normal thing to do is to modularize your code. The extreme of this is to use class modules. Modularization makes maintenance easier and it makes code more reusable. However, there is a cost. Every time a call is executed, a number of stack operations are performed. Any variables that are passed must use the stack, either ByVal or ByRef, and a number of hidden control structures must be accessed. This overhead can become significant in a high-performance serial communications application. Again, this suggestion no longer applies to VB.NET. Default object properties have been removed from the .NET framework.

Examples of "reduced-modularization" are the HexTerm and SerialMonitor applets in the last section. When I first coded them, I modified the ShowData subroutine from the VBTerm example to process and display received data. However, I found that the call/return overhead of this routine causes serious performance problems at higher serial rates. I moved the code that had been in the ShowData routine to the OnComm routine. This change resulted in measurably improved performance.

The HexTerm and SerialMonitor OnComm routines have one critical fragment of code that is comparatively slow. That fragment is:

```
For I = 1 To Len(Buffer)
    Temp = Hex$(Asc(Mid$(Buffer, I, 1))) & " "
    If Len(Temp) = 2 Then
        Temp = "0" & Temp
    End If
    temp1 = temp1 & Temp
Next I
```

The string Buffer has to be converted a character at a time to extract the ASCII value, convert to hex, and to add a leading zero (if needed). In Visual Basic, this is a fairly slow process. A DLL might be written to replace this fragment with a single function. The DLL would speed the process somewhat. Many of us do not have the inclination to write DLLs to speed code. If all else fails, the alternative of implementing slow code in a DLL should be considered. I have included DLLs named RGUTIL16.DLL and RGUTIL32.DLL. Each has a single function called BuffToHex. The BuffToHex function replaces the For/Next loop. Compared to the For/Next loop, the BuffToHex function requires about 10-20% of the time to execute. Refer to the code on the CD ROM for an example of this.

The 32-bit version of SerialMonitor in the previous section uses arrays of type Byte to buffer data. The byte array is required to handle binary data. But, it has the salutary effect of speeding the formatting of data for display. See the highlighted code in OnComm for the code required to handle binary data.

I made one more change that is worth noting to improve performance in the OnComm routine. The Select Case structure executes more slowly than the comparable If/Then/Else structure. Select Case improves readability — again, at the cost of performance.

You may have noticed that I use the Visual Basic "&" operator to append data to a string, not the "+" operator. Not only does this improve readability somewhat but it can also avoid some of the issues that you might encounter using Variant strings. VB.NET introduces some new issues that involve strings. VB.NET strings are "immutable." What this means is that the string cannot be changed. Any code that creates a new string by concatenation (the & operator) actually creates a completely new string and copies the new data while destroying the old string. This can cause a significant performance hit. While VB.NET offers a new String object that provides ways to improve performance, the unfortunate fact is that when strings are used to store data for serial communications (for the most part) we have to continue to use the same syntax that we did for earlier versions. So this is an area for concern.

What process requires the most time when displaying data? Why, it's adding text to the text box. Visual Basic memory management makes this process even more time-consumptive the longer the program runs. As memory becomes fragmented, the process of adding text to a text box slows. This can affect long-term performance of a program so it should be recognized.

While we're discussing text boxes, I need to mention one important fact. There are a couple of ways to add text to a text box. You might use:

```
TextName.Text = TextName.Text & NewText
```

Do not do this. It is very slow!

All of the code in this book uses this method:

```
TextName.SelStart = Len(TextName.Text)
TextName.SelText = NewText
```

This is much faster. Sometimes you can avoid the need to position the insertion point. If the user cannot enter text in the text box manually, you can remove the assignment to SelStart. The equivalent syntax for VB.NET is:

```
TextName.SelectionStart = TextName.Text.Length
TextName.SelectedText = sometext
```

You should use With/End With blocks when you refer to an object multiple times within an area of code. This will improve performance because the object reference needs to be qualified only once within the block. Without the With/End With block, VB will have to re-qualify the object reference for every statement that uses it. Do not jump into or out of a With/End With block.

When discussing performance and code structure, I need to mention one Visual Basic (VB6 and earlier) function that **should not be used**. That function is **IIF**. This function is not contained in the VBRUNxx.DLL (or corresponding VB runtime DLL). Rather it is in MSAFINX.DLL so there is substantial extra overhead when this function is used. Also, the function often must perform more evaluations than the corresponding If/Then/Else structure. I never use the IIF function.

One thing that you may need to do when evaluating the performance of your application is to measure elapsed time. The software facilities under Windows are somewhat limited. A standard Windows timer, such as is used by the Timer function, the Windows API GetTickCount and GetCurrentTime and the Visual Basic Timer control, has a resolution of 55 mS (though this varies with OS – Windows 2K does not have this limitation). This is not sufficient for any other than gross timing. A more satisfactory time function is provided by the multimedia subsystem function, TimeGetTime. TimeGetTime has a resolution of 1 mS. Here is the 16-bit declaration:

```
Declare Function TimeGetTime& Lib "MMSYSTEM.DLL" ()
```

Here is the 32-bit declaration:

```
Declare Function timeGetTime Lib "winmm.dll" Alias _
    "TimeGetTime" () As Long
```

The Long that is returned represents the number of milliseconds since the current Windows session started. It rolls over about every 49 days, so most uses will not have to account for the possibility of rollover. See Chapter 3 for some comments on delay timing.

Use the Debug window to display information at runtime. It is Visual Basic's most useful debug tool. However, realize that the Debug window display is fairly slow so its use can affect performance. You usually should run any performance evaluations using compiled code, especially in VB 5/6 where the compiler optimizations also can influence performance.

The MSCOMM and other communications custom controls also offer opportunities for optimization. Let's go over these next.

The InBufferSize and OutBufferSize properties are used to specify the communications buffers that the communications driver (COMM.DRV or Serial.vxd) maintains for each open serial port. You might think that these sizes do not matter too much. However, they do influence performance.

First, InBufferSize should be sufficiently large enough to allow serial receive data to be buffered while your application is out of context, i.e., when other applications that are being multitasked are executing. You should size this property conservatively. Allow a larger buffer than you feel should be needed. For example, InBufferSize = 8192 buffers about 4.2 seconds of data at 19.2k bps. This would be sufficient for many applications and environments but it might not be enough for some. Certainly, if you use a serial rate of 115k bps then it might be prudent to allocate InBufferSize equal to 30744. I like to use multiples of 1024. This is the largest multiple that can be used in MSCOMM without resorting to the API. It is about 2.6 seconds of buffering.

Second, OutBufferSize should be selected based on the device to which you are connecting. If this is a modem, you will want to size OutBufferSize slightly smaller than the typical modem buffer. This will reduce somewhat the amount of flow control that is invoked by the modem. It also will keep any status window, such as that presented by commercial controls when doing a file transfer, reflecting the actual status of the file transfer more accurately. I use 1024 for OutBufferSize because most modems have a 1.5K buffer.

If you are not using a file transfer protocol but are sending large amounts of data, your code design needs to recognize that the buffer size available for send may be less than that which needs to be sent. Your code may need to take this into consideration. For example,

```
Dim Temp As Integer
Do
    Temp = Comm1.OutBufferSize - Comm1.OutBufferCount
    If Temp >= Len(Buffer$) Then
        Comm1.Output = Buffer$
        Exit Do
    ElseIf Temp > 0 Then
        Comm1.Output = Left$(Buffer$, Temp)
        Buffer$ = Mid$(Buffer$, Temp + 1)
    End If
    DoEvents
Loop
```

This code fragment sends only that part of the Buffer$ that will fit in the transmit buffer without overflow. This sort of routine never hurts; it accommodates any OutBufferSize and data rate. And, it accounts for transmit delay caused by flow control.

The InputLen property specifies the maximum number of characters that will be input when data is input from the control. Unless there is some overriding reason, InputLen should be 0, the default. Then all available data will be read. This reduces the chance of a receive buffer overflow by freeing the entire receive buffer.

On 16-bit versions of MSCOMM, the Interval property should be set to 55. This means that OnComm receive events can be generated every 55 milliseconds. The default Interval of 1000 (1 second) is much too long and can result in data loss at higher speeds.

The NullDiscard property usually should be left at the default, False. Set this property to True only if you have to overcome a hardware or system problem where nulls are received but are not data. Some systems send a null character to "mark time," e.g., to keep a link open but they have no other meaning.

RThreshold can be a very important property. If you use OnComm receive processing, this specifies how many characters must be received before an OnComm receive event is generated. A common mistake is to set this property to a number greater than 1, assuming that it will reduce the number of OnComm receive events and thereby reduce the load caused by processing receive data. However, RThreshold = 1 only assures that a receive event will be generated when data is received. It does not specify how often such events are generated. In fact, (with Interval = 55), at 9600 bps an OnComm receive event will be generated for every 53 receive characters. This calculation assumes continuous data. But, setting RThreshold to 1 assures that a receive event will be generated as needed if data is not back-to-back. Under certain circumstances, it is possible to miss data if RThreshold is greater than 1. Recommendation: set RThreshold = 1 to enable OnComm receive.

Unless there is an overriding reason, the CDTimeout, CTSTimeout, and DSRTimeout properties should not be used. Leave them at the default 0 to eliminate spurious OnComm events.

Enable flow control when your system supports it. The most effective flow control is hardware (RTS/CTS). Remember to enable the same flow control in your modem that you enable in your program. Also, remember that you cannot use software flow control (Xon/Xoff) if you are using most error-corrected file transfer protocols.

MSCOMM sometimes generates an erroneous parity error OnComm event when parity detection is disabled. The Settings property is used to enable parity detection. The ParityReplace property can be set to "" to disable parity detection. However, occasionally, when data is already present and the port is first opened, an MSCOMM_ER_RXPARITY event will be generated even though no parity checking is enabled   often add code to the OnComm event to double-check the Settings or ParityReplace properties so that a misleading error is not displayed.

VB5/6 offers the function StrConv to convert to and from strings **and** arrays of Type Byte. These functions are **much** faster than looping through a string and/or array to process one character at a time. Even faster are string and array functions in a function library like the commercial product, MicroDexterity's Stamina. See Appendix A for contact information.

Visual Studio .NET (2002, 2003 and 2005) changes performance issues substantially. One of the indictments made by critics of .NET is that it is bloated, and that (all things being equal) performance is poorer than for earlier versions of Visual Studio. To some extent, this is true, and applications that rely on high-speed serial communications may be seen to suffer as a result. Also, there are new features that are make .NET designs much more powerful – these features, where threading is a good example, increase the opportunity to cause trouble if misapplied.

However, proper design will mitigate the problems using .NET, and the new features that are available make it a powerful alternative for the future. Here are some quick suggestions:

- In .NET Strings are easily misused. Strings are immutable objects. That is, any manipulation of the string (concatenation, removal or the copying of a substring, for example) causes a new string to be created, the operation then is performed, and the old string is destroyed. These operations are expensive, and reduced performance can be the result.

  Perform operations on strings with caution. Look out for opportunities to avoid or to postpone such manipulation.

  Use the StringBuilder object when possible. It speeds character concatenation substantially. Realize that StringBuilder is not a panacea. It still may be necessary to search for substrings, or to add or remove characters contained within it. Any such operation may result in the creation of an actual String – thus, such operations may incur the performance costs that would have been seen if a String had been used initially.

- Threading (free threads and asynchronous delegates) offer much. Proper use may improve both perceived performance and actual performance. However, when threads are used there is a cost. Creation, management, and destruction of threads and associated resources is a non-trivial process. Use threads needed, and avoid their use unless the design actually benefits.

  The requirements of cross-thread communication must be understood.

  A free-thread (worker thread) cannot directly manipulate elements in the User Interface. The UI is STAThread (Single Apartment Threaded). If a free thread attempts to access the UI directly, the program may fail immediately, and certainly will fail eventually. If the free-thread raises an event that will result in an update to the UI, a common "event" in serial communications applications, the event code **cannot** do something as simple as displaying data in a textbox or changing a progress bar. Code in the event subroutine executes in the same thread context as the free thread that generated it. To avoid cross-thread conflict, code in the event subroutine must call a delegate routine to perform the actual changes to elements of the UI. This delegate routine (called with either BeginInvoke or Invoke symantics) transfers execution of the UI update from the free thread to that of the STAThread. Examples of these techniques are in Chapters 5 and 9.

  Threads may share objects or data. Sharing between threads must be controlled to avoid the possibility of conflict. One thread might attempt to change the content of a shared object while another thread is using that object. This can lead to hard to find bugs. Sharing of data and objects must use any one of the several threading constructs that .NET provides for the purpose. Study SyncLock, Monitor, and the other synchronization mechanisims before implementing your own threaded architecture.

- Use the `System.Text.Encoding.ASCII.GetString` function to convert from an array of type Byte to an ASCII string. Note, this conversion will fail if the array contains values higher than 127 (&H7F).

Lastly, let's talk about system optimization.

It's an old adage but it is still true. You get what you pay for. Faster hardware is better. All other things being equal, a 4 GHz Pentium4 will be much better and faster (but not 40X as fast) than a 100 MHz Pentium. Fortunately, Moore's law has allowed us to improve processor performance at a steady rate and reduced cost for many years. When in doubt – get a new computer!

More system memory is better. For example, if you have a Windows 95 system with 8 MB of memory and you boost the memory to 16 MB, in some cases you can double the throughput of a communications application. The reason is that Windows may have to page data and code to disk on the 8 MB system whereas this slow form of memory management may not be needed on the 16 MB system. If you can increase the size of the RAM cache memory, do so. Some systems come with a 64K cache that can be increased to 256K. This can improve performance substantially. The same argument can be made for Windows 98 and all versions of Windows NT, except that the minimum starting point is higher. Windows 2K/XP demands at least 128 MB. When I wrote the 3rd Edition of this book I used 256 MB on my Windows 2K system (a Dell 800 MHz Pentium 3). Most of my work for the 4th Edition has been done under Windows XP using a Dell 3 GHz system equipped with 512MB of RAM, though more recently I have upgraded to 1 GB of RAM. This illustrates that hardware performance does not stand still. None-the-less, we may have to support legacy systems that are less capable.

If your application updates the display frequently or uses significant amounts of graphics then a high-performance graphics accelerator may help.

Use the fastest hard drive that you can. Keep sufficient space on the drive free so that Windows can do any paging to disk that it may need to do. More disk cache is not always better, so you may have to experiment to optimize the size of the cache. Frequently defragment the hard drive.

Do not forget that your application may not be the only one that is being executed, so these suggestions will apply if anything else is running on the computer. For example, you may need to reduce the priority of background printing to assure that your application gets as much time as possible. This is especially true under Windows 3.1x. If you multi-task DOS applications, it is possible to almost kill the performance of Windows applications.

11.4   Telephone Line Simulators

If you are using modems and have to develop a serious modem application, you will have to test it many, many times. This can involve hundreds or thousands of telephone calls. Many of us need to use our telephones for real work so tying them up testing modems can be a drag. If your telephone usage is metered then modem testing also can be very expensive. That's where telephone line simulators come in.

A telephone line simulator acts like a simple telephone central office. It provides a two-wire talk path (signal path), so called "talk battery" or loop current (this is used by the telephone system to detect when a line goes offhook), and a ring generator to signal another telephone set (or modem) that is connected to the simulator.

Simple telephone line simulators, also called ring-down generators, detect when a line goes offhook. As soon as a line goes offhook, the other line is rung. When that line is answered (it also goes off hook), ringing is terminated and the two lines are electrically connected. See the Resources appendix for some sources.

More complex telephone line simulators have these same functions but do much more. They furnish dial tone to a line that is offhook but not terminated. They accept dial signals, usually both tone and pulse. They furnish call progress tones such as BUSY when the dialed line is already offhook and ringback to the calling line when ringing is sent to the other line. This type of simulator is very useful because it allows you to test re-dial on busy and other exceptions to normal dialing. It is commonly available in two and four line systems. I use one from Teltone Corporation, a model TLS-5 four-line simulator. I have listed several vendors of these intermediate-complexity simulators in the Resources appendix.

Even more sophisticated (also, read: expensive) telephone line simulators are available. These form complete telephone system simulators. In addition to normal dialing and progress tones functions, they allow you to simulate telephone line impairments such as noise, variable receive levels, frequency and delay distortion. My favorite line simulator in this genre is the Teltone TLS-5D. It is comparatively inexpensive and has most of the features that might be needed. This sort of simulator is invaluable for anyone who designs modems but is also useful for anyone who develops critical applications that rely on modem communications. See the Resources appendix for other options.

Also available are other options such as other call-progress tones, programmable telephone numbers, call-waiting signals, and Caller ID.

There are ISDN simulators that provide all ISDN housekeeping data, 2 64k bps data channels, and so forth. These can be a great benefit if developing ISDN applications.

11.5   Communications Hardware Debugging

Many serial communications problems result from two devices attempting to use the same IRQ, e.g., if you install a serial mouse on Com1 and a modem on Com3. Unless your PC supports IRQ sharing (the ability to have multiple ports using the same IRQ), this could cause your mouse or modem to lose functionality. MCA and EISA machines support IRQ sharing. In addition, some serial I/O cards support IRQ sharing even if your machine is not specifically configured to do so. Windows 3.1 fully supports such serial I/O cards.

If you have serial devices on both COM1 and COM3, or COM2 and COM4, and your machine architecture does not support IRQ sharing then you need to reassign the IRQ for COM3 or COM4.

To properly reassign the IRQ:

1.   Determine the valid IRQ settings for the serial device by referring to the manufacturer's documentation. Most serial devices such as I/O cards or internal modems can use either of several valid IRQs. For example, a modem on COM3 may be able to use IRQ5 if IRQ4 is not available.

2.   Determine which IRQs are available on your PC.

     You can use the Microsoft Diagnostic (MSD) utility. MSD is included with Windows version 3.1 and Windows 95/98. You generally can use the Control Panel / System / Hardware / Device Manager / Ports / Properties / (port in question) / Properties / Resources tab to determine the IRQ's under Windows 2K/XP.

3.   Point both your serial device and Windows to the IRQ that you have determined is valid and available. Set your serial device to the alternate IRQ by following the directions in the manufacturer's documentation. Point Windows to the alternate IRQ by using Control Panel to place an entry in your SYSTEM.INI file.

NOTE: Under Windows 3.x, your serial device setting must match the COMxIRQ setting in your SYSTEM.INI file. If these two settings do not match, the COM port will not function properly.

To use MSD to Determine Available IRQs, first quit Windows. Next, at the MS-DOS command prompt, type "MSD" (without the quotation marks) and hit enter. An initial screen appears that provides information about MSD. Choose the OK button to view the MSD main menu. Next, at the MSD main menu screen, choose the IRQ Status button. The IRQ Status screen shows which IRQs are currently assigned (not available). Note each IRQ number where the entry in the corresponding Detected column reads "No". If "No" is shown in the Detected column then the corresponding IRQ is available. If you do not find a "No" entry in the Detected column then there are no available IRQs and you should not attempt to reassign IRQs. Last, use Control Panel to assign one of the available IRQs to the COM port you want to change. You must also change the hardware setting to the same IRQ. Follow the manufacturer's instructions.

Do you have an internal modem and a serial I/O card? You might see one of the following symptoms:

•   The modem dials and rings but does not connect.

•   The system reboots when the modem attempts to dial out.

•   The system reboots when the modem should have connected.

•   The cursor does not work in Window Terminal.

If your internal modem uses a COM port that is addressed by your serial I/O card, you can experience IRQ conflicts. To correct this problem, disable the COM port setting on the serial I/O card. For example, if your internal modem is set to use COM2 and you have a serial I/O card that recognizes COM2, you may need to disable COM2 on the serial I/O card for the modem to work correctly.

COM4 PROBLEMS

If your communications software does not recognize a serial device on COM port 3 or 4, you may have a system BIOS that was manufactured before those COM ports became standard. You should upgrade your BIOS or, better choice, upgrade your motherboard.

The COM4 default address of 02E8 conflicts with some peripheral devices, including the 8514/A, Ultra (ATI), and S3 (Orchid Fahrenheit 1280 STB WIND/X, Diamond Stealth VRAM) display adapters, and certain network adapters. Do not re-address COM4 in this situation. Contact your hardware manufacturer for information about how to reset the other adapter's default address.

To identify an IRQ conflict, first remove any serial hardware you have installed on your system (mouse, network card, fax board, modem, and so forth) and then restart Windows. If Windows now runs in 386 enhanced-mode with no problems, the problem may be caused by an IRQ conflict. Reinstall each separate piece of serial hardware, one at a time, restarting Windows between each addition. This isolates the hardware that is causing the IRQ conflict. When the problem recurs, you can assume it is caused by the hardware you have just reinstalled. Reassign the IRQ for the piece of hardware in question.

Windows 95/98 provides substantial built-in IRQ debugging facilities. For this purpose, use Settings/Control Panel/System. Select the Device Manager tab and choose the serial port in question from the tree display. Conflicts will be listed if any are detected.

Suppose that you are experiencing data overrun errors and code optimization does not help. What more might you look for?

There are factors that can affect interrupt latency in the enhanced-mode of Windows and, to a lesser extent, Windows 95/98/NT. Overrun errors can occur more often if more than one virtual machine (VM) is running, i.e., when standard MS-DOS applications are running under Windows.

The source of this problem is the enhanced-mode Windows virtual machine architecture. There is a certain amount of overhead associated with virtualizing interrupts and device ports. However, the largest problem in this area is interrupt latency caused by transitions between virtual machines, i.e., between DOS tasks.

Enhanced-mode Windows performs preemptive multitasking between virtual machines. Several times each second, Windows performs a "task switch" from one VM to another. If COMMAND.COM or an MS-DOS application is running under Windows then two VMs are present in the system, one for the MS-DOS application and one for Windows itself and its applications. If a serial port interrupt occurs while the MS-DOS VM is active, the interrupt cannot be processed because the COMM driver cannot be called until the Windows VM becomes active again. Normally this will be handled quickly enough but there are times when the switch does not occur fast enough.

The following three items are factors that affect interrupt latency and task switches in enhanced mode:

- MS-DOS spends much of its time in a state that prevents Windows from performing a task switch. Almost every MS-DOS call places MS-DOS into this state. Therefore, task switches cannot occur during file I/O, directory manipulation, screen I/O, getting or setting the system time, and so on. Even running COMMAND.COM in a window incurs MS-DOS calls to blink the cursor.

- Large amounts of file I/O in an MS-DOS application can cause an application to spend a great deal of time inside MS-DOS. Floppy disk file I/O has the greatest impact. If too much time is spent in MS-DOS, a task switch will not occur soon enough for the serial port interrupt to be processed correctly.

- Interrupts cannot be processed when interrupts are disabled. There are various times in both Windows and MS-DOS when interrupts are disabled. Also, Windows provides expanded memory (EMS) emulation for banking Windows applications into and out of conventional memory. Interrupts are disabled during the EMS task switch. These times are generally very short but when they occur in conjunction with task switch latency, they can combine to cause the problem.

- Higher-priority interrupts are processed before lower-priority interrupts. This often is not a problem; however, difficulties have been seen occasionally when the system uses a serial mouse. The mouse has a higher interrupt priority than the serial port. If the mouse is using one of the serial ports, communication with the mouse is relatively slow. Therefore, if the mouse is very active, mouse processing takes a high priority and a relatively long time to complete. This can cause the system to miss a serial port interrupt. In all of these cases, a faster machine performs better because the time spent in MS-DOS or with interrupts disabled is less and, therefore, more time is available to process the serial interrupts.

- Eliminating unnecessary TSRs (Terminate and Stay Resident programs) and MS-DOS drivers may help; decreasing the number of applications and VMs running simultaneously can only help matters.

Windows 2K/XP (and later) should remove the concerns about MS-DOS. Most 16-bit code has been removed from these 32-bit OS's. This should make these systems much more stable and should provide better performance under all normal conditions.

11.6   "Plug and Play" Problems

When you start Windows NT (and other OS derived from NT, such as Windows XP), NTDETECT searches for the pointing device (usually a mouse). In the course of this process, data is sent to the serial (COM) ports. If a serial mouse **apparently** is detected, Windows NT disables the port so a device driver for the mouse can load instead. If a port is not detected, Windows disables the port. A disabled COM port does not display any information in Control Panel Device Manager.

Common serial devices that will cause a problem are GPS receivers and industrial scales. However, any device that has the potential to output data without external intervention may cause this problem. When the PC boots, these devices may cause a mouse driver to be installed, thus disabling use of that port for conventional serial IO.

The solution is not simple. Easiest to suggest, though hardest to implement in practice, is to disconnect or power-down the device until after Windows has started. A more satisfactory solution is to disable serial mouse detection on startup.

How to Disable Detection of Devices on Serial Ports

The information in this article applies to all NT based Windows OS.

- Make a backup copy of the Boot.ini file.

- Remove the **hidden, system, and read-only** attributes from the Boot.ini file.

- Using a text editor (such as Notepad) open the Boot.ini file.

- Add the /NoSerialMice option to the end of each entry in the [operating systems] section of Boot.ini. See the example below for more information.

- Save Boot.ini and quit Notepad.

- Restore the hidden, system, and read-only attributes to the Boot.ini file.

- Shutdown and restart Windows.

- The following is a sample of the Boot.ini file (Windows XP):

```
[boot loader]
timeout=30
default=multi(0)disk(0)rdisk(0)partition(2)\WINDOWS
[operating systems]
multi(0)disk(0)rdisk(0)partition(2)\WINDOWS="Microsoft
Windows XP Professional" /fastdetect /NoSerialMice
```

NoSerialMice Syntax:

/NoSerialMice   - Disables the detection of serial mice on all COM ports.

/NoSerialMice:COMx   - Disables the detection of serial mice on COM x, where x is the number of the port.

/NoSerialMice:COMx,y,z - Disables the detection of serial mice on COM x, y and z.

NOTE: The /NoSerialMice option is not case sensitive.

11.7   ACPI Problems

ACPI (Advanced Configuration and Power Interface) is an open industry specification co-developed by **Hewlett-Packard**, **Intel**, **Microsoft**, **Phoenix**, and **Toshiba**.

ACPI establishes industry-standard interfaces for OS-directed configuration and power management on laptops, desktops, and servers.

ACPI evolves the existing collection of power management BIOS code; Advanced Power Management (APM) application programming interfaces (APIs, PNPBIOS APIs, and Multiprocessor Specification (MPS) tables into a power management and configuration interface specification.

The specification enables new power management technology to evolve independently in operating systems and hardware while ensuring that they continue to work together.

ACPI (Advanced Configuration and Power Interface) signals hardware to switch on, off, or go into sleep mode.

OK, so what can go wrong? Occasionally, power saving modes may cause erratic operation of a serial port. When a computer enters hibernation, the result may be a reduction of the maximum serial port speed to 9600 bps which may result in data loss. If this happens, then it may make sense to disable ACPI for that serial port. This requires editing the Windows registry, so it is not a step to be taken lightly. A better avenue may be to use the computer BIOS setup program to disable all serial port power saving modes. Still, you may find that you need to take these steps.

How to disable a serial port so that it will not be controlled by ACPI:

- Start RegEdit.exe. NOTE: Create a backup of the registry first. Incorrect changes may crash your system and/or force a complete reinstall of the operating system.

- Search for the registry-key: HKEY_LOCAL_MACHINE\Enum\ACPI*PNP0501. The sub keys found should be like 00000001 and 00000002. By default 00000001 should be COM1 and 00000002 COM2 but not always. Open a sub key and look for the keys "PORTNAME", "FRIENDLYNAME" and "DeviceDesc". If your sure you have the comport you need, look at the key "ConfigFlags", set at "00 00 00 00".

- Change this setting into "02 00 00 00".

- Close RegEdit and reboot your computer.

This process removes the serial port from the hardware list and ACPI no longer handles it.

11.8   USB Serial Port Problems

USB serial port adapters have become ubiquitous. They are used by many GPS receiver manufacturers, and are the popular way to provide serial ports for notebook and tablet PCs and to expand the port capacity of desktop machines.

However, there is a dark side to their use. These adapters use a software device driver to emulate the hardware functionality normally provided by the UART in a conventional serial port. As the cost of the USB adapters has declined, so has the quality associated with the software driver that is furnished with the adapter.

One common problem that is seen is that an application (typically the one that we are developing) opens the UAB adapter serial port. Subsequently, the application has the need to close then to re-open that port. Not infrequently the re-open attempt fails. The USB serial port device driver notifies Windows that the port still is open, and cannot be re-opened. When this happens, there is little that the application can do to restore normal access to the port. Often, it is necessary to remove and to reinsert the USB adapter, and occasionally it is necessary to do a full reboot of the Windows operating system.

The fault here lies with the device manufacturer. However, because the cost of the hardware is so low, there is little incentive for the manufacturer to provide a more robust device driver.

The only really satisfactory solution is to use USB adapters from manufacturers who stand behind their products. Unfortunately, this means that the hardware cost increases with this increase in reliability.

11.9   Serial Port Emulation for Pocket PCs and Windows CE devices

Compact device development can be simplified by using the device emulators that are furnished as part of the SDKs (Software Development Kits) that are provided by Microsoft and other vendors. These emulators allow you to deploy and debug code in a purely software environment that executes on the desktop development computer. Thus, it may be possible to do initial design, development and debugging before deploying to a physical Pocket PC or Windows CE hardware.

Unfortunatly, the emulators provided for Pocket PC 2000, Pocket PC 2002, and Pocket PC 2003, along with the comparable emulators for the same generation Windows CE devices did not provide proper serial port emulation. It was possible to map a desktop PC serial port to the emulator, and **at times** this mapping would allow the PC serial port to be accessed from software executing under the emulator. However, often this fails. These device emulators actually execute x86 code that is compiled for their use. This x86 come may exhibit slightly different behavior than the SH3, MIPS, Intel and ARM processors that are used by various PocketPC or Windows CE devices.

Another issue that should be addressed is that software device emulators do not provide an accurate simulation of the actual device performance. Often applications that use serial communications have some performance requirments that have to be met. These applications will have to be tuned for real-life operation using actual hardware. Thus, I recommend that applications be deployed to actual hardware for debugging and testing early in the development cycle. This will make the overall development cycle much more satisfactory.

Windows Mobile 5 device emulation is **much** more roubust than that for earlier devices. The device emulators execute the same code that the target device uses (such as XScale processors that execute ARM code). Instead of having two code bases, one for debug on the emulator, and a second for debug on the actual device, exactly the same code are used. **Note:** my earlier comments about device performance still apply. Windows Mobile 5 device emulation does not provide performance comparable to that of the target device, and, of course, emulation performance is largely governed by the speed and other characteristics of the host PC and operating system.

Appendix A   Resources

Listed here are good sources of Visual Basic programming and technical books, serial communications software vendors, software utility vendors, multiple modem boards and serial port hardware vendors, catalogs, and technical magazines.

I will not list most common modem vendors. There are too many; technical specifications and availability information is widespread.

Books

Here is a short list of books that I have found to be useful. Please do not assume that every valuable book is listed. I have only listed the ones that I have actually read. You can check the Books link on my homepage (see the Introduction for the URL) for an up-to-date list.

USB Complete: Everything You Need to Develop Custom USB Peripherals, Third Edition, ISBN 1931448027 by Jan Axelson, 2005. Here are the programming secrets for the Universal Serial Bus. USB provides a single, easy-to-use interface for multiple peripherals. But this ease of use has come at the cost of greater complexity for the developers who design USB peripherals and write the program code that communicates with them. This edition includes programming using Visual Studio .NET.

Jan shows how to write applications that communicate with USB devices using Visual Basic (or any programming language that can call Win32 API functions). There may be no need to write low-level device drivers; you can use the drivers included with Windows 98/2000. This book has original examples, hands-on guidance, and practical tips. *USB Complete* is an independent guide that doesn't promote one manufacturer's products. It will help you decide whether or not USB is the right choice for a peripheral. You'll find tips on selecting the controller chip that matches your design's needs and enables you to get your peripheral up and running. An example project shows how to develop a custom USB peripheral from start to finish, including device firmware and application programming.

Visual Basic Programmer's Guide to the Win32 API, by Daniel Appleman, Sams, ISBN: 0672315904. This book concentrates on the 32-bit API. It includes a CD-ROM with example programs, custom controls, and other support material. Web page: www.desaware.com. Email: support@desaware.com. Desaware Inc, 1100 E. Hamilton Ave. Suite 4, Campbell, CA 95008, (408) 377-4770, fax: (408) 371-3530

Developing ActiveX Components with Visual Basic 6.0, by Daniel Appleman, Sams. ISBN: 1-56276-510-8. A very good reference for learning to design ActiveX EXEs, DLLs. and custom controls using Visual Basic.

Serial Port Complete, by Jan Axelson, Lakeview Research, ISBN: 096508192-3. Programming and circuits for RS-232 and RS-485 links and networks. Includes sample applications for Visual Basic, Basic Stamp, and 8052 BASIC. Includes a disk.

Parallel Port Complete, by Jan Axelson, Lakeview Research, ISBN: 096508191-5. Programming, interfacing, and using the PC's parallel printer port. Includes EPP, ECP, and IEEE-1284. Source code in Visual Basic. Includes a disk.

Automating Science and Engineering Laboratories with Visual Basic, by Mark Russo and Martin Echols, John Wylie and Sons, ISBN: 0471254932. It is written by two Bristol-Myers-Squibb engineers who are active in lab automation and have presented courses on this topic at national and international meetings. This book is divided into 26 chapters with 3 appendices. Its 345 pages allow the topic to be covered in depth and to make an important contribution to lab automation literature. It is an excellent resource applying Visual Basic to laboratory operations.

Hardcore Visual Basic, Version 5.0, by Bruce McKinney, Microsoft Press, ISBN: 1-57231-422-2. Push the envelope of Visual Basic 5.0. Covers AddressOf, Implements, Enum, and more. Includes a CD ROM.

Advanced Peripherals, Data Communications - Local Area Networks - UARTS, National Semiconductor, 2900 Semiconductor Drive, PO Box 58090, Santa Clara, CA 95052, 408-721-5000. This data book describes the ins and outs of UARTS.

Magazines

There are dozens of magazines that you may find to be of value. I have listed those that I read regularly, though I have not included all. There may be a few here that you have not encountered before. Several of these are free to "qualified" subscribers (marked with an **) so do not be afraid that you will be spending too much money.

"**Visual Studio Magazine**", Fawcette Technical Publications. For subscriptions call 303-661-1816, 800-848-5523 or mail Visual Studio Magazine, PO Box 58872, Boulder CO. You can contact them on the Internet at www.visualstudio.com.

"Scientific Computing and Automation**", Gordon Publications, 301Gibralter Drive, Box 650, Morris Plains, NJ 07950, 201-292-5100.

"Test & Measurement World**", Chaners Publishing CO., 8773 S. Ridgeline Blvd., Highlands Ranch, CO 80126, 617-558-4671, Email tmw@cahners.com

"**Nuts & Volts Magazine**", T & L Publications Inc, 430 Princeland Court, Corona, CA 91719, 909-371-8497 or 800-783-4624, Internet at http://www.nutsvolts.com.

"**Circuit Cellar Ink**", Circuit Cellar Inc, PO Box 698, Holmes, PA 19043, 860-875-2751, Internet at http://www.circellar.com.

"**I&CS, Instrumentation & Control Systems**", Chilton Company, I&CS, Box 2025, Radnor, PA 19089, 610-964-4405, Internet at http://www.chiltonco.com/ics.

**"Telecommunications", Horizon House Publications Inc, 685 Canton Street, Norwood, MA 02062, 215-788-4402, Internet at http://www.telecoms-mag.com/tcs.html.

**"GPS World", Advanstar Communications, Subscriptions: PO Box 6148, Duluth, MN 55806, 218-723-9477 or 800-346-0085, Internet at http://www.gpsworld.com.

Catalogs

I have listed a number of catalogs here. Some may be familiar and others may be less so. I have listed pertinent items found in each catalog. Some of these catalogs list only products from one company. However, most offer a variety of manufacturers (some OEM named).

"Vbxtras", Xtras Inc, 1905 Powers Ferry Road, Suite 100, Atlanta, GA 30339, 770-952-6356 or 800-987-7280, Internet at http://www.xtras.com. This catalog probably lists the largest number of Visual Basic tools and add-ons of any publication (that I have seen). A user of the product writes each product description, so you get a minimum of marketing hype..

"Components Paradise", Programmer's Paradise, 1163 Shrewsbury Ave., Shrewsbury, NJ 07702, 800-445-7899, Internet at http://www.pparadise.com. A variety of Visual Basic tools and add-ons are listed.

"DatacomDirect Catalog", PE Patton Electronics Co, 7622 Rickenbacker Drive, Gaithersburg, MD 20879, 301-975-1000. This catalog lists short-range modems, fiber optic modems, RS-232, RS-422, and RS-485 devices, test adapters, breakout boxes, switches, and other datacom devices.

"Datacom/Networking Cookbook", TELEBYTE Technology Inc, 270 Pulaski Road, Greenlawn, NY 11740, 516-426-3232. This catalog lists short-range modems, fiber optic modems, RS-232, RS-422, and RS-485 devices, test adapters, breakout boxes, switches, and other datacom devices. This catalog includes several serial data analyzers, both hardware- and software-based.

"B&B Electronics Manufacturing Company", 707 Dayton Road, PO Box 1040, Ottawa, IL 61350, 815-433-5100, at www.bb-elec.com. This catalog lists short-range modems, fiber optic modems, RS-232, RS-422, and RS-485 devices, USB adapters, test adapters, breakout boxes, switches, and other datacom devices.

"Jensen Tools", Jensen Tools Inc, 7815 S 46th Street, Phoenix, AZ 85044, 602-968-6231 or 800-426-1194. This catalog lists a variety of tools, including telecom test tools and telephone line simulators.

"Provantage", Provantage Corporation, 7249 Whipple Avenue NW, North Canton, OH 44720, 800-336-1166, Internet at http://www.provantage.com. This catalog lists a variety of programming tools. More importantly, they sell BreakOut-II, an inexpensive software Serial Line Analyzer.

"**Global Computer Supplies**", 2318 East Del Amo Blvd., Dept. GF, Compton, CA 90220, 310-603-2266 or 800-845-6625. Lists a variety of computer products including modems and other data communications devices.

"**MISCO**", One MISCO Plaza, Holmdel, NJ 07733, 800-876-4726. Lists a variety of computer products including modems and other data communications devices.

"**BLACK BOX Catalog**", Black Box Corporation, 1000 Park Drive, Lawrence, PA 15055, 412-746-5500 or 800-552-6816. This catalog lists a very large number of conventional dial modems, ISDN, short-range modems, fiber optic modems, RS-232, RS-422, and RS-485 devices, test adapters, breakout boxes, switches, and other datacom devices.

"**Hello Direct**", Hello Direct, 5893 Rue Ferrari Drive, San Jose, CA 95138, 800-444-3556. Lots of telephone accessories including some to share telephone, fax, and data modems, digital PBX interfaces for data modems, and Caller ID.

"**Data Comm Warehouse**", 1720 Oak Street, PO Box 301, Lakewood, NJ 08701, 800-328-2261, Internet at http://www.warehouse.com. This catalog lists a variety of modems and other data communications devices.

"**Telecom Design Solutions**", Teltone Corporation, 22121 20th Avenue SE, Bothell, WA 98021, 800-427-7862. This catalog lists several telephone line and ISDN simulators. Teltone telephone line simulators are widely used.

"**C&S Sales**", C&S Sales Inc, 150 W Carpenter Ave., Wheeling, IL 60090, 708-541-0710 or 800-292-7711. This catalog has a variety of electronic test equipment, including instruments that have RS-232 serial interfaces.

"**EXTECH Instruments**", 335 Bear Hill Road, Waltham, MA 02154, 617-890-7440, Internet at www.extech.com. This catalog has a variety of electronic test equipment, including instruments that have RS-232 serial interfaces.

"**Tucker Electronics**", PO Box 551419, Dallas, TX 75355, 800-527-4642, Internet at http://www.tucker.com. This catalog has a variety of electronic test equipment, including instruments that have RS-232 serial interfaces.

"**Davis Instruments**", 4701 Mount Hope Drive, Baltimore, MD 21215, 800-368-2516. This catalog has a variety of electronic instruments for industrial test and monitoring, including instruments that have RS-232 serial interfaces.

"**JAMECO Electronic Components/Computer Products**", Jameco, 1355 Shoreway Road, Belmont, CA 94002, 415-592-8097 or 800-831-4242, Internet at http://www.jameco.com. This catalog has a variety of electronic test equipment, including instruments that have RS-232 serial interfaces

"**CyberResearch PC Systems Handbook**", CyberResearch Inc, 25 Business Park Drive, Branford, CT 06405, 203-483-8815 or 800-341-2525, Internet at http://www.cyberresearch.com. This catalog has a variety of data acquisition systems and industrial computer components, including serial communications accessories.

"**Communication SOURCE-BOOK**", Industrial Computer Source PO Box 910557, San Diego, CA 92191, 619-677-0898 or 800-459-7442, Internet at http://www.industry.net/indcompsrc. This catalog, and others from Industrial Computer Source, contains a variety of serial communications devices and accessories.

"**T & M Catalog**", Agilent Technologies. Request catalogs via the web at www.agilent.com. Phone 800-452-4844 or 877-4AGILENT. There are regional sales offices around the world. This catalog shows all of Agilent's (HP's) test instrumentation, including telecom and datacom test equipment.

"**Data Acquisition Control & Test I/O Cards, Catalog**" Access I/O Products Inc 9400 Activity Road, San Diego, CA 92126, 619-693-9005 or 800-326-1649, Internet at http://www.acces-usa.com. This catalog has a series of data acquisition cards and accessories, including serial adapters.

"**Home Automation Products Catalog**", Home Controls Incorporated, 7626 Miramar Road, Suite 3300, San Diego, CA 92126, 619-693-8887 or 800-266-8765. This catalog carries a series of devices for home automation, using the X-10 networking system. X-10 can be controlled via RS-232 serial interfaces.

"**Tiger Software**", 8700 W Flagler St 4th Floor, Miami, FL 33174, 800-888-4437. This catalog carries computer hardware and software, including modems and X-10 networking components for home automation.

"**TECHNI-TOOL**", 5 Apollo Road, Box 368, Plymouth Meeting, PA 19462, 610-941-2400, Internet at http://www.techni-tool.com. This catalog carries a variety of tools and test instruments, including RS-232 breakout boxes and testers.

"**VIKING Electronics Inc**", 1531 Industrial Street, Hudson, WI 54016, 715-386-8861, Internet at http://www.vikingelectronics.com. Telephone systems, specialized telephones, and ring-down generators for demonstration and testing are featured.

Marlin P. Jones and Assoc Inc Box 12685, Lakepark, FL 33403, 800-652-6733 or 561-848-8236. Internet at www.mpja.com. Inexpensive electronic kits for serial and parallel data acquisition. Books, tools and accessories for enthusiasts

Electronix Express (a division of RSR Electronics Inc), 365 Blair Road, Avenet, NJ 07001, 800-972-2225 or 732-281-8020, Internet at www.elexp.com. They are a mail order and web provider of inexpensive electronic test equipment (including the MAS-345 DVM with RS-232 interface, for which I provide example data logger code on the CD ROM), kits, books, and supplies.

The **x10.com** web catalog sells what its name implies, lots of X10-based networking products. It is hard to beat their price and selection. Go to www.x10.com.

Serial Port Hardware And Serial Devices

Many of the catalogs listed previously have serial port hardware. Here is a listing of some more specialized vendors.

Socket Communications, 37400 Central Court, Newark, California 94560, 510-744-2700. Internet at www.socketcom.com. Socket Communications sells a variety of communications adapters for Compact Flash and PC Card (PCMCIA), marketed most commonly for use with Pocket PC and other Windows CE devices.

National Instruments, 6504 Bridge Point Parkway, Austin, TX 78730, 800-433-3488, Internet at http://www.natinst.com. National Instruments sells a variety of data acquisition hardware and software, and RS-232 and RS-485 serial port boards.

CmC, Connecticut microComputer Inc, PO Box 186, Brookfield, CT 06804, 800-426-2872. CmC sells a variety of serial port hardware, including addressable RS-232 and RS-485 networking.

DuTec, 4801 James McDivitt Road, PO Box 964, Jackson, MI 49204, 800-248-1632. Sells a nice RS-232 to RS-485 converter that is fully isolated and that requires no special handshaking. It uses built-in hardware to detect transmit data and to enable and disable the transmit driver.

ICS DataCom, 473 Los Coches Street, Milpitas, CA 95035, 800-952-4499. ICS sells a small RS-232 to RS-484 adapter that plugs directly into the PC serial port connector.

MCC, Micro Computer Control, PO Box 275, Hopewell, NJ 08525, 609-466-1751, Internet at http://www.mcc-us.com. MCC sells a RS-232 to I2C adapter called IPort. I2C is a special serial format that is used by some micro controllers. The IPort adapter turns your PC serial port into an I2C Bus Master and Slave.

FiberPlex Incorporated, 10840-412 Guilford Road, Annapolis Jct. MD 20701. 301-604-0100. Email: prsales@fiberplex.com. FiberPlex sells a fiber optic, high-speed modem/isolator, and a number of other fiber optic products.

Phoenix Digital, 7650 E Evans Road, Bldg A, Scottsdale, AZ 85260, 02-483-7393. Phoenix Digital sells industrial communications products, including a fiber optic modem useful at distances up to six miles.

Western Telematic Inc, 5 Sterling, Irvine, CA 92718, 14-586-9950 or 800-854-7226. WTI sells a variety of serial devices such as code activated switches, and RS-232 to RS-485 converters.

Prairie Digital Inc, 846 Seventeenth Street, Prairie du Sac, WI 53578, 8-643-8599. Prairie Digital sells an inexpensive serial port data acquisition and control system. It does digital and analog I/O and has special features for stepper motor control, all quite inexpensively.

Telecom Analysis Systems Inc, 34 Industrial Way East, Eatontown, NJ 07224. 908-544-8700, Internet at www.taskit.com. TAS sells several different test systems for modems, ISDN terminal adapters, and cellular modems. These are perhaps the most powerful telephone line simulators that are available.

Software Interphase Inc, 82 Cucumber Hill Rd., Foster, RI 02825. 401-397-2340, Internet at sinterphas@aol.com. Software Interphase sells a line of serial networkable modules for control of external devices. Varieties are offered for relays and motor speed control. Windows DLLs are included.

Measurement & Control Products Inc, 415 Madison Avenue, 22 Floor, New York, NY 10164. 212-229-2141. Measurement & Control Products sells a very nice hand-held RS-232 data monitor. It handles asynchronous and synchronous data from 300 bps to 64k bps.

MESA Systems Co., 119 Herbert Street, Framingham NJ 01701. 508-820-1561. MESA sells a remote digital transmitter interface module with an RS-485 interface. It measures analog sensors and converts them to engineering units for transmission to a PC. The DTIM is temperature compensated and is mounted in a weather proof housing.

Cambridge Electronics Laboratories, 20 Chester Street, Somerville, MA 02144. 617-629-2805. Cambridge sells QuickLink, an inexpensive one-way ring down unit that can be used as a simple telephone line simulator.

Consultronics, Limited, 160 Drumlin Circle, Concord, Ontario. 905-738-3741 or 800-267-7235. Consultronics sells a variety of telecommunications test equipment, including the TCS 700. The TCS 700 is a complete system for testing modems that connect to the PSTN or leased analog lines. It consists of a bi-directional impairment simulator, a dual port exchange simulator providing two connections for PSTN modems, two local loop simulators, two PCM / ADPCM simulators and Microsoft Windows software that performs through port file transfer testing and allows the user to set up the TCS 700.

Dries Associates Inc, 7457 Elmwood Avenue, Middleton, WI 53562. 608-831-5542. Dries sells a RS-232 port multiplexer that allows you to connect from 4 to 256 serial devices to a PC.

Lava Computer Manufacturing Inc, 28A Dansk Court, Rexdale, ON M9W 5V8. 416-674-5962. Lava sells a line of high performance serial boards that use 16650 and 16750 UARTs for higher performance than is available from 16550 based ports. These benefit from a high-performance driver, like Pacific CommWare's TurboComm 95.

Telulex Inc, 2455 Old Middlefield Way S., Mountain View, CA 94043. 415-938-0240, Internet at http://www.telulex.com. The Telulex Modes SG-100 Signal Generator is a very flexible 0-20MHz signal generator. It can be controlled and programmed via a RS-232 serial port.

TELATEMP Corporation, PO Box 5160, Fullerton, CA 92838, 800-321-5160. TELATEMP sells a model TK-1 Transit TermLogger that is used to log temperatures on in-transit shipments. It has a serial port and logged data can be downloaded to a PC.

DGH, PO Box 5638 Manchester, NH 03108. 603-622-0452. DGH sells a line of data acquisition and control modules that can be linked serially and interfaced to a PC.

SiliconSoft, 5131 Moorpark Ave., #303, San Jose, CA 95129, 800-969-4411. SiliconSoft sells a very inexpensive series of serial "dongles" for analog data acquisition.

Worthington Data Solutions, 232 Swift St., Santa Cruz, CA 95060. 800-345-4220 or 408-458-9964. Internet at www.barcodehq.co. Worthington sells a variety of barcode readers including, but not limited to, serial readers at reasonable prices.

Wireless Communications

See also, Bluetooth and WiFi serial.

Radiotronix, 905 Messenger Lane, Moore, OK 73160. 405-794-7730. Internet at http://radiotronix.com. Radiotronix designs and sell a range of standard and custom wireless serial hardware. Serial rates up to 115200 bps are available, and designs are FCC approved.

Adcon Telemetry Inc, 1001 Yamoto Road, Suite 305, Boca Raton, FL 33431. 561-989-5309 or 800-360-5309. Internet at http://www.adcon.com. Adcon sells a very small, license free transceiver that can send serial data at distances up to one mile, with data rates up to 19200 bps. It has built-in networking protocols for extended applications.

Point Six Inc, 391 Codell Dr., Lexington, KY 40509. 859-266-3606. www.pointsix.com. Point Six sells a set of 418 MHz wireless modems with point-to-point and multidrop capability.

AERO-Repco Systems Inc, 2400 Sand Lake Road, Orlando, FL 32809, 800-950-5633, Internet at http://www.aerotrom-repco.com. ARS sells the NLR series of wireless modems that provides rates up to 38400 bps and ranges up to 1.5 miles. No license is required. They provide point-to-point or multipoint operation and have a built-in voice channel. ARS also has conventional radio modems with ranges up to 15 miles.

Proxim Inc, 295 North Bernardo Avenue, Mountain View, CA 94043, 415-960-1630. Proxim sells the ProxLINK wireless RS-232 modem. It supports ranges up to 2 miles with a long-range antenna.

MICRILOR Inc, 17 Lakeside Park, Wakefield, MA 01880. 617-246-0103. Wireless modems with speeds up to 64k bps and ranges up to 1000 ft.

O'Neill Connectivities Inc, 800-624-5296. OCI furnishes LAWNII wireless RS-232 systems with ranges of up to 3000 feet.

MONICOR Wireless Products, 2964 NW 60th Street, Fort Lauderdale, FL 33309. 954-979-1907. Email: monicor@ix.netcom.com. Monicor sells a series of radio modems that provide speeds up to 19200 bps and ranges up to one mile.

Comrad, Communications Research & Development Corporation, 7210 Georgetown Road, Suite 300, Indianapolis, IN 46268. 317-290-9107. COMRAD sells 900 MHz Wireless Data Link. It supports speeds up to 38400 bps and distances up to 1500 feet.

GRE America Inc, 425 Harbor Blvd., Belmont, CA 94002. 415-591-1400 or 800-233-5973. GRE sells a line of spread spectrum transceivers that serial speeds up to 38400 bps. They can be used in point-to-point and multipoint applications.

Solid State Electronics Corporation, 18646 Parthenia Street, Northridge, CA 91324. 818-993-8257. SSE sells a variety of radio modems that provide speeds up to 9600 bps and ranges up to 10 miles.

EST, Electronic Systems Technology Inc, 415 North Quay Street, Kennewick, WA 99336. 509-735-9092, Internet at http://www.esteem.com. EST sells a series of wireless modems that support speeds up to 19200 bps. The modems have Hayes AT command set emulation.

Magellan Systems Corporation, 960 Overland Court, San Dimas, CA 91773. 909-394-5000. Magellan sells an Inmarsat satellite telephone that has data capability. The notebook computer sized unit costs about $5 per minute to operate and gives you world-wide communications.

Other satellite telephones are available with similar specifications. **NEC** (Nippon Electric Company) sells one, as do others.

Almost Wireless

X-10, with sources listed in the Catalogs section, is a lower-performance control network that is commonly used in home automation. Other, higher performance serial networking systems are listed here.

Intellon Corporation, 5100 West Silver Springs Blvd., Ocala, FL 34482. 904-237-7416, Internet at http://www.intelon.com. Intellon provides hardware and software for developing CEBus-compliant (EIA-600 standard), spread spectrum communications networks. Like X-10, CEBus is often used in home automation. However, it also is used in industrial automation networks. Both wired, over various media, and wireless communications are supported.

Echelon Corporation, 4015 Miranda Ave., Palo Alto, CA 94304. 415-855-7400 or 800-258-4LON, Internet at http://www.lonworks.echelon.com. Echelon provides hardware and software for developing LonWorks ™ spread spectrum communications networks. LonWorks is used in home automation. However, it also is widely used in industrial automation networks. Both wired, over various media, and wireless communications are supported.

GPS

There are many Global Positioning System satellite receiver manufacturers, such as Trimble, Garmin, Magellan, Eagle, Motorola, etc. However, the best source for information on most of this is:

Navtech Seminars and GPS Supply, 2775 S. Quincy Street, Suite 610, Arlington, VA 22206. 703-931-0500 or 800-628-0885, Internet at http://www.navtechgps.com. Navtech offers a number of GPS receivers that have serial interfaces. Most of these can be used with the GPS applet that is included in this book.

Multiport Boards

SEALEVEL Communications and I/O. Sealevel, 155 Technology Place, PO Box 830, Liberty, SC 29657. 864-843-4343, Internet at http://www.sealevel.com. Sealevel manufactures a number of Windows serial port boards, including multi-port boards.

Digi International, 11001 Bren Road East, Minnetonka, MN 55343. 612-912-3444 or 800-344-4273, Internet at http://www.digi.com. Digi makes a series of multiport boards.

Comtrol Corporation, 900 Long Lake Road, Suite 210, Saint Paul, MN 55112. e: 612-631-7654 or 800-926-6876. Comtrol sells the RocketPort series of multiport boards. At this writing the RocketPort boards have drivers for Windows NT but not Windows 95/98.

Maxpeed Corporation, 1120 Chess Drive, Foster City, CA 94404. 415-345-5447, Internet at http://www,maxpeed.com. Maxpeed sells a variety of multiport boards.

CONSENSYS, (no address), 905-940-2900 or 800-388-1896, Internet at http://www.consensys.com. CONSENSYS sells the ChiliPORTS series of intelligent multiport boards.

Central Data, 1602 Newton Drive, Champaign, IL 61821. 217-359-8010 or 800-482-0398, Internet at http://www.cd.com. Central Data sells a series of PCI bus multiport boards.

Equinox Systems Inc, One Equinox Way, Sunrise, FL 33351. 305-746-9000 or 800-275-3500, Internet at http://www.equinox.com. Equinox sells a series of multiport boards. They also sell modem-pooling software.

Quatech Inc, Phone 330-655-9000 or toll free at 800-553-1170, fax 330-655-9010, email sales@quatech.com, and on the web at www.quatech.com. Boards feature 2, 4, or 8 independent RS-232 ports with Speeds up to 921.6 kbps. Boards use 16750 UARTs with 64-byte FIFOs. Full modem control and hardware and software flow control, with DB-9, DB-25, or RJ-11 connectors. A Surge suppression package is available. Boards have Windows 95/98/Me/NT/2000/XP support. Quatech also manufactures RS-422/485 multiport boards, Compact Flash serial adapters (one and two port) and USB serial adapters.

Inside Out Networks, 7004 Bee Caves Rd, Bldg 3, Ste. 200, Austin, Texas 78746. Phone 512-06-0600; fax 512-306-0694, and on the web at www.ionetworks.com. Inside Out Networks specializes in Enterprise USB Connectivity, makes simple-to-install, high-performance solutions for attaching serial, parallel, and modem devices to PCs, thin clients, and servers utilizing USB technology. Products included RS-232 and RS-422/485 single and multiport adapters.

Bluetooth and WiFi Serial

In this section I will provide information on Bluetooth devices that are intended to work with standard serial APIs. Thus, they will be compatible with the software techniques described in this book. In general, these devices are intended to be "RS-232 serial cable replacements" to provide a wireless solution that requires no special software.

Free2Move, The F2M01 provides Ad Hoc connectivity will give you the serial link when as the communicating devices are within range of each other. It integrates a Bluetooth module in very dense packing. A class 1 Bluetooth solution is used offering a nominal range of approximately 10-100m. No external drivers are needed to use the plug. A Windows application is included that can be used to configure the plug to suit your requirements. You can get information on the web at www.free2move.se. In the US you may purchase from www.expansys.us.

The **Socket Cordless Serial Adapter** with *Bluetooth* Wireless Technology is a small adapter allows any device with a standard 9-pin serial port to communicate wirelessly. It will communicate with another Cordless Serial Adapter or any *Bluetooth* enabled device. You don't need to install any drivers on the host device. Included in the package is the Socket Cordless Serial Adapter Configuration Utility, an easy-to-use Windows application which lets you to reprogram many of the default settings on the Cordless Serial Adapter. You can custom configure settings to match your individual needs. The utility runs under Windows 95, 98, Me, 2000 or XP. Also included is 9-pin female/female null modem connector. You can also reprogram your Cordless Serial Adapter manually through AT commands via a terminal device. Socket's AT Command Set provides even more extensive options for custom configurations. The Cordless Serial Adapter communicates over the *Bluetooth* Serial Port Profile. The adapter can be powered from the included AC power plug or from the host device using pin 9 of the adapter. Contact Socket Communications on the web at www.socketcom.com.

Lantronix, 15353 Barranca Parkway, Irvine, CA 92618 USA sells both standalone WiFi Device Servers and embedded WiFi Device Servers that provide serial connectivity, supporting RS-232, RS-422 and RS-485 interfaces. Contact them on the web at www.lantronix.com. **Many** oems use the Lantronix embedded device server as the heart of their product. These manufacturers are too numerous to mention separately.

Software

TechArts, 829 E. Molloy Road, Syracuse, NY 13211. 315-455-1003 to 800-455-9853, Internet at http://www.techarts.com. TechArts sells Net2Com outbound modem pooling for Windows NT. This has support for Windows 3.x and Windows 95 client redirectors via Int14.

Cherry Hill Software, 7 Vermont View Drive, Suite 18, Watervliet, NY 12189. 518-786-3153. HiCom/9 driver. Supports up to nine serial ports on Windows 3.x systems.

Pacific CommWare Inc 2895 Highway 66, Ashland, OR 97520. 541-482-2744, Internet at http://www.turbocom.com. Pacific CommWare sells communications drivers for Windows 3.x and Windows 95/98. These support multiport boards and high-performance serial ports. Also available is TurboCommander Pro for Windows 3.x. TurboCommander Pro provides software for analysis and debugging of serial communications.

AggSoftware, on the web at www.aggsoft.com. Agg Software sells a variety of software products that can be vital for debugging serial software problems including the outstanding Advanced Serial Port Monitor (see the CD ROM for a demo version). Other products are Virtual Null modem and Serial Port Logger. More specialized are CNC Syntax Editor and DNC Precision for use with numerical control machines.

Sax Software Corporation, 950 Patterson St. , Eugene, OR 97401, 800-645-3729 or 541-344-2235, Internet at http://www.saxsoft.com or www.sax.net. Sax sells Sax Comm Objects. These are custom controls, and the Professional version includes Sax Basic Engine, a powerful macro-BASIC language that you can use in your Visual Basic applications. Sax also sells Sax.net Communications, a native .NET serial communication component.

PRUDENS Inc, 100 Borris Circle, #302, Streamwood, IL 60107, Internet at prudens@pcweb.dpliv.com. Prudens sells a variety of Windows 95/98 utilities and developer tools for Windows 95/98. ComSpy 95 is a utility that can help debug communications applications. A shareware version can be downloaded from http://www.pcmag.com or from various forums on CompuServe, including WINSHARE.

Dean Software Design, PO Box 13032, Mill Creek, WA 98082-1032. Email: 75240.65@compuserve.com. Dean Software sells InfoSpy, a Windows 3.x/Windows 95/98 spy utility that includes a communications message spy.

SoftCircuits, P.O. Box 1355, West Jordan, UT 84084-8355, (801) 282-0646, Internet at http://www.softcircuits.com. SoftCircuits does custom programming and provides a number of free programming utilities and source code on their Web site.

Mabry Software, 503 316th Street Northwest, Stanwood, WA 98292. 360-629-9278 or 800-996-2279, Internet at http://www.mabry.com, e-mail at mabry@mabry.com. Mabry sells dozens of custom controls of a utilitarian nature. Of special interest may be their Internet controls. Some of these are RAS (Dialer), NetTime, FTP, Mail (SNMP, POP, and IMAP). .

Tetradyne Software Inc, 2542 S. Bascom Ave, Suite #206, Campbell, CA 95008. 408-377-6367, Internet at http://www.tetradyne.com. Tetradyne (formerly WinStar Technologies) sells DriverX for I/O port, memory access, and interrupt handling under Windows 95/98 and Windows NT/2K/XP. They also sell other developer tools for device driver development for Windows 32-bit OS'.

Sysinternals, on the Internet at www.sysinternals.com. Freeware PortMon.exe (see the Debugging chapter).

Logisoft Inc, 5565 Preston Oaks Rd #294, Dallas, TX 75240. Phone 972.385.1669, fax 972.385.1698. Internet at http://www.logisoft-inc.com, email logisoft@logisoft-inc.com. Logisoft sells X/Pager, a TAPI enabled OCX for alphanumeric paging using modems and the Internet.

MicroDexterity Inc, PO Box 5372, Plymouth, MI. 48170-5372. Internet at www.microdexterity.com, email at info@mdxi.com. Phone 888-891-0700 or 313-453-5872. MicroDexterity sells Stamina, a complete function library that includes hundreds of useful functions for calculating CRCs and manipulating string and byte array data.

Frontline Test System Inc, PO Box 7507, Charlottesville, VA., 22906-7507. Phone 804-984-4500. Internet at www.fte.com, email at sales@fte.com. FTS sells SerialTest, a fine software-based serial data analyzer.

Franson Technology AB, Arkovagen 45, 121 55 Johanneshov, Sweden. Franson sells a variety of powerful communications software products for programmers, including: SerialTools .NET/ActiveX, BlueTools (Bluetooth), GPSTools .NET/ActiveX, GPSGate for Windows and Pocket PC, and CoordTrans. Contact franson at www.franson.com. Demo versions are available for download. You also can contact me directly for special license pricing on Franson products at: dick_grier@msn.com.

Multiple Modem Boards

MultiTech Systems Inc, 2205 Woodale Drive, Mounds View, MN 55112. 612-785-3500 or 800-328-9717, Internet at http://www.multitech.com. The MultiModem ISI provides four V.34, 33.6k bps modems on one board.

Ariel Corporation, 2540 Route 130, Cranbury, NJ 08512. 609-860-2900, Internet at http://www.ariel.com. Ariel CTI-modem has up to 24, V.34 modems on one board.

SECOND WAVE Inc Building 13, 2525 Wallingwood Drive, Austin, TX 78746. 512-329-9283. Email: secdwave@flash.net. Second Wave sells the Comm Blaster multi-modem PCI board, with up to eight V.34 modems on one board.

Xircom Inc, Corporate Headquarters, 2300 Corporate Center Dr. ,Thousand Oaks, CA 91320. 805 376-9300, Internet at http://www.zircom.com. Xircom sells Xircom Inc MPM-4 and MPM-8 boards. These provide four or eight V.34 modems on one board.

PLC Software

See **LUCA** for one commercial product that provides PLC (Programmable Logic Controller) communications support. Here are some other resources.

Automated Solutions. 1415 Fulton Rd. #205-A12, Santa Rosa, CA 95403, (707) 578-5882 (800) 410-4632. Internet at www.automatedsolutions.com.

Parijat Controlware. 1425 Blalock Suite 201, Houston, TX 77055, (713)935-0900. Intenet at www.parijat.com.

ConsoliTech Corp. 800-727-7959 or 317-258-3544. Internet at www.consolitech.com.

CimQuest (InGear). 518 Kimberton Rd. Suite 325 Phoenixville, PA 19460. 610-935-8282. Internet at www.ingear.cimquest.com.

Appendix B   VT-100 Terminal Emulator

I have included a folder on the CD ROM that provides a complete VT-100 terminal emulation. It is written in VB6, and uses MSComm32.ocx for the serial connection. Previous editions of this book included code written for 16-bit versions of Visual Basic; however, I did not find it to be very satisfactory.

I do not have space in this text to go into detail on its operation. However, it should provide a good starting point if you have to include this sort of functionality in your applications. An even more satisfactory solution might be to use a commercial communications add-on that provides this emulation (See Chapter 7).

Appendix C   NMEA-0183 Protocol

There are several proprietary serial protocols used to communicate with GPS receivers. I am aware of only one standard protocol, NMEA-0183. I have reproduced a subset of it here. This protocol is reduced to code for receive data in the GPS applet in Chapter 4.

The NMEA protocol is fully defined in "NMEA 0183, Version 2.0." Copies can be obtained from the National Marine Electronics Association, PO Box 50040, Mobile, AL 36605. Another source is the Radio Technical Commission for Maritime Services' "RTCM Recommended Standards for Differential Navstar GPS Service, Version 2.0, RTCM Special Committee No. 104." Copies can be obtained from RTMC, PO Box 19087, Washington, DC 20036.

Recommended Minimum Specific GPS/TRANSIT Data (RMC)

$GPRMC,<1>,<2>,<3>,<4>,<5>,<6>,<7>,<8>,<9>,<10>,<11>*hh<CF><LF>

Item	Description
<1>	UTC time of position fix, hhmmss format
<2>	Status, A = valid position, V = NAV receiver warning
<3>	Latitude, ddmm.mmm format (leading zeros are sent)
<4>	Latitude hemisphere, N or S
<5>	Longitude, dddmm.mmm format (leading zeros are sent)
<6>	Longitude hemisphere, E or W
<7>	Speed over ground, 0.0 to 999.9 knots
<8>	Course of ground, 000.0 to 359.9 degrees, true (leading zeros are sent)
<9>	UTC date of position fix, ddmmyy format
<10>	Magnetic variation, 000.0 to 180.0 degrees (leading zeros are sent)
<11>	Magnetic variation direction, E or W (westerly adds to true course)

Global Positioning System Fix Data (GGA)

$GPGGA,<1>,<2>,<3>,<4>,<5>,<6>,<7>,<8>,<9>,M,<10>,M,<11>,<12>*hh<CR><LF>

Item	Description
<1>	UTC time of position fix, hhmmss format
<2>	Latitude, ddmm.mmm format (leading zeros are sent)
<3>	Latitude hemisphere, N or S
<4>	Longitude, dddmm.mmm format (leading zeros are sent)
<5>	Longitude hemisphere, E or W
<6>	GPS quality indication, 0 = fix not available, 1 = non-differential GPS fix available, 2 = differential GPS (DGPS) fix available
<7>	Number of satellites in use, 00 to 08 (leading zeros are sent)
<8>	Horizontal dilution of precision, 1.0 to 99.9
<9>	Antenna height above/below mean sea level, -9999.9 to 99999.9 meters
<10>	Geoidal height, -999.9 to 9999.9 meters
<11>	Differential GPS (RTCM-SC104) data age, number of seconds since last valid RTMC transmission (null if non-DGPS)
<12>	Differential Reference Station ID, 0000 to 1023 (leading zeros are sent, null if non-DGPS)

There are several other potentially useful frames. The description of these will be left to the complete documents.

Appendix D   PC Serial Port Description

The basic PC serial port can be treated as though it were just the UART that is used. There are other components, of course. However, those other components are not material when it comes to the software interface to the port. So, what I'll describe here are the UART registers and the meaning of each function that uses each register.

You can refer to the National Semiconductor Advanced Peripherals data book (listed in Appendix A) for more information. Other UART vendors publish comparable data books.

Register Address 8250										
0 DLAB = 0	**0 DLAB = 0**	**1 DLAB = 0**	**2**	**3**	**4**	**5**	**6**	**7**	**0 DLAB = 1**	**1 DLAB = 1**
Receive Buffer Register Read Only	Transmit Holding Register Write Only	Interrupt Enable Register	Interrupt Ident. Register Read Only	Line Control Register	Modem Control Register	Line Status Register	Modem Status Register	Scratch Register	Divisor Latch (LSB)	Divisor Latch (MSB)
Bit #										
0	Data Bit 0 (Note 1)	Data Bit 0	Receive Data Available	"0" if Interrupt Pending	Word Length Select Bit 0	Data Terminal Ready (DTR)	Data Ready (DR)	Delta Clear To Send (DCTS)	Bit 0	Bit 0	Bit 8
1	Data Bit 1	Data Bit 1	Transmit Holding Register Empty	Interrupt ID Bit (0)	Word Length Select Bit 1	Request To Send (RTS)	Overrun Error	Delta Data Set Ready (DDSR)	Bit 1	Bit 1	Bit 9
2	Data Bit 2	Data Bit 2	Receiver Line Status	Interrupt ID Bit (1)	Number of Stop Bits	Out 1	Parity Error	Trailing Edge Ring Indicate	Bit 2	Bit 2	Bit 10
3	Data Bit 3	Data Bit 3	Modem Status	0	Parity Enable	Out 2	Framing Error	Delta Data Carrier Detect (DDCD)	Bit 3	Bit 3	Bit 11
4	Data Bit 4	Data Bit 4	0	0	Even Parity Select	Loop	Break Interrupt	Clear To Send (CTS)	Bit 4	Bit 4	Bit 12
5	Data Bit 5	Data Bit 5	0	0	Stick Parity	0	Transmit Holding Register	Data Set Ready (DSR)	Bit 5	Bit 5	Bit 13
6	Data Bit 6	Data Bit 6	0	0	Set Break	0	Transmit Empty	Ring Indicate (RI)	Bit 6	Bit 6	Bit 14
7	Data Bit 7	Data Bit 7	0	0	Divisor Latch Access Bit (DLAB)	0	0	Data Carrier Detect (DCD)	Bit 7	Bit 7	Bit 15

Note 1: Bit 0 is the least significant bit. It is the first bit to be transmitted or received.

	Register Address 16450/16550											
Bit #	**0 DLAB = 0**	**0 DLAB = 0**	**1 DLAB = 0**	**2**	**2**	**3**	**4**	**5**	**6**	**7**	**0 DLAB = 1**	**1 DLAB = 1**
	Receive Buffer Register Read Only	Transmit Holding Register Write Only	Interrupt Enable Register	Interrupt Ident. Register Read Only	FIFO Control Register Write Only	Line Control Register	Modem Control Register	Line Status Register	Modem Status Register	Scratch Register	Divisor Latch (LSB)	Divisor Latch (MSB)
0	Data Bit 0 (Note 1)	Data Bit 0	Receive Data Available	"0" if Interrupt Pending	FIFO Enable	Word Length Select Bit 0	Data Terminal Ready (DTR)	Data Ready (DR)	Delta Clear To Send (DCTS)	Bit 0	Bit 0	Bit 8
1	Data Bit 1	Data Bit 1	Transmit Holding Register Empty	Interrupt ID Bit (0)	RCVR FIFO Reset	Word Length Select Bit 1	Request To Send (RTS)	Overrun Error	Delta Data Set Ready (DDSR)	Bit 1	Bit 1	Bit 9
2	Data Bit 2	Data Bit 2	Receiver Line Status	Interrupt ID Bit (1)	XMIT FIFO Reset	Number of Stop Bits	Out 1	Parity Error	Trailing Edge Ring Indicate	Bit 2	Bit 2	Bit 10
3	Data Bit 3	Data Bit 3	Modem Status	Interrupt ID Bit (2)	DMA Mode Select	Parity Enable	Out 2	Framing Error	Delta Data Carrier Detect (DDCD)	Bit 3	Bit 3	Bit 11
4	Data Bit 4	Data Bit 4	0	0	Reserved	Even Parity Select	Loop	Break Interrupt	Clear To Send (CTS)	Bit 4	Bit 4	Bit 12
5	Data Bit 5	Data Bit 5	0	0	Reserved	Stick Parity	0	Transmit Holding Register	Data Set Ready (DSR)	Bit 5	Bit 5	Bit 13
6	Data Bit 6	Data Bit 6	0	FIFOs Enabled (Note 2)	RCVR Trigger (LSB)	Set Break	0	Transmit Empty	Ring Indicate (RI)	Bit 6	Bit 6	Bit 14
7	Data Bit 7	Data Bit 7	0	FIFOs Enabled (Note 2)	RCVR Trigger (MSB)	Divisor Latch Access Bit (DLAB)	0	0	Data Carrier Detect (DCD)	Bit 7	Bit 7	Bit 15

Note 1: Bit 0 is the least significant bit. It is the first bit to be transmitted or received. Note 2: These bits are always 0 in the 16450 mode.

Appendix E   ASCII Control Codes

Hexadecimal	Decimal	Key	Name	Description
00	0	^@	NUL	Null
01	1	^A	SOH	Start of Header
02	2	^B	STX	Start of Text
03	3	^C	ETX	End of Text
04	4	^D	EOT	End of Transmission
05	5	^E	ENQ	Enquiry
06	6	^F	ACK	Acknowledge
07	7	^G	BEL	Bell
08	8	^H	BS	Backspace
09	9	^I	HT	Horizontal Tab
0A	10	^J	LF	Line Feed
0B	11	^K	VT	Vertical Tab
0C	12	^L	FF	Form Feed
0D	13	^M	CR	Carriage Return
0E	14	^N	SO	Shift Out
0F	15	^O	SI	Shift In
10	16	^P	DLE	Data Link Escape
11	17	^Q	DC1	Device Control 1
12	18	^R	DC2	Device Control 2
13	19	^S	DC3	Device Control 3

Hexadecimal	Decimal	Key	Name	Description
14	20	^T	DC4	Device Control 4
15	21	^U	NAK	Negative Acknowledge
16	22	^V	SYN	Synchronous Idle
17	23	^W	ETB	End Transmission Block
18	24	^X	CAN	Cancel
19	25	^Y	EM	End of Medium
1A	26	^Z	SUB	Substitute
1B	27	^[	ESC	Escape
1C	28	^\	FS	File Separator
1D	29	^]	GS	Group Separator
1E	30	^^	RS	Record Separator
1F	31	^_	US	Unit Separator

The Key column refers to the combination of CTRL key (^) and the specified key to generate a control code.

The Visual Basic manuals describe the remainder of the ASCII/ANSI character set.

Appendix F   Basic Modem AT Command Set

The commands listed here are used by the majority of Hayes compatible modems. However, other commands are less standardized. Be sure to consult your modem's manual to be certain about a specific command.

Command	Description
+++	Returns the modem to command mode from online mode. Online mode is when your modem is connected to another modem. Carrier detect will be true. In order for your modem to recognize the escape sequence, you cannot send any characters to the modem for at least 1 second before and after the escape sequence. This command is generally used to get the modems attention so you can send an ATH command to force it to hang up the line. The modem responds with OK when it goes into command mode. You can then issue any AT command.
ATO	Go offhook and start Originate mode handshake, or return to online mode, if already connected.
A/	Repeat the last command issued to the modem. Do not terminate with a carriage return.
ATA	Go offhook and start Answer mode handshake.
ATDx	Dial a telephone number. Replace x with P or T for pulse or tone dialing, respectively. Append all digits next. The P or T can be omitted. If not specified, the default dialing method will be used. A comma will insert a 2-second delay before dialing the next digit.
ATEn	Modem echo commands. ATE or ATE0 disables echo, ATE1 enables echo.
ATHn	Hookswitch control. ATH or ATH0 causes the modem to go on hook (hang up), while ATH1 causes the modem to go offhook.
ATMn	Speaker control. ATM or ATM0 turns the speaker off. ATM1 enables the speaker while not connected. ATM2 enables the speaker at all times.
ATSr=n	Set the modem S-Register. The r specifies which register, and the n specifies the value to be used.

Command	Description
ATSr?	Read the modem S-Register. The r specified which register is read.
ATVn	Select a modem result format. ATV or ATV0 select numeric result codes. ATV1 selects full (verbose) responses.
ATXn	Select result set. ATX or ATX0 sets the minimum result set. ATX1 selects a more complete result set. ATXn, where n > 0, also enables other modem functions, such as extended tones detection. Refer to your modem manual for a complete description.
ATZ	Reset modem to defaults. This is equivalent to powering the modem off then back on. The defaults may be the factory setting or they may be ones previously set using the AT&W command.
AT&Fn	Reset modem to a factory profile. Refer to your modem manual for a description of the profile for each n.
AT&W	Save the current modem settings as the power on defaults. They will be used subsequently, until changed by another AT command.
AT&Cn	Select the state of DCD (carrier detect) when the modem is connected. AT&C0 sets DCD true at all times. AT&C1 causes DCD to reflect the modem connect state, true when connected and false when not connected.
AT&Dn	Specifies how the modem responds to lowering DTR. AT&D0 causes the modem to ignore DTR. AT&D1 is not used. AT&D2 causes the modem to disconnect and return to command mode, when DTR is lowered (set to false). AT&D1 may or may not be used. Check your modem manual.

Appendix G   Telocator Alphanumeric Protocol (TAP)

Appendix G has been moved to the CD ROM.

Index